International Conference on Quantum Phenomena, Quantum Control and Quantum Optics (Quantum Fest 2016)

Journal of Physics: Conference Series Volume 839

Mexico City, Mexico
17 – 21 October 2016

ISBN: 978-1-5108-4097-3
ISSN: 1742-6588

Printed from e-media with permission by:

Curran Associates, Inc.
57 Morehouse Lane
Red Hook, NY 12571

proceedings.com

Some format issues inherent in the e-media version may also appear in this print version.

Printed by Curran Associates, Inc. (2017)

For permission requests, please contact the Institute of Physics
at the address below.

Institute of Physics
Dirac House, Temple Back
Bristol BS1 6BE UK

Phone: 44 1 17 929 7481
Fax: 44 1 17 920 0979

techtracking@iop.org

Additional copies of this publication are available from:

Curran Associates, Inc.
57 Morehouse Lane
Red Hook, NY 12571 USA
Phone: 845-758-0400
Fax: 845-758-2633
Email: curran@proceedings.com
Web: www.proceedings.com

Table of contents

Volume 839

Quantum Fest 2016 International Conference on Quantum Phenomena, Quantum Control and Quantum Optics

17–21 October 2016, Mexico City, Mexico

Accepted papers received: 2 May 2017
Published online: 26 May 2017

Preface

Papers

IOP Conf. Series: Journal of Physics: Conf. Series **839** (2017) 011001 doi:10.1088/1742-6596/839/1/011001

Preface

The *Quantum Fest* is a periodic annual festival on Quantum Phenomena, Quantum Control and Geometry of Quantum States, organized by the Center for Research and Advanced Studies (Cinvestav by its acronym in Spanish) and Unidad Profesional Interdisciplinaria en Ingeniería y Tecnologías Avanzadas del Instituto Politécnico Nacional (UPIITA-IPN) in México City, Mexico. The aim of this meeting is to bring together students and researchers which are engaged in the subjects of the festival, from both theoretical and experimental approaches, in order to get lively discussions and to enable a closer contact between them.

The *Quantum Fest* was celebrated for the first time in the Physics Department of Cinvestav (2010), since then it has been hosted in Cinvestav, UPIITA-IPN and the Tecnológico de Monterrey, Campus Estado de México (ITESM-CEM).

The *Quantum Fest 2016* is the seventh edition of the festival, it took place from October 17 to 21 in the *Sala de Usos Múltiples, Edificio I* of UPIITA-IPN, and was addressed to join the celebration of the first eighty years of the Instituto Politécnico Nacional as well as the first twenty years of the Unidad Profesional Interdisciplinaria en Ingeniería y Tecnologías Avanzadas del Instituto Politécnico Nacional. We would like to thank the willing of the UPIITA-IPN to offer its facilities as the venue of the festival; all its help provided to simplify the logistics and organization of the conference has been welcomed and is acknowledged.

The topics addressed at the short courses of the Quantum Fest 2016 were time asymmetric quantum mechanics, quantum resonances, models of quantum field theory in metamaterials, singular potentials and self-adjoint extensions, nonclassical states of light, Hardy functions and Hilbert space operators.

The Lecturers of Quantum Fest 2016 were:

Manuel Gadella (Valladolid University, Spain)

Maribel Loaiza (Department of Mathematics, Cinvestav, Mexico)

Luis Miguel Nieto (Valladolid University, Spain)

Oscar Rosas-Ortiz (Physics Department, Cinvestav, Mexico)

Besides the courses lectured by Profs. Gadella, Loaiza, Nieto and Rosas-Ortiz, there were 10 plenary presentations, 10 contributed talks and 31 posters. This time, more than 50 participants from Instituto Politécnico Nacional, Cinvestav, Universidad Nacional Autónoma de México, Tecnológico de Monterrey and Universidad Politécnica del Valle de México came together to attend the courses, talks and poster sessions delivered during the festival.

We would like to emphasize that the plenary as well as the contribution talks are presented only by invitation of the festival Organizing Committee.

The Plenary Speakers of Quantum Fest 2016 were:

Bijan K. Bagchi (Department of Applied Mathematics, University of Calcutta, India)

Rafael Baquero (Physics Department, Cinvestav, Mexico)

Rodolfo Id Betan (National University of Rosario, Argentina)

Octavio Castaños (ICN-UNAM, Mexico)

Véronique Hussin (Department of Mathematics and Statistics, University of Montréal, Canada)

Bogdan Mielnik (Physics Department, Cinvestav, Mexico)

Mikhail S. Plyushchay (Physics Department, University of Santiago de Chile, Chile)
Vladimir Rabinovich (ESIME-IPN, Mexico)
José Recamier (ICF-UNAM, Mexico)
Dieter Schuch (Goethe-Universitaet Frankfurt, Germany)

The informal and lively discussions at Sala de Usos Múltiples of UPIITA during either the coffee breaks or the poster sessions, have been a breeding ground for scientific collaboration between the participants of the conference. We are certain that this is crucial to strength the rapprochement between students and researchers.

This year, the academic recognition as the best poster expositor was awarded to Kevin Zelaya for his work entitled "Time-dependent exactly solvable potentials generated by Darboux transformations". The figure in the cover of this volume has been borrowed from the cartel presented by Mr. Zelaya.

The members of the 2016 Organizing Committee were:

Sara Cruz y Cruz (UPIITA-IPN)
Zurika Blanco (Physics Department, Cinvestav)
Zulema Gress (UPIITA-IPN)

The Quantum Fest 2016 would not have been possible without the valuable financial support of the sponsor institutions: Instituto Politécnico Nacional, UPIITA-IPN and Cinvestav.

Our thanks to Dr. Alejandro García Guerra, Director de la UPIITA-IPN for endless support and host during the festival.

Special thanks to Mmes Berenice León, Rebeca Octaviano and Miriam Lomelí, Secretaries of the festival, for invaluable help in many aspects of the organization, including the design of the posters and distribution of announcements, hotel and flight reservations, and the scheduling of the event. Berenice, Rebeca and Miriam: the Bar Quantum is very grateful to you.

The publication of this volume of Journal of Physics: Conference Series has been supported by the Instituto Politécnico Nacional in the framework of the celebration of its 80th Anniversary. We would like to express our gratitude to Lic. Hilda Castro, Subdirectora Administrativa de la UPIITA-IPN for facilitating the administrative duties related with the publication of these Proceedings.

The present volume includes only the positively refereed contributions that were submitted by the participants of the festival.

Finally, we announce that the Quantum Fest will take place every two years from now on. The eight festival will be celebrated in the facilities of the Physics Department of Cinvestav.

Sara Cruz y Cruz
Instituto Politécnico Nacional, UPIITA

México City, April 2017

Peer review statement

All papers published in this volume of *Journal of Physics: Conference Series* have been peer reviewed through processes administered by the proceedings Editors. Reviews were conducted by expert referees to the professional and scientific standards expected of a proceedings journal published by IOP Publishing.

IOP Conf. Series: Journal of Physics: Conf. Series **839** (2017) 012001 doi:10.1088/1742-6596/839/1/012001

Mathematical Foundations of Time Asymmetric Quantum Mechanics

M Gadella

Departamento de Física Teórica, Atómica y Optica. Facultad de Ciencias, Universidad de Valladolid, Paseo Belén 7. 47011 Valladolid, Spain

E-mail: manuelgadella1@gmail.com

Abstract. We review the mathematical tools that are suitable for a formulation of time asymmetry in quantum mechanics. In particular, Hardy functions on a half plane and rigged Hilbert spaces constructed with a subclass of Hardy functions. This time asymmetry often appears in quantum scattering and, in particular, in resonance scattering. We review the construction of Gamow vectors, often considered Gamow states for resonances. A brief summary of the fundamental ideas of time asymmetric quantum mechanics is presented in a last section.

1. Introduction

In the framework of standard non-relativistic quantum mechanics, the time evolution governed by a self adjoint Hamiltonian H is given by a group of unitary operators on a Hilbert space, depending on the parameter time t. In this case, t may reach all possible real values, i.e., $-\infty < t < \infty$. However, not all quantum dynamical processes are governed by a single Hamiltonian. Let us think in scattering, which intuitively and roughly speaking works as follows: Imagine a quantum particle. In the remote past, its quantum state is prepared as free. This means that its time evolution is given by some sort of free Hamiltonian H_0. At some spatial localization, the particle enters into an interaction region, where is subject of some forces that we shall assume that come from the existence of a potential V. Then, inside the interaction region, time evolution of its state is determined by a Hamiltonian $H = H_0 + V$, which will transform this state. Eventually, the particle will abandon the interaction region and evolve again with H_0 to be detected in the far future. In this case, time evolution is given by a Hamiltonian pair $\{H_0, H\}$. Although it is not necessary, we gain in intuition if we think on the interaction region as a bounded domain in space.

A particularly interesting situation will occur when the particle spends in the interaction a time which is much larger than the time it would stay if the interaction would be switched off. In this case, we say that a *metastable state* or *quantum resonance* has been produced. Thus, we may think on resonances as the result of a capture of some particle by a center of forces and its release or emission. The whole process is called *resonance scattering*, while the process of emission is the *decay*.

In atomic or nuclear physics, it is obvious the existence of quantum unstable states. A quantum particle (electron, alpha, etc.) is emitted spontaneously from an atom or nucleus and this is the observed situation. We may describe the situation in terms of resonance scattering, in which we ignore the capture process and identify the unstable state with the quantum resonance. Here, we are solely interested in the process of decay.

In fact, capture and decay are usually quite different process, which require different conditions. While the capture requires some conditions in the preparation of the incoming state, the emission or decay is spontaneous. Therefore, capture (or preparation of the metastable state) and decay are not mutually symmetric and they are not time reversal of each other.

This type of asymmetry can also be observed in other quantum scattering processes. There are quantum scattering processes, which are very common but instead their time reversal processes are very improbable. A typical example has been discussed in [1].

All these situations give an idea of time asymmetry in quantum mechanics. Related notions are *quantum irreversibility*, which describes the non-invariance with respect to the time reversal operation and the existence of a *time arrow*, which distinguishes between past and future in a unique way [2].

Since we usually observe decay and not capture, we need a formalism that describes the decay of quantum unstable particles or resonances. A comment is in order here. In the past physicists have distinguished between resonances and decaying states [3]. Resonances were characterized by a bump in the cross section with width Γ, which was the measurable quantity. Then, a lifetime can be defined as $\tau := \hbar/\Gamma$. Decaying states were defined by its mean life. Measurements of mean life and cross section are independent, so that resonances and decaying states could not be fully identified until precision in both types of measurements were good enough to establish the accuracy of the formula $\tau = \hbar/\Gamma$. In the relativistic case this is not possible in general since the mean lives of decaying particles are times below the minimal time interval which is possible to measure. Yet, lifetimes of relativistic particles are often determined after the measurement of the cross section by the above formula $\tau := \hbar/\Gamma$. It is sometimes claimed that this relation can be fixed in the context of time asymmetric quantum mechanics [4].

Resonances may also be reasonably modeled by an interaction between a bound state and an external field, as is in the celebrated Friedrichs model [5–7]. Yet, the Friedrichs model may be described as a resonance scattering, in which scattering matrix, Møller wave operators, etc. are well defined [8].

Along this paper, we are going to review a mathematical formalism for the theory of quantum resonances, which are typical irreversible quantum processes, based in the use of Hardy functions on a half plane. Hardy functions are complex analytic functions, which properties we review in the sequel. *Hardy functions seem to be the correct mathematical tool to describe irreversibility in Quantum mechanics.* For other descriptions, which do not exclude Hardy functions, see [9, 10].

2. Rigged Hilbert spaces

A rigged Hilbert space or Gelfand triplet is a tern of vector spaces [11–15]

$$\Phi \subset \mathcal{H} \subset \Phi^\times \,, \tag{1}$$

where:

i) $\mathcal{H}$ is an infinite dimensional (separable) Hilbert space.

ii) Φ is a *dense* subspace of $\mathcal{H}$. Dense means that for any vector $\varphi \in \mathcal{H}$, there is a sequence of vectors, $\{\varphi_n\}$ in Φ such that $\varphi_n \longmapsto \varphi$ in terms of the norm topology on $\mathcal{H}$. This means that any $\varphi \in \mathcal{H}$ can be approached by vectors in Φ with arbitrary precision. In addition, Φ has its own topology, which is *stronger* than the topology Φ inherits from $\mathcal{H}$. This means, in particular, that all convergent sequences in Φ are also convergent sequences in $\mathcal{H}$, but the converse is not true.

iii) In order to define $\Phi^\times$, let us consider the set of mappings $F : \Phi \longmapsto \mathbb{C}$, where $\mathbb{C}$ is the field of complex numbers, such that: a) F is antilinear on Φ, i.e., for any $\varphi, \psi \in \Phi$ and any pair $\alpha, \beta \in \mathbb{C}$, we have that $F(\alpha\varphi + \beta\psi) = \alpha^* F(\varphi) + \beta^* F(\psi)$, where the star denotes complex conjugation, and b) F is continuous, which in particular means that if $\varphi_n \longmapsto \varphi$ in Φ, then, $F(\varphi_n) \longmapsto F(\varphi)$ in $\mathbb{C}$. These mappings form a linear space that we denote as $\Phi^\times$. $F(\varphi)$ is the complex number resulting of applying F to the vector φ. For our purposes, we should use the Dirac notation: $F(\varphi) = \langle \varphi | F \rangle$.

Then, to any $\varphi \in \mathcal{H}$, we may associate a unique $F_\varphi \in \Phi^\times$, defined as $\langle \psi | F_\varphi \rangle := \langle \psi | \varphi \rangle$, for all $\psi \in \Phi$. After identification of F_φ with φ (that mathematicians often call an *abus de langage*, like that, in French), we conclude that $\mathcal{H} \subset \Phi^\times$. We do not want to enter in the discussion of the possible topologies on $\Phi^\times$.

Rigged Hilbert spaces have been used to (among other purposes):

i) Give a rigorous setting to the Dirac formulation of Quantum Mechanics [12, 14–16].

ii) Give a precise meaning to Gamow vectors or vector states for resonances [17–19].

iii) With the use of rigged Hilbert space of Hardy functions, provide a mathematical support for the time asymmetric quantum mechanics [20–23].

Let us briefly discuss some features related with rigged Hilbert spaces (RHS). Let A be a linear operator on $\mathcal{H}$ reduced by Φ. This means that for any $\varphi \in \Phi$, we have that $A\varphi \in \Phi$, or $A\Phi \subset \Phi$, i.e., Φ is *invariant* under the action of A. Then, A may be extended to a linear operator on $\Phi^\times$ by means of the *duality formula*:

$$\langle A\varphi | F \rangle = \langle \varphi | AF \rangle \,, \qquad \forall\, \varphi \in \Phi \,, \quad \forall\, F \in \Phi^\times \,. \tag{2}$$

It has been proved that if A is a self adjoint operator on an infinite dimensional (separable) Hilbert space $\mathcal{H}$, there is always a dense subspace Φ, in the domain of A (the space of vectors in which A acts) such that Φ reduces A [13]. This is one of the ingredients towards a rigorous Dirac formulation of quantum mechanics.

From a RHS, we may construct infinitely many others which are, in some sense, equivalent to the original one. Let U be an arbitrary unitary operator on $\mathcal{H}$, or furthermore, U might be a unitary operator between to Hilbert spaces $\mathcal{H}$ and $\mathcal{G}$, i.e., $U : \mathcal{H} \longmapsto \mathcal{G}$. We know that unitary mappings transport topological properties from $\mathcal{H}$ into $\mathcal{G}$. Moreover, there are some other important properties which are preserved relative to operators. In particular, if A is self adjoint in $\mathcal{H}$, its transformed by U on $\mathcal{G}$, UAU^{-1} is self adjoint on $\mathcal{G}$. Thus, if we have a RHS as in (1) and $U : \mathcal{H} \longmapsto \mathcal{G}$ is unitary, the triplet

$$U\Phi \subset \mathcal{G} \subset (U\Phi)^\times \tag{3}$$

is a new RHS with the same properties than the original one. The topology on Φ is transported to $U\Phi$ by U, so that Φ and $U\Phi$ have the same topological properties [18].

3. Hardy functions on a half plane

Let us consider the open upper half plane, $\mathbb{C}^+$, of the complex plane, defined as the set of complex numbers with positive imaginary axis, i.e., $\mathbb{C}^+ := \{z \in \mathbb{C} \mid \operatorname{Im} z > 0\}$. By $\mathbb{C}$ we always mean the field of complex numbers. A *Hardy function* [18, 19, 24–26] $f(z)$ on the upper half plane is a complex analytic function on $\mathbb{C}^+$ such that

$$\sup_{\alpha>0} \int_{-\infty}^{\infty} |f(x+i\alpha)|^2 \, dx < K < \infty \,. \tag{4}$$

This means that for any positive value of α, the function $f(x+i\alpha)$ is square integrable and that all integrals in (4) for all values of $\alpha > 0$ are bounded by the same finite constant K. One important consequence is that the function on the real line $\mathbb{R}$ defined as $f(x) := \lim_{\alpha\to 0} f(x+i\alpha)$ is also square integrable and

$$\int_{-\infty}^{\infty} |f(x)|^2 \, dx < K. \tag{5}$$

It is customary to call $f(x)$ the function of the boundary values of $f(z)$. The function $f(x)$ is *uniquely determined* almost elsewhere, which means that it may not be defined on a set of zero Lebesgue measure. The converse of this result is true: *if we know* that a square integrable function on the real line, $f(x)$, gives the boundary values of a Hardy function on the upper half plane, $f(z)$, then we can recover all values of $f(z)$ for $z \in \mathbb{C}^+$. Due to a theorem by Titchmarsh [24, 25], for all $z \in \mathbb{C}^+$, we have that

$$f(z) = \frac{1}{2\pi i} \int_{-\infty}^{\infty} \frac{f(x)\, dx}{x - z}, \tag{6}$$

where $f(x)$ is the boundary function of $f(x)$. The proof of this result relies in the Cauchy theorem.

Let us call $\mathcal{H}^2_+$ the set of Hardy functions on the upper half plane. These functions have the following properties:

i) We may identify any function $f(z)$ in $\mathcal{H}^2_+$ with its boundary function $f(x)$. Since $f(x)$ is square integrable, then $f(x) \in L^2(\mathbb{R})$, where $L^2(\mathbb{R})$ is the Hilbert space of all complex square integrable functions on the real line. Thus, $\mathcal{H}^2_+ \subset L^2(\mathbb{R})$.

ii) $\mathcal{H}^2_+$ is a linear space, which is a subspace of $L^2(\mathbb{R})$.

iii) According to the *Titchmarsh theorem*, we may recover the values of a function in $\mathcal{H}^2_+$ from its boundary function, *provided that we know that it is the boundary function of a Hardy function on the upper half plane*. Then, how we can say that a square integrable function is the boundary function of one in $\mathcal{H}^2_+$? The answer was given by Paley and Wiener in a celebrated result [24–26]: The Fourier transform of a square integrable function $f(x)$ on the real line is given by

$$\mathcal{F}(f) := \widehat{f}(k) := \frac{1}{\sqrt{2\pi}} \int_{-\infty}^{\infty} f(x)\, e^{-ikx}\, dx \tag{7}$$

and it is also a square integrable function with the same norm. We say that $f(x)$ is supported on an interval Δ of the real line, either finite or infinite, if its is zero outside Δ.

The *Paley-Wiener theorem* establishes that $f(x) \in \mathcal{H}^2_+$ if and only if $f(x)$ is the Fourier transform of a function supported on the negative semi axis $\mathbb{R}^- \equiv (-\infty, 0]$.

Thus, in order to recognize whether a function, $f(x)$, is in $\mathcal{H}^2_+$, we take its inverse Fourier transform. If the function vanish outside $\mathbb{R}^-$, then it is in $\mathcal{H}^2_+$, otherwise it is not.

iv) The Paley-Wiener theorem is a quite useful tool to construct *all* functions in $\mathcal{H}^2_+$. We only need to consider the space $L^2(\mathbb{R}^-)$ of all square integrable functions on $\mathbb{R}^-$ and take their Fourier transforms. Then, we write that

$$\mathcal{F}[L^2(\mathbb{R}^-)] = \mathcal{H}^2_+ . \tag{8}$$

The Fourier transform is a unitary operation on a Hilbert space. Since $L^2(\mathbb{R}^-)$ is a subspace of $L^2(\mathbb{R})$, which is also a Hilbert space, then $\mathcal{H}^2_+$ is a Hilbert subspace of $L^2(\mathbb{R})$.

v) The values of a Hardy function on the upper half plane not only may be recovered by its boundary function on the real line, but also from its boundary values on the positive semi axis $\mathbb{R}^+ \equiv [0, \infty)$ as shown by a result by van Winter [27]. This discussion requires the Mellin transform. Let $f(x)$ be a square integrable function on $\mathbb{R}^+$. Its Mellin transform is given by

$$\mathcal{M}(f)(s) = f_M(s) := \frac{1}{(2\pi)^{1/2}} \int_0^{\infty} f(x)\, x^{is-1/2}\, dx\,, \qquad s \in \mathbb{R}\,. \tag{9}$$

IOP Conf. Series: Journal of Physics: Conf. Series **839** (2017) 012001 doi:10.1088/1742-6596/839/1/012001

The Mellin transform of a square integrable function is also square integrable. The van Winter theorem states that a function $f(x) \in L^2(\mathbb{R}^+)$ is the boundary function on the positive semi axis $\mathbb{R}^+$ of a Hardy function on the upper half plane if and only if its Mellin transform $f_M(s)$ satisfies

$$\int_{-\infty}^{\infty} (1 + e^{2\pi s})|f_M(s)|^2 \, ds < \infty \,. \tag{10}$$

The values of $f(x)$ for $z = re^{i\theta}$, $0 < r \leq \infty$ and $0 < \theta \leq 2\pi$ are given by:

$$f(z) = \frac{1}{(2\pi)^{1/2}} \int_{\infty}^{\infty} f_M(s) \, (re^{i\theta})^{-is-1/2} \, ds \,. \tag{11}$$

Hardy functions in the lower half plane $\mathbb{C}^- := \{z \in \mathbb{C} \mid \operatorname{Im} z < 0\}$ are defined analogously. The space of Hardy functions in the lower half plane is denoted by $\mathcal{H}^2_-$. Functions in $\mathcal{H}^2_-$ have similar properties than functions in $\mathcal{H}^2_+$ with some minor differences. In particular, (6) needs of a minus sign right after the equal sign and in iii.) and iv.) we have to replace Fourier transforms supported in the negative semi axis by Fourier transforms of functions on the positive semi axis. There are some additional properties concerning functions in $\mathcal{H}^2_\pm$:

vi) The spaces $\mathcal{H}^2_\pm$ are subspaces of $L^2(\mathbb{R})$, which are also Hilbert spaces. Moreover, as a consequence of the Paley-Wiener theorem, each function in $\mathcal{H}^2_\pm$ is orthogonal to each function in $\mathcal{H}^2_\mp$:

$$\int_{-\infty}^{\infty} f^*_\pm(x) \, g_\mp(x) \, dx = 0 \,, \qquad f_\pm(x) \in \mathcal{H}^2_\pm \,, \quad g_\mp(x) \in \mathcal{H}^2_\mp \,. \tag{12}$$

In addition:

$$L^2(\mathbb{R}) = \mathcal{H}^2_+ \oplus \mathcal{H}^2_- \,, \tag{13}$$

where the sign $\oplus$ means orthogonal direct sum.

vii) The complex conjugate $f^*(x)$ of a function $f(x) \in \mathcal{H}^2_\pm$ is in $\mathcal{H}^2_\mp$, $f^*(x) \in \mathcal{H}^2_\mp$. Furthermore, $[f(z^*)]^* = f(z)$, where the star always denotes complex conjugation.

viii.) Any function $f(z)$ in $\mathcal{H}^2_\pm$ has the following asymptotic behavior for large values of $|z|$:

$$|f(z)| \approx |z|^{-1/2} \,. \tag{14}$$

These are the most relevant properties of Hardy functions on a half plane. Next, we are going to refine the space of Hardy functions.

3.1. Smooth Hardy functions

The Schwartz space $\mathcal{S}$ is the set of all complex functions of a real variable, $f(x)$ satisfying the following properties:

i) Any function $f(x) \in \mathcal{S}$ is continuous and has continuous derivatives of any order at all points.

ii) Any function $f(x) \in \mathcal{S}$, as well as any of its derivatives, goes to zero at the infinity faster than the inverse of any polynomial. i.e.,

$$\lim_{|x|\to\infty} x^n \, \frac{d^m f(x)}{dx^m} = 0 \,, \qquad n, m = 0, 1, 2, \ldots \,. \tag{15}$$

Functions in $\mathcal{S}$ are called the *Schwartz functions* and have the following properties [28]:

i) $\mathcal{S}$ is a vector space over the field of complex numbers.

ii) Any function in $\mathcal{S}$ is square integrable, so that $\mathcal{S} \subset L^2(\mathbb{R})$. Furthermore, $\mathcal{S}$ is dense in $L^2(\mathbb{R})$.

iii) $\mathcal{S}$ can be endowed with a metric (there exists a distance between vectors) topology, which is *stronger* than the norm topology inherited from $L^2(\mathbb{R})$. This means that the triplet of spaces

$$\mathcal{S} \subset L^2(\mathbb{R}) \subset \mathcal{S}^\times \tag{16}$$

is a RHS. The space $\mathcal{S}^\times$ is equivalent to the space of tempered distributions, with the only (very minor) difference that elements on $\mathcal{S}^\times$ are *antilinear* mappings on $\mathcal{S}$, while tempered distributions are *linear* mappings on $\mathcal{S}$.

iv) Let $[a, b]$ any interval in the real line $\mathbb{R}$, either finite or infinite. Let $\mathcal{S}[a, b]$ the space of functions in $\mathcal{S}$ *supported* on $[a, b]$. One may prove that

$$S[a, b] \subset L^2[a, b] \subset S^\times[a, b]\,, \tag{17}$$

where $L^2[a, b]$ is the Hilbert space of complex square integrable functions on the interval $[a, b]$, is again a RHS. Recall that $[a, b]$ could be either finite or infinite, as for instance $[0, \infty)$. In particular, $S[a, b]$ *is always dense in* $L^2[a, b]$.

v) The Fourier transform of a Schwartz function is also a Schwartz function. Furthermore, the Fourier transform is an one to one onto mapping from $\mathcal{S}$ onto itself that preserve the topological structre on $\mathcal{S}$.

Now, we are in the situation to construct the spaces of smooth Hardy functions. For that purpose, we shall use the Paley-Wiener theorem as an essential ingredient.

Let us consider the Schwartz spaces $\mathcal{S}(\mathbb{R}^\mp)$ and the space of Fourier transforms of functions in $\mathcal{S}(\mathbb{R}^\mp)$, $\mathcal{F}[\mathcal{S}(\mathbb{R}^\mp)]$. These spaces have the following properties:

i) After the Paley-Wiener theorem, $\mathcal{F}[\mathcal{S}(\mathbb{R}^\mp)] \subset \mathcal{H}^2_\pm$.

ii) The Fourier transform of a Schwartz function is also another Schwartz function and this operation is bijective (one to one and onto), so that

$$\mathcal{F}[\mathcal{S}(\mathbb{R}^\mp)] \equiv \mathcal{S} \cap \mathcal{H}^2_\pm\,. \tag{18}$$

iii) The spaces $\mathcal{S}^\mp$ are dense in $L^2(\mathbb{R}^\mp)$. Since the Fourier transform is unitary, then, $\mathcal{S} \cap \mathcal{H}^2_\pm$ is dense in $\mathcal{H}^2_\pm$ with the norm topology. In addition due to the above comments on unitary mappings on RHS, the triplets:

$$\mathcal{S} \cap \mathcal{H}^2_\pm \subset \mathcal{H}^2_\pm \subset (\mathcal{S} \cap \mathcal{H}^2_\pm)^\times \tag{19}$$

are well defined RHS.

iv) Let us consider the space of restrictions to $\mathbb{R}^+$ of functions in $\mathcal{H}_\pm$ that we shall denote here as $\mathcal{H}^2_\pm|_{\mathbb{R}^+}$. After the van Winter theorem, there exists one to one onto mappings (and therefore invertible) $\theta_\pm$:

$$\theta_\pm : \mathcal{H}^2_\pm \longmapsto \mathcal{H}^2_\pm\Big|_{\mathbb{R}^+}\,, \qquad \theta_\pm : \mathcal{S} \cap \mathcal{H}^2_\pm \longmapsto \mathcal{S} \cap \mathcal{H}^2_\pm\Big|_{\mathbb{R}^+}\,. \tag{20}$$

Van Winter has also proved that $\mathcal{H}^2_\pm|_{\mathbb{R}^+}$ is dense in $L^2(\mathbb{R}^+)$. From here, we can also prove that $\mathcal{S} \cap \mathcal{H}^2_\pm|_{\mathbb{R}^+}$ is also dense in $L^2(\mathbb{R}^+)$. The topology on $\mathcal{H}^2_\pm|_{\mathbb{R}^+}$ can be transported by $\theta_\pm$ to $\mathcal{S} \cap \mathcal{H}^2_\pm\Big|_{\mathbb{R}^+}$, so that the triplets

$$\mathcal{S} \cap \mathcal{H}^2_\pm\Big|_{\mathbb{R}^+} \subset L^2(\mathbb{R}^+) \subset (\mathcal{S} \cap \mathcal{H}^2_\pm\Big|_{\mathbb{R}^+})^\times \tag{21}$$

are new RHS.

It is important to point out that, although functions in $\mathcal{S} \cap \mathcal{H}^2_\pm|_{\mathbb{R}^+}$ can be uniquely extended to the negative semi axis $\mathbb{R}^-$ (and to a half plane), *as functions in* $\mathcal{S} \cap \mathcal{H}^2_\pm\Big|_{\mathbb{R}^+}$, *we are considering their values on the positive semi axis* $\mathbb{R}^+$ *only.*

4. Gamow states and their mathematical construction

Once the metastable or decaying state or resonance (all these names represent the same object) has been prepared, it starts to decay. We may use an origin of times $t = 0$ at which the preparation of the decaying state is complete and starts to decay [3] (Although we shall stick to this notion of the origin of times, this determination is somehow ambiguous [29]). Then, assume that the state is represented at $t = 0$ by the state vector ψ. The non-decay probability at time $t > 0$ is

$$P(t) = |\langle\psi|e^{-itH}\psi\rangle|^2 , \tag{22}$$

where H is the interacting or total Hamiltonian, where $\{H_0, H\}$ is the Hamiltonian pair, responsible for the resonance scattering.

Assume that ψ is a normalized vector in a Hilbert space (in the subspace of scattering states) and that the Hamiltonian H is semi bounded, i.e., that its spectrum has a lower bound. This is a condition that have most of known quantum Hamiltonians. Then, $P(t)$ goes to zero as $t \longmapsto \infty$ [30]. The vector ψ denotes a decaying state if the non-decay probability is exponential function of the type $e^{-\alpha t}$, $\alpha > 0$. This is because the observed decay rate is exponential [3].

However, under the above conditions imposed to ψ and H, $P(t)$ cannot be exponential. It could be approximately exponential for almost all times within the range of observation. However, it is far from being exponential for very short (Zeno era) and very long (Khalfin region) times [30]. Both deviations of the exponential regime have been reported to be found experimentally [31, 32], but they are difficult to detect, the Zeno era because it is too short and the Khalfin region because it remains very little amount of undecayed material. Thus within a reasonable degree of accuracy, we may assume that resonances decay exponentially at all times. But then, *this exponential decay cannot be produced by a normalizable state vector*. The vector state that decays exponentially for all $t > 0$ is called the (decaying) Gamow state and it cannot be normalized with the usual L^2 norm.

It is well know that (non relativistic) resonances are very often associated to pairs of poles of the analytic continuation, $S(p)$, to the complex plane of the scattering operator (or scattering matrix), in the momentum representation, located symmetrically with respect to the negative imaginary axis. If we shift to the energy representation, the lower half plane is transformed into the second sheet of the two sheeted Riemann surface corresponding to the transformation $p = \sqrt{2mE}$. Then, resonances are characterized by pairs of poles of the analytic continuation of $S(E)$ to the second sheet, at values $z_R = E_R - i\Gamma/2$ and its complex conjugate, z_R^*. Here, E_R is the resonant energy and Γ the width, which gives the mean life [3]. There are some other mathematical and physical definitions of resonances [3] that not always coincide [30]. The study of the relations of between these definitions needs to be completed.

We are now in the position of defining the *decaying Gamow vector*, ψ^D, as an eigenvector of the total Hamiltonian H with the complex eigenvalue $z_R = E_R - i\Gamma/2$, i.e., $H\psi^D = z_R\,\psi^D$ and the *growing Gamow vector*, ψ^G, as an eigenvector of H with eigenvalue $z_R^* = E_R + i\Gamma/2$, $H\psi^G = z_R^*\,\psi^G$. Gamow vectors ψ^D and ψ^G cannot be normalizable as they are eigenvectors of a self adjoint Hamiltonian with complex eigenvalues. This means that Gamow vectors do not belong to the Hilbert space in which H is defined as a self adjoint operator. The advantage of this definition is that the decaying Gamow vector decays exponentially for all positive values of time (we should clarify this point later):

$$e^{-itH}\psi^D = e^{-itE_R}\,e^{-t\Gamma/2}\,\psi^D . \tag{23}$$

The fact is that Gamow vectors belong to the antiduals of two, in principle different, RHS. Now, we proceed to explain the idea of their mathematical construction.

In order to do it, we reduce the hypothesis to a minimum, further generalizations can be constructed without serious difficulties. Assume that H_0 and H have simple continuous spectrum equal to $\mathbb{R}^+ = [0,\infty)$, so that both have only scattering states. Assume that these operators are defined on certain Hilbert space $\mathcal{H}$. Then, according to a spectral theorem [28] pp. 226-227, there exists a unitary operator $U : \mathcal{H} \longmapsto L^2(\mathbb{R}^+)$ such that H is transformed into the multiplication operator on $L^2(\mathbb{R}^+)$. To understand this, let us consider $\psi(E) \in L^2(\mathbb{R}^+)$ such that $E\,\psi(E) \in L^2(\mathbb{R}^+)$ (we use the argument E to mean energy). On the space of functions with this property, let us define the multiplication operator $\mathcal{E}$ as $\mathcal{E}\psi(E) := E\psi(E)$. Then, $UHU^{-1} = \mathcal{E}$. Note that U *diagonalizes* H by this operation.

Concerning $\mathcal{E}$, the following properties are relevant:

i) The operator $\mathcal{E}$ leaves the spaces $\mathcal{S}\cap\mathcal{H}^2_\pm|_{\mathbb{R}^+}$ invariant, so that for any function $\psi_\pm(E) \in \mathcal{S}\cap\mathcal{H}^2_\pm|_{\mathbb{R}^+}$, we have that $\mathcal{E}\psi_\pm(E) = E\psi_\pm(E) \in \mathcal{S}\cap\mathcal{H}^2_\pm|_{\mathbb{R}^+}$.

ii) The operator $\mathcal{E}$ can be extended to the dual $(\mathcal{S}\cap\mathcal{H}^2_\pm|_{\mathbb{R}^+})^\times$ by means of the duality formula: If $\psi_\pm(E) \in \mathcal{S}\cap\mathcal{H}^2_\pm|_{\mathbb{R}^+}$ and $F_\pm \in (\mathcal{S}\cap\mathcal{H}^2_\pm|_{\mathbb{R}^+})^\times$, then,

$$\langle \mathcal{E}\,\psi_\pm(E)|F_\pm\rangle = \langle \psi_\pm(E)|\mathcal{E}\,F_\pm\rangle\,, \tag{24}$$

so that $\mathcal{E}$ acts on all vectors in $(\mathcal{S}\cap\mathcal{H}^2_\pm|_{\mathbb{R}^+})^\times$.

Next, we construct a new RHS as follows: Let U be the unitary operator which *diagonalizes* H as above ($UHU^{-1} = \mathcal{E}$). If $\Phi_\pm := U^{-1}[\mathcal{S}\cap\mathcal{H}^2_\pm|_{\mathbb{R}^+}]$, we consider these new RHS given by $\Phi_\pm \subset L^2(\mathbb{R}^+) \subset \Phi_\pm^\times$. Then,

$$UH\Phi_\pm = UHU^{-1}\,[\mathcal{S}\cap\mathcal{H}^2_\pm|_{\mathbb{R}^+}] = \mathcal{E}[\mathcal{S}\cap\mathcal{H}^2_\pm|_{\mathbb{R}^+}] \subset \mathcal{S}\cap\mathcal{H}^2_\pm|_{\mathbb{R}^+}$$

$$\Longrightarrow H\Phi_\pm \subset U^{-1}\,[\mathcal{S}\cap\mathcal{H}^2_\pm|_{\mathbb{R}^+}] = \Phi_\pm\,, \tag{25}$$

so that $\Phi_\pm$ reduce H ($H\Phi_\pm \subset \Phi_\pm$). By the duality formula, $\langle H\varphi_\pm|F_\pm\rangle = \langle\varphi_\pm|H\,F_\pm\rangle$, valid for all $\varphi \in \Phi_\pm$ and all $F_\pm \in \Phi_\pm^\times$, we extend H into the antiduals $\Phi_\pm^\times$.

Then, we may obtain some rigorous results that we list in the sequel. These results have been proved in some references like [18, 19]:

i) Assume that $z_R = E_R - i\Gamma/2$ is a resonance pole of the S matrix associated to the Hamiltonian pair $\{H_0, H\}$. Then, there exists a unique vector (which is a generalized function) $\delta_{z_R} \in (\mathcal{S}\cap\mathcal{H}^2_+|_{\mathbb{R}^+})^\times$ such that $\mathcal{E}\delta_{z_R} = z_R\,\delta_{z_R}$, i.e., δ_{z_R} is an eigenvector of $\mathcal{E}$ with eigenvalue z_R. This eigenvector lies in the dual $(\mathcal{S}\cap\mathcal{H}^2_+|_{\mathbb{R}^+})^\times$ and cannot be in $L^2(\mathbb{R}^+)$ due to the Hermiticity of $\mathcal{E}$. Analogously, there exists $\delta_{z_R^*} \in (\mathcal{S}\cap\mathcal{H}^2_-|_{\mathbb{R}^+})^\times$ with $\mathcal{E}\delta_{z_R^*} = z_R^*\,\delta_{z_R^*}$. The star always means complex conjugation.

ii) Let us define the *decaying Gamow vector* as $\psi^D := U^{-1}\,\delta_{z_R} \in \Phi_+^\times$. Then,

$$\mathcal{E}\delta_{z_R} = z_R\,\delta_{z_R} \Longrightarrow U^{-1}\mathcal{E}UU^{-1}\,\delta_{z_R} = z_R\,U^{-1}\delta_{z_R} \Longrightarrow H\psi^D = z_R\,\psi^D\,. \tag{26}$$

The latter identity makes sense in $\Phi_+^\times$. Analogously, we define the *growing Gamow vector* as $\psi^G := U^{-1}\,\delta_{z_R^*} \in \Phi_-^\times$. Then, we have that $H\psi^G = z_R^*\,\psi^G$, equation valid in $\Phi_-^\times$. Thus Gamow vectors are defined as the eigenvectors of the total Hamiltonian H with eigenvalues given by resonance poles of the S matrix in the energy representation.

iii) Our structures have been constructed using Hardy spaces on a half plane. Hardy spaces on a half plane split the unitary group given by $e^{-it\mathcal{E}}$, $t \in \mathbb{R}$ into two semigroups. In fact [18, 19]:

If $t \geq 0$, $e^{it\mathcal{E}}\mathcal{H}^2_+ \subset \mathcal{H}^2_+$ and also $e^{it\mathcal{E}}[\mathcal{S}\cap\mathcal{H}^2_+|_{\mathbb{R}^+}] \subset \mathcal{S}\cap\mathcal{H}^2_+|_{\mathbb{R}^+}$. However, for any $t_0 < 0$, there exists a function $\phi_+(E) \in \mathcal{S}\cap\mathcal{H}^2_+|_{\mathbb{R}^+}$, such that $e^{it_0\mathcal{E}}\phi_+(E) \notin \mathcal{S}\cap\mathcal{H}^2_+|_{\mathbb{R}^+}$.

If $t \leq 0$, $e^{it\mathcal{E}}\mathcal{H}^2_- \subset \mathcal{H}^2_-$ and also $e^{it\mathcal{E}}[\mathcal{S}\cap\mathcal{H}^2_-|_{\mathbb{R}^+}] \subset \mathcal{S}\cap\mathcal{H}^2_-|_{\mathbb{R}^+}$. However, for any $t_0 > 0$, there exists a function $\phi_-(E) \in \mathcal{S}\cap\mathcal{H}^2_-|_{\mathbb{R}^+}$, such that $e^{it_0\mathcal{E}}\phi_-(E) \notin \mathcal{S}\cap\mathcal{H}^2_-|_{\mathbb{R}^+}$.

vi) These ideas can be immediately carried to $\Phi_\pm$, since $U^{-1}\, e^{it\mathcal{E}}\, U = e^{itH}$. Thus:

If $t \geq 0$, $e^{itH}\Phi_+ \subset \Phi_+$. Furthermore for any $t_0 < 0$, there exists a vector $\varphi_+ \in \Phi_+$ such that $e^{it_0 H}\varphi_+ \notin \Phi_+$.

If $t \leq 0$, $e^{itH}\Phi_- \subset \Phi_-$. Furthermore for any $t_0 > 0$, there exists a vector $\varphi_- \in \Phi_-$ such that $e^{it_0 H}\varphi_- \notin \Phi_-$.

v) Let $\Phi \subset \mathcal{H} \subset \Phi^\times$ be a RHS and U a unitary operator on $\mathcal{H}$ such that $U^\dagger \Phi \subset \Phi$, where $U^\dagger$ is the adjoint of U. Then, U can be extended to an operator on $\Phi^\times$ by duality: If $\varphi \in \Phi$ and $F \in \Phi^\times$ are chosen arbitrarily, then, $\langle U^\dagger \varphi | F\rangle = \langle \varphi | UF\rangle$. This equation defines the action of U on $\Phi^\times$ in a unique way. In consequence:

$$e^{-itH}\Phi_+^\times \subset \Phi_+^\times \,, \text{if and only if } t > 0\,; \qquad e^{-itH}\Phi_-^\times \subset \Phi_-^\times \,, \text{if and only if } t < 0\,. \tag{27}$$

Due to the properties of Hardy functions the evolution group e^{-itH} has split into two semigroups, one for positive values of time and the other for negative values of time.

On the other hand, we may reconstruct our spaces in such a way that relations (27) be valid for all values of time [33].

vi) Now, it comes the essential property of Gamow states, which has been proved in [18, 19]:

For $t \geq 0$, the time evolution of the decaying Gamow state is given by

$$e^{-itH}\,\psi^D = e^{-itE_R}\, e^{-t\Gamma/2}\,\psi^D\,. \tag{28}$$

This means that the decaying Gamow state *has an exact exponential decay for* $t \geq 0$. Due to the use of Hardy functions in the definition of our spaces, time evolution is not defined for ψ^D at times $t < 0$.

For $t \leq 0$, the time evolution of the growing Gamow state is given by

$$e^{-itH}\,\psi^G = e^{-itE_R}\, e^{+t\Gamma/2}\,\psi^G\,. \tag{29}$$

This means that the decaying Gamow state *has an exact exponential grow for* $t \leq 0$ or an *exact exponential decay to the past.* As for ψ^D, time evolution is not defined for ψ^G at times $t > 0$.

vii) The growing Gamow vector as well as all *growing* process that takes place in $\Phi_-^\times$ for $t < 0$ should not be confused with the process of capture, preparation or creation of the resonance. In fact, it is the time reversal of the decaying process: something that *describes the same in the reverse direction of time.* In particular, if T is the time operator, $T\Phi_\pm^\times = \Phi_\mp^\times$ and $T\psi^D = \psi^G$ and also $T\psi^G = \psi^G$. Nevertheless, another interpretation is possible. This interpretation is the founding stone of Time Asymmetric Quantum Mechanics.

5. Time Asymmetric Quantum mechanics

As we have just pointed out in the previous section, we may reinterpret $\Phi_+^\times$ in a different way than being the time reversal of $\Phi_-^\times$. This interpretation was proposed by A. Bohm and collaborators in a series of papers [4, 20–23]. Thus, *the notion of time asymmetric quantum mechanics (TAQM) comes from the idea according to which the processes described by vectors in the space* $\Phi_-^\times$ *are not related with the time reversal of the decay as described by vectors in* $\Phi_+^\times$.

We have already mentioned in the Introduction that irreversibility in quantum mechanics include resonances but it goes beyond than resonances. In particular, it must consider many scattering situations in which the probability of the process in one direction of time is much smaller than the probability of its time reversal [1, 2].

In order to introduce a formulation of TAQM, we need to add to the standard formulation of Quantum Mechanics a new axiom to the existing ones. This new axiom is called the **Hardy space axiom**. Since TAQM manifest itself on scattering processes, resonate or not, it is reasonable to present it in this context and using a causality principle. This causality principle states than in any scattering experiment, first comes the preparation and then the registration.

We may assume that in our scattering experiment, the scattering Møller wave operators exists [34]. Let us denote these operators Ω_{IN} and Ω_{OUT} for the incoming and outgoing Møller operator respectively. The relation between the free and perturbed scattering states is given by these operators [34], so that we define

$$\Phi_{IN} := \Omega_{IN}^{-1}\,\Phi_- \qquad ; \qquad \Phi_{OUT} := \Omega_{OUT}^{-1}\,\Phi_+ \,. \tag{30}$$

Under the simplest conditions (no bound states, no singular spectrum, if any of these elements were present, we have to use the Hilbert space of scattering states [18]), Φ_{IN} and Φ_{OUT} are *dense* in the Hilbert space $\mathcal{H}$. These spaces define two new RHS [18, 19]

$$\Phi_{IN} \subset \mathcal{H} \subset \Phi_{IN}^\times \qquad ; \qquad \Phi_{OUT} \subset \mathcal{H} \subset \Phi_{OUT}^\times \,. \tag{31}$$

Due to the construction of the RHS (31) and the properties of the Møller operators, we have the following result:

$$\text{For } t<0\,, \quad e^{-itH_0}\,\Phi_{IN}^\times \subset \Phi_{IN}^\times\,; \quad ; \quad \text{for } t>0\,, \quad e^{itH_0}\,\Phi_{OUT} \subset \Phi_{OUT}\,. \tag{32}$$

To prove the second relation in (32), take $t > 0$ and $\phi_{out} \in \Phi_{OUT}$. We have:

$$\phi_{out}(t) = e^{itH_0}\phi_{out} = \Omega_{OUT}^{-1}\,e^{itH}\Omega_{OUT}\,\Omega_{OUT}^{-1}\,\phi_+ = \Omega_{OUT}^{-1}\,e^{itH}\,\phi_+ \in \Phi_{OUT}\,, \quad t>0\,. \tag{33}$$

To prove the first relation in (32) we proceed analogously with Φ_{IN} and then, apply the duality formula $\langle e^{itH_0}\varphi|\psi_{in}\rangle = \langle\varphi|e^{-itH_0}\psi_{in}\rangle$ valid for all $\varphi \in \Phi_{IN}$.

In order to introduce the *Hardy space axiom*, we need to do it on the preparation and registration in an scattering experiment separately.

i) *Preparation.* Before the scattering, we have to prepare states of particles to be scattered. This is done with some sort of device called the *preparation apparatus*, which produce the *incoming state* ψ_{in}, which after preparation and before entering into the interaction region, evolves freely.

Then, for incoming states the *Hardy space axiom* is stated as follows:

The incoming state ψ_{in} *belongs to a space of incoming free states,* $\Phi_{IN}^\times$

Due to the use of Hardy functions in the construction of Φ_{IN}, and therefore of $\Phi_{IN}^\times$, time evolution is defined for times $t < 0$ only, on all vectors in $\Phi_{IN}^\times$. This, for any $t < 0$, $\psi_{in}(t) = e^{-itH_0}\psi_{in}(0)$, where $\psi_{in}(0) := \psi_{in}$.

ii) *Registration.* The unstable quantum state decays after the particle leaves the interaction region. Decaying products (we use this word for general scattering experiments, not only for resonances) are registered in a registration apparatus. This registration apparatus lies outside the interaction region so that it detects free states. As a matter of fact, the registration apparatus detects the projection into the region it covers of the total outgoing state. This projection is given by a vector $\phi_{out}(0) = \phi_{out}$. And this vector is *observed*, so that we *should look at the operator* $|\phi_{out}\rangle\langle\phi_{out}|$ *as an observable.*

Then, for the registration part the *Hardy space axiom* is stated as follows:

The vector ϕ_{out} *that defines the observable* $|\phi_{out}\rangle\langle\phi_{out}|$, *lies in* Φ_{OUT}, $\phi_{out} \in \Phi_{OUT}$.

Thus, according to (32), $\phi_{out}(t) = e^{itH_0}\phi_{out}$, $t > 0$. Observables must evolve according to the Heisenberg rule of time evolution, something which is obvious here, since

$$|\phi_{out}(t)\rangle\langle\phi_{out}(t)| = e^{itH_0}\,|\phi_{out}\rangle\langle\phi_{out}|\,e^{-itH_0}\,. \tag{34}$$

Another way of seeing that the registered observable should evolve in time as an observable comes from the Born probability of measuring $|\phi_{out}(t)\rangle\langle\phi_{out}(t)|$ in the *arbitrary* (pure) state given by the density operator $\rho(t) := |\psi(t)\rangle\langle\psi(t)|$ given by ($t > 0$, $\psi(0) = \psi$):

$$\mathcal{P}_{\rho(t)}(|\phi_{out}\rangle\langle\phi_{out}|) = |\langle\phi_{out}|\,\psi_{in}(t)\rangle|^2 = \mathrm{Tr}\{\,|\phi_{out}\rangle\langle\phi_{out}|[e^{-iH_0t}|\psi\rangle\langle\psi|e^{iH_0t}]\}$$
$$= \mathrm{Tr}\{e^{itH_0}|\phi_{out}\rangle\langle\phi_{out}|e^{-iH_0t}\} = \mathcal{P}_{\rho}(|\phi_{out}(t)\rangle\langle\phi_{out}(t)|)\,. \qquad (35)$$

Thus, $|\phi_{out}\rangle\langle\phi_{out}|$ evolves for $t > 0$ as an observable. Thus, we have complete consistency.

In consequence: states are vectors $\psi_{in} \in \Phi_{IN}^{\times}$ or density operators $|\psi_{in}\rangle\langle\psi_{in}|$, the latter acting as operators from Φ_{IN} into $\Phi_{IN}^{\times}$ (this space is commonly denoted as $\mathcal{L}(\Phi_{IN}, \Phi_{IN}^{\times})$). Observables are operators $|\phi_{out}\rangle\langle\phi_{out}|$ on Φ_{OUT}. The time reverse operation transforms $\Phi_{\pm}^{\times}$ into $\Phi_{\mp}^{\times}$ and from here one deduces that it transforms $\Phi_{IN}^{\times}$ onto $\Phi_{OUT}^{\times}$ and vice versa. However, it is not true that it transforms $\Phi_{OUT}^{\times}$ onto either Φ_{IN} or $\mathcal{L}(\Phi_{IN}, \Phi_{IN}^{\times})$. Also, time reversal of observables are not states.

Finally, let us re-state the **Hardy class axiom** as:

We take as space of in-states $\Phi_{IN}^{\times}$. Any vector of $\Phi_{IN}^{\times}$ can be represented by a smooth Hardy function on the lower half plane. Analogously, we take as space of observables the operators formed by dyads $|\phi_{out}\rangle\langle\phi_{out}|$ of vectors ϕ_{out} in Φ_{OUT}. These vectors can be represented by smooth Hardy functions on the lower half plane.

In conclusion, we have summarized the essential mathematical formalism that yields to a presentation of the TAQM. We have skipped many details, in particular those concerning to scattering and resonance scattering. The final description of TAQM, as given in the present paper, is very short so that we encourage the interested reader a careful study of the original papers listed in our references.

Acknowledgements

I wish to acknowledge to Sara Cruz y Cruz and Oscar Rosas Ortiz and the whole team in UPIITA for the excellent organization of the Quantum Fest 2016. In addition, I would like to thank all them for the far beyond excellent hospitality we have received in Mexico.

References

[1] Lee TD 1981 *Particle Physics and Introduction to Field Theory* (London: Harwood Academic Publishers), pp 286-290
[2] Bohm A and Harshman N L 1998 Quantum theory in the rigged Hilbert space. Irreversibility from causality, in *Irreversibility and Causality*, Springer Lecture Notes in Physics **504** (Berlin: Springer)
[3] Bohm A 2001 *Quantum Mechanics: Foundations and Applications* 3rd Ed. (Berlin: Springer)
[4] Bohm A, Erman F and Uncu H 2011 *Turk J Phys* **35** 209
[5] Friedrichs KO 1948 *Commun Pure Appl Math* **1** 361
[6] Horwitz L P and Marchand P 1971 *Rocky Mountain J Math* **1** 225
[7] Gadella M and Pronko G P 2011 Fort Phys **59** 795
[8] Exner P 1985 *Open Quantum Systems and Feynman Integrals* (Reidel: Dordrecht)
[9] Castagnino M et al 2001 *J Phys A: Math Gen* **34** 10067
[10] Gadella M 2015 *Found Phys* **45** 177
[11] Gelfand I M 1964 *Generalized Functions, vol 4* (New York: Academic)
[12] Bohm A 1978 *The Rigged Hilbert Space and Quantum Mechanics*, Springer Lecture Notes in Physics **78** (New York: Springer)
[13] Roberts J E 1966 *Commun Math Phys* **3** 98
[14] Antoine J P 1969 *J Math Phys* **10** 53
[15] Melsheimer O 1974 *J Math Phys* **15** 902
[16] Gadella M and Gómez F 2002 *Found Phys* **32** 815
[17] Bohm A 1981 *J Math Phys* **22** 2813

[18] Bohm A and Gadella M 1989 *Dirac Kets, Gamow Vectors and Gelfand triplets*, Springer Lecture Notes in Physics **348** (Berlin: Springer)
[19] Civitarese O and Gadella M 2004 *Phys. Rep* **396** 41
[20] Bohm A 1999 *Phys Rev A* **60** 861
[21] Bohm A, Loewe M and van de Ven B 2003 *Fortschr Phys* **51** 551
[22] Bohm A, Kaldass H and Wickramasekara S 2003 *Fortschr Phys* **51** 569
[23] Bohm A, Gadella M and Kielanowski P 2011 *SIGMA* **7** 086
[24] Koosis P 1998 *Introduction to H^p Spaces* (Cambridge: Cambridge University Press)
[25] Duren PL 1970 *Theory of H^p Spaces* (New York: Academic)
[26] Paley R, Wiener N 1934 *Fourier Transforms in the Complex Domain* (Providence: American Mathematical Society)
[27] Van Winter C 1971 *Trans Am Math Soc* **162** 103
[28] Reed M and Simon B 1972 *Functional Analysis* (New York: Academic)
[29] Bohm A and Bryant P 2011 *Int J Theor Phys* **50** 2094
[30] Fonda L, Ghirardi GC and Rimini A 1978 *Rep Prog Phys* **41** 587
[31] Fischer MC, Gutiérrez-Medina B and Raizen MG 2001 *Phys Rev Lett* **87** 040402
[32] Rothe C, Hintschich S L and Monkman A P 2006 *Phys Rev Lett* **96** 163601
[33] Gadella M and Laura R 2001 *Int J Quan Chem* **81** 307
[34] Amrein W O, Jauch J and Sinha B 1977 *Scattering Theory in Quantum Mechanics* (New York: Benjamin)

IOP Conf. Series: Journal of Physics: Conf. Series **839** (2017) 012002 doi:10.1088/1742-6596/839/1/012002

A Short Introduction to Hilbert Space Theory

M Loaiza

Department of Mathematics, Cinvestav, Av. Instituto Politécnico Nacional 2508, México City, Mexico

E-mail: mloaiza@math.cinvestav.mx

Abstract. We present a short introduction to Hilbert spaces and linear operators defined on them.

1. Introduction

Hilbert spaces are the closest generalization to infinite dimensional spaces of the Euclidean spaces. These notes were written for students wishing a basic introduction to Hilbert space theory but who have no knowledge of Banach spaces. First, we consider a normed space and we see that if the space is finite dimensional all the norms defined on it generate the same topology, so that the convergence of sequences does not depend on the norm that is used. Later we consider linear transformations defined in a normed space and we see that all of them are continuous if the space is finite dimensional. Continuous linear transformations are called bounded operators and, by introducing a norm in the space of all bounded operators, we convert it into a Banach space under certain conditions. Once we have clear the concepts of convergence of a sequence and completeness of the space, we define Hilbert spaces and consider some of their properties. Mainly we focus on the Pythagorean theorem, the Cauchy-Schwarz inequality, the Parallelogram Identity and in the introduction of the concept of basis for a Hilbert space. Furthermore, we show that every Hilbert space of dimension n is isomorphic to $\mathbb{C}^n$ and that every separable Hilbert space is isomorphic to ℓ^2, the space of all square summable sequences.

One of the main theorems related to Hilbert spaces is the Riesz Representation Theorem, which characterizes the continuous linear transformations defined in a Hilbert space with values in $\mathbb{C}$. This theorem allows us to define the adjoint operator of a bounded operator. Finally we study the spectrum of an operator and we see some of the properties of the operator in terms of its spectrum.

The results included here are classical and can be found in many books of functional analysis.

2. Metric spaces and continuous functions

Let X be a set, *a metric (or distance)* d, defined in X, is a function $d : X \times X \to \mathbb{R}_+ = \{x \in \mathbb{R} \,|\, x \geq 0\}$ such that

(i) $d(x, y) = d(y, x)$,

(ii) $d(x, z) \leq d(x, y) + d(y, z)$,

(iii) $d(x, y) = 0$ if and only if $x = y$.

IOP Conf. Series: Journal of Physics: Conf. Series **839** (2017) 012002 doi:10.1088/1742-6596/839/1/012002

The classical examples of metrics defined in $\mathbb{R}^n$ are the following: for $a = (a_1, \ldots, a_n), b = (b_1, \ldots, b_n)$

$$d_1(a,b) = \max\{|a_i - b_i|, i = 1, \ldots, n\},$$
$$d_2(a,b) = \sum_{k=1}^{n} |a_i - b_i|,$$
$$d_3(a,b) = \left(\sum_{k=1}^{n} (a_i - b_i)^2 \right)^{1/2}.$$

Every set is a metric space with the *discrete metric*:

$$d(x,y) = \begin{cases} 1 & x \neq y, \\ 0 & x = y. \end{cases}$$

Let (X,d) be a metric space, $x \in X$ and $r > 0$. The *open ball* centered at the point x and with radius r is the set $B_r(x) = B(x,r) = \{y \in X : d(x,y) < r\}$. The following figure shows $B_1(0)$ in $\mathbb{R}^2$ with the metrics d_1, d_2 and d_3 respectively.

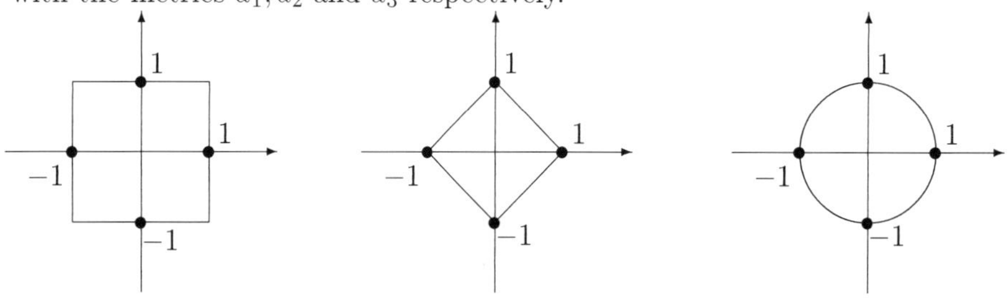

Open sets in a metric space (X,d) are defined as follows. A set $U \subset X$ is called open if, for each $x \in U$ there exists a radius $r > 0$, depending on x, such that $B_r(x) \subset U$. Observe that

(i) The union of open sets is an open set.
(ii) The intersection of a finite collection of open sets is an open set.
(iii) X and $\emptyset$ are open sets.

A family τ of subsets of X with properties i), ii) and iii) is called a *topology* for X.

For example, if X has the discrete metric then, any subset of X is open since

$$B_r(x) = \begin{cases} \{x\} \text{ if } r < 1, \\ X \text{ if } r \geq 1. \end{cases}$$

Different metrics defined on a set can produce the same topology, for example, in $\mathbb{R}^2$, the metrics d_1, d_2 and d_3 generate the same topology.

Let (x_n) be a sequence of points in a metric space (X,d). We say that the sequence (x_n) *converges* to x, and write $x_n \to x$, if given $\epsilon > 0$ there exists $N > 0$ such that for $n \geq N$, $d(x_n, x) < \epsilon$. A function $f : (X, d_X) \to (Y, d_Y)$ is called *continuous at the point* $x \in X$ if given any sequence (x_n) such that $x_n \to x$ then $f(x_n) \to f(x)$.

A sequence (x_n) in a metric space (X,d) is called a *Cauchy sequence* if given $\epsilon > 0$ there exists N such that for $n, m > N, d(x_n, x_m) < \epsilon$. Any convergent sequence is a Cauchy sequence and since the elements of a Cauchy sequence are getting closer and closer when n tends to ∞, it is natural to think that every Cauchy sequence is convergent. However, the following

IOP Conf. Series: Journal of Physics: Conf. Series **839** (2017) 012002 doi:10.1088/1742-6596/839/1/012002

example shows that this is not the case. Let P be the set of polynomials in $[0,1]$ with the metric $d(p,q) = \sup_{x\in[0,1]} |p(x) - q(x)|$ and consider the sequence $p_n(x) = \sum_{k=0}^{n} \frac{x^k}{k!}$. Note that $p_n(x) \to e^x$, which is not a polynomial anymore. Another classical example of a Cauchy sequence that is not convergent is obtained by considering the vector space $\mathbb{Q}$, consisting of all rational numbers and an irrational point $x \in \mathbb{R}\setminus\mathbb{Q}$. The decimal expansion of x, $x = a.a_1a_2a_3...$ is the limit of the sequence of rational numbers $b_n = a.a_1 \ldots a_n$. Thus, (b_n) is a Cauchy sequence that does not converge in $\mathbb{Q}$. Metric spaces such that every Cauchy sequence is convergent are called *complete spaces.*

Among all metric spaces those that have the structure of a vector space are quite important. Most of the examples given above are vector spaces. Let X be a vector space over $\mathbb{C}$ or $\mathbb{R}$. A norm $\| \ \|$ defined in X is a function $\| \ \| : X \to \mathbb{R}_+$, such that

(i) $\|x\| \geq 0, \|x\| = 0$ if and only if $x = 0$,

(ii) $\|\lambda x\| = |\lambda|\|x\|$,

(iii) $\|x + y\| \leq \|x\| + \|y\|$.

A normed space has a direct relation between its algebraic structure and its topological structure. A normed space which is a complete metric space with the distance $d(x,y) = \|x - y\|$ is called a *Banach space.* The most natural examples of Banach spaces are $\mathbb{C}^n, \mathbb{R}^n$ with the Euclidean norm. Actually, they are Banach spaces with any norm defined in them. This property is not exclusive of these spaces, every finite dimensional normed space is a Banach space (see [3], Theorem 2.4-2, for details).

As an example of an infinite dimensional Banach space we have $C([0,1])$, the set of all complex-valued continuous functions defined on $[0,1]$ with the uniform norm, $\|f\|_\infty = \sup_{x\in[0,1]} |f(x)|$.

Consider now the set ℓ^2, consisting of all complex valued sequences (a_n) such that $\sum_{n=1}^{\infty} |a_n|^2 < \infty$, with the norm

$$\|(a_n)\| = \sqrt{\sum_{n=1}^{\infty} |a_n|^2}.$$

This space is a vector space with the following operations: if $a = (a_n), b = (b_n) \in \ell^2$ and $\lambda \in \mathbb{C}$ then

$$a + b = (a_n + b_n),$$
$$\lambda a = (\lambda a_n).$$

The space ℓ^2 is a Banach space that will play a very important role in the characterization of separable Hilbert spaces, in the sense we define later on.

Let (x_n) be a sequence in a normed space X, the series $\sum_{n=1}^{\infty} x_n$ is said to be *convergent* if the sequence (s_n), where $s_n = x_1 + x_2 + \cdots + x_n$, is convergent. The series $\sum_{n=1}^{\infty} x_n$ is said to be *absolutely convergent* if $\sum_{n=1}^{\infty} \|x_n\|$ converges. The following theorem, which proof follows easily, characterizes Banach spaces.

Theorem 2.1 *A normed space is a Banach space if and only if every absolutely convergent sequence is convergent.*

3. Hilbert spaces

For simplicity from now on we only consider vector spaces over $\mathbb{C}$. Let V be a vector space, an *inner product* is a function $\langle , \rangle : V \times V \to \mathbb{C}$ such that for $u, v, w \in V$ and $\lambda \in \mathbb{C}$,

(i) $\langle u, v\rangle = \overline{\langle v, u\rangle}$;
(ii) $\langle \lambda u, v\rangle = \lambda\langle u, v\rangle$;
(iii) $\langle u+v, w\rangle = \langle u, w\rangle + \langle v, w\rangle$;
(iv) $\langle u, u\rangle \geq 0$ and $\langle u, u\rangle = 0$ if and only if $u = 0$.

One of the most important inequalities in an inner product space is the *Cauchy-Schwarz inequality*. It states the following, for any u, v in an inner product space

$$|\langle u, v\rangle| \leq \|u\|\|v\|.$$

Using last inequality it is easy to prove that $\|u\| = \sqrt{\langle u, u\rangle}$ is a norm in V.

A *Hilbert space* is a vector space with inner product $\langle\, ,\, \rangle$ such that it is a Banach space with the norm induced by the inner product.

The simplest example of a Hilbert space is $\mathbb{C}^n$ with the inner product:

$$\langle x, y\rangle = \sum_{i-1}^{n} x_i\overline{y_i},$$

where $x = (x_1, \ldots, x_n)$ and $y = (y_1, \ldots, y_n)$. The metric induced by this inner product is the usual Euclidean metric in $\mathbb{C}^n$.

The space ℓ^2 is an infinite dimensional vector space that is a Hilbert space with the product:

$$\langle a, b\rangle = \sum_n a_n\overline{b_n}, \text{ where } a = (a_1, a_2, \ldots), b = (b_1, b_2, \ldots).$$

It is easy to prove that if $\|\ \|$ is a norm induced by an inner product then,

$$2\|u\|^2 + 2\|v\|^2 = \|u+v\|^2 + \|u-v\|^2.$$

The latter equation is called the *Parallelogram Identity* and is an important tool to determine if a Banach space is a Hilbert space or not. For example, the space of all continuous functions defined on $[0, 1]$ is not a Hilbert space since it does not hold the Parallelogram Identity. Indeed, let $f(x) = x$, $g(x) = 1$, then $\|f+g\| = 2$, $\|f-g\| = 1$, $\|f\| = 1$, $\|g\| = 1$.

If a norm holds the Parallelogram Identity this norm defines an inner product in the space by the *polarization identity*

$$\langle x, y\rangle = \frac{1}{4}(\|x+y\|^2 - \|x-y\|^2 + i\|x+iy\|^2 - i\|x-iy\|^2).$$

Two points x and y in an inner product space are *orthogonal* if $\langle x, y\rangle = 0$. A set of vectors $\{x_\alpha\}_{\alpha\in I}$ is *orthonormal* if $\|x_\alpha\| = 1$ for each $\alpha \in I$, and $\langle x_\alpha, x_\beta\rangle = 0$ if $\alpha \neq \beta$. Observe that, if x, y are orthogonal then,

$$\|x+y\|^2 = \|x\|^2 + \|y\|^2,$$

which is the Pythagorean Theorem.

An orthonormal set in a Hilbert space $\mathcal{H}$ is called a *basis* of $\mathcal{H}$ if the set of all linear combinations of its elements is dense in $\mathcal{H}$. It means that we can approximate each element of $\mathcal{H}$ by linear combinations of the elements of its basis. If $\mathcal{H}$ is not a finite dimensional vector space, a basis in the sense given above is never a basis of $\mathcal{H}$ as a vector space or Hamel basis (see [1] exercise 16, Chapter 1, Section 4). Such as in the case of vector spaces a Hilbert space always has a basis.

IOP Conf. Series: Journal of Physics: Conf. Series **839** (2017) 012002 doi:10.1088/1742-6596/839/1/012002

The space $\mathbb{C}^n$ is a finite dimensional Hilbert space. A basis of $\mathbb{C}^n$ is $\{e_m\}_{m=1}^n$ where

$$e_m = (0, \ldots, 0, 1, 0, \ldots, 0).$$

In every vector space one can write an element of the space as a linear combination of the elements of the basis. If $\mathcal{H}$ is a Hilbert space which basis is not finite, we can write each element in terms of the elements of the basis.

In the following theorem, which proof can be found in [6], Theorems 4.14 and 4.15, the series $\sum_{n=1}^{\infty} a_n e_n$ has the usual meaning

$$\sum_{n=1}^{\infty} a_n e_n = \lim_{n\to\infty} \sum_{n=1}^{m} a_n e_n.$$

We consider a Hilbert space that has a numerable basis.

Theorem 3.1 *If $\{e_n\}_{n\geq 1}$ is a basis of a Hilbert space $\mathcal{H}$ then,*

(i) Each element $a \in \mathcal{H}$ can be written uniquely as

$$a = \sum_{n=1}^{\infty} a_n e_n, \; a_n = \langle a, e_n \rangle.$$

(ii) The inner product of two elements $a = \sum_{n=1}^{\infty} a_n e_n, b = \sum_{n=1}^{\infty} b_n e_n$, is given by

$$\langle a, b \rangle = \sum_{n=1}^{\infty} a_n \overline{b_n}$$

and

$$\|a\|^2 = \sum_{n=1}^{\infty} |a_n|^2$$

For simplicity, in last theorem we considered a Hilbert space with a numerable basis. The general version of this theorem can be found in [1], Chapter 1, Lemma 4.13.

A Hilbert space with numerable or finite basis is called *separable*. On the other hand, a metric space is called separable if it has a numerable dense set. Since any Hilbert space is a metric space we could get confused with these two definitions of separability. However, a Hilbert space is separable if and only if, it is separable as metric space. It is easy to see that the set of all linear combinations of elements of the basis of $\mathcal{H}$, whose coefficients are complex numbers with rational components, is a numerable dense set in $\mathcal{H}$.

Let $(\mathcal{H}, \langle\, , \,\rangle_{\mathcal{H}})$ and $(\mathcal{K}, \langle\, , \,\rangle_{\mathcal{K}})$ be two Hilbert spaces. A linear operator $T : \mathcal{H}$ onto $\mathcal{K}$ is called an isomorphism if, for each $x, y \in \mathcal{H}$

$$\langle x, y \rangle_{\mathcal{H}} = \langle T(x), T(y) \rangle_{\mathcal{H}}.$$

Theorem 3.2 *Let $\mathcal{H}$ be a separable Hilbert space. Then,*

(i) If the dimension of the space is n then, $\mathcal{H}$ is isomorphic to $\mathbb{C}^n$.

(ii) If the space has a numerable basis, then $\mathcal{H}$ is isomorphic to ℓ^2.

Proof. We only consider the infinite dimensional case. Let $\{e_n\}_{n\in\mathbb{N}}$ be an orthonormal basis for $\mathcal{H}$. Define $T:\mathcal{H}\to\ell^2$ by

$$T(y) = (\langle y, e_n\rangle)_{n\in\mathbb{N}}, \text{ if } y = \sum_{n=1}^{\infty}\langle y, e_n\rangle e_n.$$

Then,

$$\begin{aligned}\langle T(x), T(y)\rangle &= \sum_{n=1}^{\infty}\langle x, e_n\rangle\overline{\langle y, e_n\rangle} = \sum_{n=1}^{\infty}\langle x, \langle y, e_n\rangle e_n\rangle \\ &= \langle x, \sum_{n=1}^{\infty}\langle y, e_n\rangle e_n\rangle = \langle x, y\rangle.\end{aligned}$$

Thus, $\mathcal{H}$ is isomorphic to ℓ^2. □

The latter theorem is very important concerning the Hilbert structure of the space. However, most of the Hilbert spaces used in applications have additional properties related to the elements of the space. For example, the Hilbert space could be the Bergman space, the Harmonic space, the Hardy space, etc. In any case the nature of the elements in the Hilbert space is very important.

4. Orthogonal projections

A set A in a metric (in general in a topological space) is *closed* if its complement is an open set. For the special case of metric spaces we can prove that a set A is closed using sequences of points in A. Indeed, A is closed if and only if for any convergent sequence (x_n) of points in A the limit of the sequence is also a point in A.

For infinite dimensional Banach spaces we can find non closed subspaces. For example, the set of all polynomials in $C[0,1]$ is a non closed subspace.

Theorem 4.1 *Any finite dimensional subspace of a Banach space is closed.*

Let $\mathcal{H}$ be a Hilbert space and $M\subset\mathcal{H}$. For a point $x\in\mathcal{H}$ the *distance* from x to M is defined by

$$d(x, M) = \inf\{d(x, m)|m\in M\}.$$

Theorem 4.2 *Let $\mathcal{H}$ be a Hilbert space and M a closed subspace of $\mathcal{H}$. Let $x\in\mathcal{H}\setminus M$, then there exists a unique point $z\in M$ such that*

$$\|x - z\| = \inf\{d(h, m)|m\in M\} = d(x, M) := d.$$

Proof. Let (y_n) be a sequence of points in M such that

$$\lim_{n\to\infty}\|x - y_m\| = d.$$

Then, using the Parallelogram Identity, we have that

$$\|y_n + y_m - 2x\|^2 + \|y_n - y_m\|^2 = 2\|y_n - x\|^2 + 2\|y_m - x\|^2.$$

This fact implies that

$$\|y_n - y_m\|^2 = 2\|y_n - x\|^2 + 2\|y_m - x\|^2 - 4\|\frac{y_n + y_m}{2} - x\|^2.$$

Then,

$$\|y_n - y_m\|^2 \leq 2\|y_n - x\|^2 + 2\|y_m - x\|^2 - 4d^2 \to 0 \text{ when } n, m \to \infty.$$

Thus, there exists $z \in M$ such that $y_n \to z$. Then,

$$d(x, M) = \lim_{n\to\infty} \|x - y_n\| = \|x - \lim_{n\to\infty} y_n\| = \|x - z\|.$$

Assume that there exists $z_1 \in M$ such that $d = \|x - z_1\|$. Then, the sequence $z, z_1, z, z_1, z, z_1, \ldots$ is such that $d = \lim_{n\to\infty} \|x - w_n\|$ where,

$$w_n = \begin{cases} z & \text{if } n \text{ is even,} \\ z_1 & \text{if } n \text{ is odd.} \end{cases}$$

Since $(x - w_n)$ converges, it is a Cauchy sequence which implies that (w_n) is a Cauchy sequence, then $z = z_1$. □

The existence of the element z in last theorem looks quite natural for normed spaces, this fact is not true for general Banach examples. The reader can find an example in [6], exercise 3.9. The reason our intuition fails is because we think of distances as Euclidean distances.

Let $M \subset \mathcal{H}$, the *orthogonal complement* of M is the set

$$M^\perp = \{h \in \mathcal{H} | \langle h, m\rangle = 0 \,\forall\, m \in M\}.$$

The orthogonal complement of M holds the following properties:

(i) $M^\perp$ is a closed subspace of $\mathcal{H}$.
(ii) If M is a closed subspace of $\mathcal{H}$ then, $(M^\perp)^\perp = M$.
(iii) $M \cap M^\perp = \{0\}$.

Theorem 4.3 *Let $\mathcal{H}$ be a Hilbert space and M a closed subspace of $\mathcal{H}$. Then, every element $x \in \mathcal{H}$ is written uniquely as*

$$x = z + w,\ z \in M \text{ and } w \in M^\perp.$$

That is, $\mathcal{H} = M \oplus M^\perp$.

Proof. Let $x \in \mathcal{H}$ and $z \in M$ such that

$$\|x - z\| = d(x, z) = d(x, M).$$

Let $w = x - z$, then using [6], Lemma 4.23, $w \in M^\perp$.

If $x = z_1 + w_1$, $z_1 \in M$ and $w_1 \in M^\perp$. Then,

$$z + w = z_1 + w_1,$$

Then

$$w - w_1 = z_1 - z \in M \cap M^\perp = \{0\}.$$

□

Using last decomposition of $\mathcal{H}$ as a direct sum of M and $M^\perp$, the *orthogonal projection* P_M from $\mathcal{H}$ onto M is defined as

$$P_M(x) = z.$$

Observe that $\|x\|^2 = \|z+w\|^2 = \|z\|^2 + \|w\|^2 \geq \|P_M(x)\|^2$, which implies that P_M is a continuous linear operator.

IOP Conf. Series: Journal of Physics: Conf. Series **839** (2017) 012002 doi:10.1088/1742-6596/839/1/012002

5. Continuous linear functionals and the Riesz representation theorem

Let $\mathcal{H}$ be a Hilbert space, a linear function $\varphi : \mathcal{H} \to \mathbb{C}$ is called a functional (or linear functional). For example, let $\mathbb{C}^n, n_0 \in \{1, 2, \ldots, n\}$. Define

$$\varphi(z_1, z_2, \ldots, z_n) = z_{n_0}.$$

Then

$$\varphi(z) = \langle z, e_{n_0} \rangle,$$

where

$$e_{n_0} = (0, \ldots, 0, 1, 0, \ldots, 0), \text{ where the } n_0 \text{ component of } e_{n_0} \text{ equals } 1.$$

The functional φ is continuous.

Theorem 5.1 *Let $\mathcal{H}$ be a Hilbert space and $\varphi : \mathcal{H} \to \mathbb{C}$ a continuous functional. Then,*

$$\dim N(\varphi)^\perp = 1,$$

where $N(\varphi) = \{h \in \mathcal{H} | \varphi(h) = 0\}$.

Proof. It is easy to see that $N(\varphi)$ is a closed subspace of $\mathcal{H}$. Since $\varphi \neq 0, N(\varphi) \neq \mathcal{H}$. Then, $N(\varphi)^\perp \neq \{0\}$ (recall that $\mathcal{H} = N(\varphi) \oplus N(\varphi)^\perp$). Take $x_1 \neq 0, x_2 \neq 0$ in $N(\varphi)^\perp$. Then, there exists a complex number $a \neq 0$ such that

$$\varphi(x_1) + a\varphi(x_2) = 0.$$

Then $\varphi(x_1 + ax_2) = 0$, which implies that $x_1 + ax_2 \in N(\varphi) \cap N(\varphi)^\perp = \{0\}$. Thus, $\dim N(\varphi)^\perp = 1$. □

The following theorem is called the Riesz Representation Theorem.

Theorem 5.2 *Let $\mathcal{H}$ be a Hilbert space and $\varphi : \mathcal{H} \to \mathbb{C}$ a continuous linear functional. Then, there exists a unique $z \in \mathcal{H}$ such that*

$$\varphi(x) = \langle x, z \rangle \, \forall x \in \mathcal{H}.$$

Proof. Assume that $\varphi \neq 0$ then, $N(\varphi)^\perp$ has dimension 1. Take z_0 in $N(\varphi)^\perp$ with $\|z_0\| = 1$. Let $x \in \mathcal{H}$ then,

$$x = x - \langle x, z_0 \rangle z_0 + \langle x, z_0 \rangle z_0.$$

Observe that $\langle x - \langle x, z_0 \rangle z_0, z_0 \rangle = \langle x, z_0 \rangle - \langle x, z_0 \rangle \langle z_0, z_0 \rangle = 0$. Thus

$$x - \langle x, z_0 \rangle z_0 \in (N(\varphi)^\perp)^\perp = N(\varphi).$$

Then,

$$\varphi(x) = \varphi(\langle x, z_0 \rangle z_0) = \langle x, z_0 \rangle \varphi(z_0) = \langle x, \overline{\varphi(z_0)} z_0 \rangle.$$

Let $z = \overline{\varphi(z_0)} z_0$. If $\varphi(x) = \langle x, z_1 \rangle$ then

$$\langle x, z - z_1 \rangle = 0 \, \forall x \in \mathcal{H}.$$

In particular $\langle z - z_1, z - z_1 \rangle = 0$ which implies that $z = z_1$. □

IOP Conf. Series: Journal of Physics: Conf. Series **839** (2017) 012002 doi:10.1088/1742-6596/839/1/012002

6. Bounded operators

Let $(X, \| \|_X), (Y, \| \|_Y)$ be two normed spaces. A linear transformation $T : X \to Y$ is called *bounded* if there exists $C > 0$ such that

$$\|T(x)\|_Y \leq C\|x\|_X, \qquad \forall x \in X. \tag{6.1}$$

For example,

(i) the identity operator $I : X \to X$ is bounded.
(ii) Let $x_0 \in [a, b]$, the function $T : C[a, b] \to \mathbb{C}$ given by $T(f) = f(x_0)$ is a bounded operator.

It is easy to prove that a linear operator T is bounded if and only if $T(M)$ is bounded for any bounded set M. The following theorem which proof can be found in any book of functional analysis, characterizes bounded operators.

Theorem 6.1 *Let $(X, \| \|_X), (Y, \| \|_Y)$ be two normed spaces and $T : X \to Y$ be a linear operator. The following sentences are equivalent.*

(i) A linear operator T is bounded.
(ii) The operator T is continuous.
(iii) The operator T is continuous at the point 0.
(iv) The operator T is continuous at some point x_0.

The set $\mathcal{B}(X, Y)$ consisting of all bounded operators from X to Y has a natural vector space structure. Furthermore, if $X = Y$, $\mathcal{B}(X) := \mathcal{B}(X, X)$ is an algebra, with the product of two elements given by its composition. Besides, the norm in this space is defined as

$$\|T\| = \inf\{C : C \text{ holds } (6.1)\}.$$

The latter norm can be calculated in many forms.

Theorem 6.2 .

(i) $\|T\| = \sup_{\|x\| \leq 1} \|T(x)\|$.
(ii) $\|T\| = \sup_{\|x\| = 1} \|T(x)\|$.
(iii) $\|T\| = \sup_{x \neq 0} \frac{\|T(x)\|}{\|x\|}$.

The space $\mathcal{B}(X, Y)$ is a Banach space if and only if Y is a Banach space. As we mentioned above, $\mathcal{B}(X)$ is an algebra. Besides, if $S, T \in \mathcal{B}(X)$ then

$$\|ST\| \leq \|S\|\|T\|,$$

which makes $\mathcal{B}(X)$ a Banach algebra if X is a Banach space. For a detailed introduction to Banach algebras see [7].

As an example of a bounded operator, consider the set $M_{n,m}(\mathbb{C})$ consisting of all $n \times m$ complex valued matrices. Each linear transformation $T_A : \mathbb{C}^m \to \mathbb{C}^n$ is given by

$$T_A(x) = Ax,$$

where A is the matrix of the linear transformation in the standard basis for $\mathbb{C}^n$ and $\mathbb{C}^m$:

$$Ax = \begin{pmatrix} a_{11} & a_{12} & \dots & a_{1m} \\ a_{21} & a_{22} & \dots & a_{2m} \\ \vdots & \vdots & \dots & \vdots \\ a_{n1} & a_{n2} & \dots & a_{nm} \end{pmatrix} \begin{pmatrix} x_1 \\ x_2 \\ \vdots \\ x_m \end{pmatrix} = \begin{pmatrix} a_{11}x_1 + a_{12}x_2 + \dots + a_{1m}x_m \\ a_{21}x_1 + a_{22}x_2 + \dots + a_{2m}x_m \\ \vdots \\ a_{n1}x_1 + a_{n2}x_2 + \dots + a_{nm}x_m \end{pmatrix}. \tag{6.2}$$

It is easy to see that T_A is a bounded operator. Thus, every linear transformation $T : \mathbb{C}^m \to \mathbb{C}^n$ is bounded. We mentioned above that any two norms defined on a finite dimensional vector space generate the same family of open sets. The following theorem is a consequence of this fact.

Theorem 6.3 *If X is a finite dimensional normed space then, any linear operator T defined in X, is bounded.*

The latter theorem is not valid for infinite dimensional normed spaces. For example, let $P[0,1]$ be the set of all polynomials, defined on the interval $[0,1]$, with the norm

$$\|p\| = \sup_{x\in[0,1]} |p(x)|.$$

And define $T : P[0,1] \to P[0,1]$ by $T(f) = f'$.

For each $n \in \mathbb{N}$, take $f_n(x) = x^n$. Observe that $\|f_n\| = 1$ but $\|T(f_n)\| = \|nx^{n-1}\| = n$, which shows that T is not bounded.

Two important examples of bounded operators defined in ℓ^2 are the shift operators:

$$S(a_1, a_2, a_3, \dots) = (0, a_1, a_2, a_3, \dots), \tag{6.3}$$

$$T(a_1, a_2, a_3, \dots) = (a_2, a_3, \dots). \tag{6.4}$$

Theorem 6.4 *Let $(\mathcal{H}, \langle , \rangle)$ be a Hilbert space and $T : \mathcal{H} \to \mathcal{H}$ be a bounded operator. Then, there exists a unique bounded operator $T^* : \mathcal{H} \to \mathcal{H}$ such that for $x, y \in \mathcal{H}$,*

$$\langle T(x), y\rangle = \langle x, T^*(y)\rangle.$$

Proof. Take $y \in \mathcal{H}$ and define

$$L(x) = \langle T(x), y\rangle.$$

Then,

$$\|L(x)\| \le \|T(x)\|\|y\| \le \|T\|\|x\|\|y\|.$$

Using the Riesz Representation Theorem, there exists a unique $h \in \mathcal{H}$ such that

$$L(x) = \langle x, h\rangle.$$

Define $T^*(y) = h$. A detailed proof of the boundedness of T^* can be found in [5], Ch. 12. □

For example, consider the shift operators S, T, defined in (6.3) and (6.4) respectively. Then,

$$\langle S(a_1, a_2, a_3, \dots), (b_1, b_2, b_3, \dots)\rangle = \sum_{n=1}^{\infty} a_n \overline{b_{n+1}} = \langle (a_1, a_2, a_3, \dots), (b_2, b_3, b_4, \dots)\rangle.$$

Thus $S^* = T$. Similar calculations show that $T^* = S$, which means $(T^*)^* = T$.

Actually the operation $T \mapsto T^*$ is an involution in $\mathcal{B}(\mathcal{H})$. That means, for $S, T \in \mathcal{B}(\mathcal{H})$

IOP Conf. Series: Journal of Physics: Conf. Series **839** (2017) 012002 doi:10.1088/1742-6596/839/1/012002

(i) $(T^*)^* = T$,
(ii) $(ST)^* = T^*S^*$,
(iii) $(S+T)^* = S^* + T^*$,
(iv) $(\lambda T)^* = \overline{\lambda} T^*$.

Besides, it is easy to see that $\|T^*\| = \|T\|$ and that $\|T^*T\| = \|T\|^2$. The last properties of the adjoint operator and the fact that $\mathcal{B}(\mathcal{H})$ is a Banach algebra imply that $\mathcal{B}(\mathcal{H})$ is a C*-algebra, see [4] for details.

An operator $T \in \mathcal{B}(\mathcal{H})$ is said to be

(i) *self adjoint or hermitian* if $T = T^*$,
(ii) *normal* if $TT^* = T^*T$,
(iii) *unitary* if $T^*T = TT^* = I$.

7. The spectrum of a linear operator

Let X be a Banach space the operator $T \in \mathcal{B}(X)$ is called *invertible* if there exists $S \in \mathcal{B}(X)$ such that

$$ST = I \text{ and } TS = I,$$

where I denotes the identity operator in X. If X is a finite dimensional normed space and $T : X \to X$ is a linear operator, the following statements are equivalent

(i) T is invertible,
(ii) T is injective,
(iii) T is surjective,
(iv) there exists $S \in \mathcal{B}(X)$ such that $ST = I$,
(v) there exists $S \in \mathcal{B}(X)$ such that $TS = I$.

The last sentences are not equivalent if X is not a finite dimensional normed space. For example, the operators S and T defined in (6.3) and (6.4) respectively satisfy $ST \neq I$ and $TS = I$.

Let $T \in \mathcal{B}(X)$, the *spectrum* of T is defined as the set

$$\sigma(T) = \{\lambda \in \mathbb{C} | T - \lambda I \text{ is not invertible}\}.$$

For example, if we have the linear transformation $T_A : \mathbb{C}^n \to \mathbb{C}^n$, given in (6.2), the spectrum of this operator is the set of all eigenvalues of the matrix A.

The following theorem generalizes the formula

$$\frac{1}{1-z} = \sum_{n=0}^{\infty} z^n, \text{ if } |z| < 1.$$

Theorem 7.1 *If $T \in \mathcal{B}(X)$ is such that $\|T\| < 1$ then, $I - T \in \mathrm{Inv}(\mathcal{B}(X))$ and*

$$(I-T)^{-1} = \sum_{n=0}^{\infty} T^n.$$

Proof. Since $\|T\| < 1$ and $\|T^n\| \leq \|T\|^n$ the series $\sum_{n=0}^{\infty} \|T^n\|$ is convergent which implies that $\sum_{n=0}^{\infty} T^n$ converges (see Theorem 2.1). Define

$$S = \sum_{n=0}^{\infty} T^n = \lim_{N \to \infty} \sum_{n=0}^{N} T^n.$$

Thus,

$$(I-T)S = S - TS = \sum_{n=0}^{\infty} T^n - \sum_{n=0}^{\infty} T^{n+1} = I.$$

Similarly $S(I-T) = I$. Thus,

$$(I-T)^{-1} = \sum_{n=0}^{\infty} T^n.$$

□

Denote by $\mathrm{Inv}(\mathcal{B}(X))$ the set of all invertible operators. Last theorem has two very important consequences.

(i) $\mathrm{Inv}(\mathcal{B}(X))$ is an open set in $\mathcal{B}(X)$.

(ii) The inversion $T \mapsto T^{-1}$ is a continuous function on $\mathrm{Inv}(\mathcal{B}(X))$.

Theorem 7.2 *The spectrum $\sigma(T)$ of an operator T, is a compact set of the complex plane and it is contained in the closed disk $\{z \in \mathbb{C} | |z| \le \|T\|\}$.*

Proof. $\mathbb{C} - \sigma(T)$ is open because $\mathrm{Inv}(\mathcal{B}(X))$ is open and the inversion is continuous. If $\lambda \in \mathbb{C}$ is such that $|\lambda| > \|T\|$ then, $\lambda I - T = \lambda(I - \frac{T}{\lambda})$ is an invertible operator, which implies that $\lambda \notin \sigma(T)$. □

One can prove that, the spectrum of any element $T \in \mathcal{B}(\mathcal{H})$ is not empty. In [7] can be found the proof of this result in the context of Banach algebras. This proof is based in the Hahn-Banach Theorem and the Liouville Theorem.

The spectrum of an operator plays a very important role in the study of the properties of this operator. The following theorem shows this fact.

Theorem 7.3 ([5]) *A normal operator $T \in \mathcal{B}(\mathcal{H})$ is*

(i) self adjoint if and only if $\sigma(T)$ lies in the real axis,

(ii) unitary if and only if $\sigma(T)$ lies in the unit circle.

Many properties of an operator such as compactness are closed related to the properties of the spectrum of the operator. Fur further reading on this topic we refer the reader to the references.

References

[1] Conway J 1990 *A Course in Functional Analysis* (New York: Springer-Verlag).
[2] Debnath L and Mikusiński P 2005 *Introduction to Hilbert Spaces with applications* (Burlington: Elsevier Academic Press).
[3] Kreysig E 1978 *Introductory Functional Analysis with Applications* (New York: John Wiley and Sons).
[4] Murphy G 1990 *C*-algebras and Operator Theory* (San Diego: Academic Press).
[5] Rudin W 1973 *Functional Analysis* (New York: MacGraw-Hill).
[6] Young N 1988 *An Introduction to Hilbert Space* (Cambridge: Cambridge University Press).
[7] Zhu K 1993 *An Introduction to Operator Algebras* (Boca Raton: CCR Press).

IOP Conf. Series: Journal of Physics: Conf. Series **839** (2017) 012003 doi:10.1088/1742-6596/839/1/012003

Richardson's Solutions in the Real– and Complex–Energy Spectrum

R M Id Betan

Instituto de Física Rosario (CONICET-UNR), Bv. 27 de Febrero 210 bis, S2000EZP Rosario. Argentina
Facultad de Ciencias Exactas, Ingeniería y Agrimensura (UNR), Av. Pellegrini 250, S2000BTP Rosario. Argentina
Instituto de Estudios Nucleares y Radiaciones Ionizantes (UNR), Riobamba y Berutti, S2000EKA Rosario. Argentina

E-mail: idbetan@ifir-conicet.gov.ar

Abstract. The constant pairing Hamiltonian holds exact solutions worked out by Richardson in the early Sixties. This exact solution of the pairing Hamiltonian regained interest at the end of the Nineties. The discrete complex-energy states had been included in the Richardson's solutions in [1]. In this contribution we reformulate the problem of determining the exact eigen-energies of the pairing Hamiltonian when the continuum is included through the single particle level density. The solutions with discrete complex-energy states is recovered by analytic continuation of the equations to the complex energy plane. This formulation may be applied to loosely bound system where the correlations with the continuum-spectrum of energy is really important. Some details are given to show how the many-body eigen-energy emerges as sum of the pair-energies.

1. Introduction

It is an usual practice, in many-body calculation, starts from a single-particle representation. Probably, the best way to define this basis, is from the eigenfunctions of the single-particle Hamiltonian obtained from the many-body mean-field [2]. For loosely-bound systems, it is expected that the eigen-energies of the single-particle Hamiltonian, will lay very close to the continuum threshold. This means that the Fermi level will be also very close to continuum or even it may lies in the continuum itself. For these kind of system, correlations with continuum spectrum of energy becomes very important and so they have to be taken into account explicitly.

The residual interaction coming from the mean-field approximation contains all possible correlations between the different configurations. This implies that the many-body basis will, rapidly increase with the number of particles in the system. Besides, the need to include the continuum throws the dimension to huge numbers. The pairing interaction gives a good approximation to the residual interaction [3,4]. It replaces the matrix diagonalization procedure by a roots finding procedure from a small set of algebraic non-linear equations.

The independent state pairing interaction admits exact solution that was worked out in 1963 by Richardson [5]; only a few years after that Bardeen, Cooper and Shrieffer [6] explained microscopically the superconductivity as the pairing of electrons in the neighborhood of the Fermi level.

The exact treatment of the pairing correlations with continuum representation in many-body systems, allows to assess the importance of the continuum-continuum correlations in systems with many particles. This is achieved by solving the Richardson equations with resonant and non-resonant configurations [7, 8].

The goal of this work is to complement [9] and [8] by giving a guided demonstration of how, the pair-energies and the Richardson's equations, emerge from the many-body eigenvalue equation.

In sections 2 and 3 we defined the model interaction and the solution for the many-body system, respectively. The complementary equations (Richardson equations) needed to find the many-body eigen-energy is given in section 4 for the box representation. The quasi-bound continuum states in the box are changed by the single particle density in section 5 and the Richardson equations are given. Section 6 gives the Richardson equations for a complex energy representation. Some discussion are made in the last section 7. Four appendixes complement this work by given some details for calculations.

2. Pairing Hamiltonian

The original derivation of the Richardson's equations in [5, 10] is very cumbersome. From a Richardson's suggestion [9], Von Delft and Braun derived Richardson's equations in a simpler and clearer way. This last approach is adopted in this work to derived the equation with continuum spectrum and its extension to the complex energy plane. Due to the difficulties of dealing explicitly with the continuum, we will work out the equations by considering the system immersed in an spherical box of dimension R. At the end of the calculations we will make the formal limit of the size of the box to infinity in order to have the final equations in the real energy representation. The final stage will consist to make analytic extension of the equations to the complex energy plane.

Let us assume that the single-particle levels of our representation has negative and positive energy ε_j, and that each state is doubly degenerate. The constant pairing Hamiltonian reads,

$$H = \sum_{j\sigma} \varepsilon_j c^\dagger_{j\sigma} c_{j\sigma} - G \sum_{ij} c^\dagger_{j+} c^\dagger_{j-} c_{j-} c_{j+}\,, \tag{1}$$

with $|j, \pm\rangle$ time-reversed states, and G the strength of the pairing interaction. The fermion operators $c^\dagger_{j\sigma}$ satisfy the usual anti-commutation relationship.

Now we introduce the pair creation operators $b^\dagger_j = c^\dagger_{j+} c^\dagger_{j-}$, which satisfy the following commutation relations $[b_j, b^\dagger_{j'}] = \delta_{jj'}(1 - 2n_j) = \delta_{jj'}(1 - 2b^\dagger_j b_j)$, with $n_j = c^\dagger_{j+}c_{j+} + c^\dagger_{j-}c_{j-}$. The last identity in the commutation follows because, as it is explained in [11], p. 38, '...the ground state of H has no pair state $(j+, j-)$ occupied by a single fermion.' The other commutations relations are nil. The operators $b^\dagger_j$ also satisfy $[b^\dagger_j b_j, b^\dagger_{j'}] = \delta_{jj'} b^\dagger_j$ and $(b^\dagger_j)^2 = 0$.

For practical purpose needed in the next section we write the Hamiltonian in terms of the operators $B^\dagger_0 = \sum_i b^\dagger_i$ and $B_0 = \sum_i b_i$,

$$H = \sum_j 2\varepsilon_j b^\dagger_j b_j - G B^\dagger_0 B_0 \tag{2}$$

3. Eigenvalue equation and many-body eigenenergy

We start by making the following anzats for the $2n$-body wave function [9], where n is the number of pairs in the system,

$$|\psi_n\rangle = \prod_{\nu=1}^{n} B^\dagger_{J_\nu}|0\rangle = \prod_{\nu=1}^{n}\left(\sum_j \frac{b^\dagger_j}{2\varepsilon_j - E_{J_\nu}}\right)|0\rangle \tag{3}$$

Our goal is to find the many-body eigen-energy $\mathcal{E}_n$ and the coefficients E_{J_ν} which define the corresponding many-body eigenfunction. To reach this goal we solve the many-body Schrödinger equation with the above wave function

$$H|\psi_n\rangle = \mathcal{E}_n|\psi_n\rangle\,, \tag{4}$$

which can be written in the following way

$$\left[H, \prod_{\nu=1}^{n} B^\dagger_{J_\nu}\right]|0\rangle = \mathcal{E}_n|\psi_n\rangle, \tag{5}$$

where we have used $H|0\rangle = 0$, i.e. the eigenvalues $\mathcal{E}_n$ of H are defined with respect to the vacuum state $|0\rangle$.

Using the identity (A.1) of the Appendix A we write the commutator $[H, \prod_\nu B^\dagger_{J_\nu}]$ in terms of the single commutator $[H, B^\dagger_J]$,

$$[H, \prod_{\nu=1}^{n} B^\dagger_{J_\nu}] = \sum_{\nu=1}^{n}\left\{\left(\prod_{\eta=1}^{\nu-1} B^\dagger_{J_\eta}\right)[H, B^\dagger_{J_\nu}]\left(\prod_{\mu=\nu+1}^{n} B^\dagger_{J_\mu}\right)\right\}. \tag{6}$$

By replacing the value of $[H, B^\dagger_J]$ calculated in the Appendix B (see Eq. B.16) we arrive at the expression

$$[H, \prod_{\nu=1}^{n} B^\dagger_{J_\nu}] = \sum_{\nu=1}^{n}\left\{\left(\prod_{\eta=1}^{\nu-1} B^\dagger_{J_\eta}\right)\left[E_{J_\nu}B^\dagger_{J_\nu} + B^\dagger_0\left(1 - G\sum_j \frac{1-2b^\dagger_j b_j}{2\varepsilon_j - E_{J_\nu}}\right)\right]\left(\prod_{\mu=\nu+1}^{n} B^\dagger_{J_\mu}\right)\right\}$$

By distributing the product and gathering the indexes when possible we get

$$\begin{aligned}[H, \prod_{\nu=1}^{n} B^\dagger_{J_\nu}] &= \sum_{\nu=1}^{n} E_{J_\nu}\prod_{\eta=1}^{n} B^\dagger_{J_\eta} \\ &+ \sum_{\nu=1}^{n}\left[1 - \sum_j \frac{G}{2\varepsilon_j - E_{J_\nu}}\right] B^\dagger_0 \prod_{\eta=1,\eta\neq\nu}^{n} B^\dagger_{J_\eta} \\ &+ \sum_{\nu=1}^{n}\left(\prod_{\eta=1}^{\nu-1} B^\dagger_{J_\eta}\right)\sum_j \frac{2GB^\dagger_0 b^\dagger_j b_j}{2\varepsilon_j - E_{J_\nu}}\left(\prod_{\mu=\nu+1}^{n} B^\dagger_{J_\mu}\right)\end{aligned} \tag{7}$$

IOP Conf. Series: Journal of Physics: Conf. Series **839** (2017) 012003 doi:10.1088/1742-6596/839/1/012003

Next we apply the above identity to the vacuum $|0\rangle$ and use $\prod_{\eta=1}^{n} B^{\dagger}_{J_\eta}|0\rangle = |\psi_n\rangle$ to obtain

$$
\begin{aligned}
[H, \prod_{\nu=1}^{n} B^{\dagger}_{J_\nu}]|0\rangle &= \sum_{\nu=1}^{n} E_{J_\nu}|\psi_n\rangle \\
&+ \sum_{\nu=1}^{n}\left[1 - \sum_j \frac{G}{2\varepsilon_j - E_{J_\nu}}\right] B^{\dagger}_0 \prod_{\eta=1,\eta\neq\nu}^{n} B^{\dagger}_{J_\eta}|0\rangle \\
&+ \sum_{\nu=1}^{n}\left(\prod_{\eta=1}^{\nu-1} B^{\dagger}_{J_\eta}\right)\sum_j \frac{2GB^{\dagger}_0 b^{\dagger}_j b_j}{2\varepsilon_j - E_{J_\nu}}\left(\prod_{\mu=\nu+1}^{n} B^{\dagger}_{J_\mu}\right)|0\rangle
\end{aligned}
\tag{8}
$$

Using Eq. (5) for the left hand side we have

$$
\begin{aligned}
\mathcal{E}_n|\psi_n\rangle &= \sum_{\nu=1}^{n} E_{J_\nu}|\psi_n\rangle \\
&+ \sum_{\nu=1}^{n}\left[1 - \sum_j \frac{G}{2\varepsilon_j - E_{J_\nu}}\right] B^{\dagger}_0 \prod_{\eta=1,\eta\neq\nu}^{n} B^{\dagger}_{J_\eta}|0\rangle \\
&+ \sum_{\nu=1}^{n}\left(\prod_{\eta=1}^{\nu-1} B^{\dagger}_{J_\eta}\right)\sum_j \frac{2GB^{\dagger}_0 b^{\dagger}_j b_j}{2\varepsilon_j - E_{J_\nu}}\left(\prod_{\mu=\nu+1}^{n} B^{\dagger}_{J_\mu}\right)|0\rangle
\end{aligned}
\tag{9}
$$

If the second and third lines of the above equation would be zero, the eigenenergy of the many-body system would be given in terms of the parameters E_{J_ν} by

$$
\mathcal{E}_n = \sum_{\nu=1}^{n} E_{J_\nu} \tag{10}
$$

In the next section we show which are the conditions that the parameters E_{J_ν} have to satisfy, in order that these 'dangerous' terms cancel out.

4. Richardson equations

In this section we are going to reduce the last two terms of the right hand of equation (9) to a more amenable expression. As a consequence, we are going to get the Richardson equations, which give the conditions that the parameters E_{J_ν} must to satisfy to validate Eq. (10).

Let us write the dangerous term as $R|0\rangle = R_1|0\rangle + R_2|0\rangle$, with

$$
R_1 = \sum_{\nu=1}^{n}\left[1 - \sum_j \frac{G}{2\varepsilon_j - E_{J_\nu}}\right] B^{\dagger}_0 \prod_{\eta=1,\eta\neq\nu}^{n} B^{\dagger}_{J_\eta} \tag{11}
$$

$$
R_2 = \sum_{\nu=1}^{n}\left(\prod_{\eta=1}^{\nu-1} B^{\dagger}_{J_\eta}\right)\sum_j \frac{2GB^{\dagger}_0 b^{\dagger}_j b_j}{2\varepsilon_j - E_{J_\nu}}\left(\prod_{\mu=\nu+1}^{n} B^{\dagger}_{J_\mu}\right) \tag{12}
$$

IOP Conf. Series: Journal of Physics: Conf. Series **839** (2017) 012003 doi:10.1088/1742-6596/839/1/012003

Next, we are going to reduce $R_2|0\rangle$ to a structure similar to $R_1|0\rangle$. In the Appendix C we show that $\sum_j 2GB_0^\dagger b_j^\dagger b_j/(2\varepsilon_j - E_{J_\nu})(\prod_{\mu=\nu+1}^{n} B_{J_\mu}^\dagger)$ can be written as

$$\sum_{\nu'=\nu+1}^{n}\left\{\left(\prod_{\eta'=\nu+1}^{\nu'-1} B_{J_{\eta'}}^\dagger\right)\sum_j \frac{2GB_0^\dagger b_j^\dagger}{(2\varepsilon_j - E_{J_\nu})(2\varepsilon_j - E_{J_{\nu'}})}\left(\prod_{\mu'=\nu'+1}^{n} B_{J_{\mu'}}^\dagger\right)\right\}.$$

Then

$$R_2|0\rangle = \sum_{\nu=1}^{n}\left(\prod_{\eta=1}^{\nu-1} B_{J_\eta}^\dagger\right)\sum_{\nu'=\nu+1}^{n}\left\{\left(\prod_{\eta'=\nu+1}^{\nu'-1} B_{J_{\eta'}}^\dagger\right)\sum_j \frac{2GB_0^\dagger b_j^\dagger}{(2\varepsilon_j - E_{J_\nu})(2\varepsilon_j - E_{J_{\nu'}})}\left(\prod_{\mu'=\nu'+1}^{n} B_{J_{\mu'}}^\dagger\right)\right\}|0\rangle.$$

In the Appendix D we worked out the expression $\sum_j \frac{2GB_0^\dagger b_j^\dagger}{(2\varepsilon_j - E_{J_\nu})(2\varepsilon_j - E_{J_{\nu'}})}$ and got

$$\sum_j \frac{2GB_0^\dagger b_j^\dagger}{(2\varepsilon_j - E_{J_\nu})(2\varepsilon_j - E_{J_{\nu'}})} = \frac{2GB_0^\dagger}{E_{J_\nu} - E_{J_{\nu'}}}(B_{J_\nu}^\dagger - B_{J_{\nu'}}^\dagger). \tag{13}$$

Therefore

$$R_2|0\rangle = \sum_{\nu=1}^{n}\left(\prod_{\eta=1}^{\nu-1} B_{J_\eta}^\dagger\right)\sum_{\nu'=\nu+1}^{n}\left\{\left(\prod_{\eta'=\nu+1}^{\nu'-1} B_{J_{\eta'}}^\dagger\right)\frac{2GB_0^\dagger}{E_{J_\nu} - E_{J_{\nu'}}}(B_{J_\nu}^\dagger - B_{J_{\nu'}}^\dagger)\left(\prod_{\mu'=\nu'+1}^{n} B_{J_{\mu'}}^\dagger\right)\right\}|0\rangle.$$

This equation can be written as

$$\begin{aligned} R_2|0\rangle &= B_0^\dagger \sum_{\nu=1}^{n}\sum_{\nu'=\nu+1}^{n}\frac{2G}{E_{J_\nu} - E_{J_{\nu'}}}\prod_{\eta=1,\eta\neq\nu'}^{n} B_{J_\eta}^\dagger|0\rangle \\ &\quad - B_0^\dagger \sum_{\nu=1}^{n}\sum_{\nu'=\nu+1}^{n}\frac{2G}{E_{J_\nu} - E_{J_{\nu'}}}\prod_{\eta=1,\eta\neq\nu}^{n} B_{J_\eta}^\dagger|0\rangle \end{aligned} \tag{14}$$

By changing the indexes of the summations in the first line using Eq. (A.2) and then changing $\nu \leftrightarrow \nu'$ we obtain

$$\begin{aligned} R_2|0\rangle &= B_0^\dagger \sum_{\nu=1}^{n}\sum_{\nu'=1}^{\nu-1}\frac{2G}{E_{J_{\nu'}} - E_{J_\nu}}\prod_{\eta=1,\eta\neq\nu}^{n} B_{J_\eta}^\dagger|0\rangle \\ &\quad - B_0^\dagger \sum_{\nu=1}^{n}\sum_{\nu'=\nu+1}^{n}\frac{2G}{E_{J_\nu} - E_{J_{\nu'}}}\prod_{\eta=1,\eta\neq\nu}^{n} B_{J_\eta}^\dagger|0\rangle \end{aligned} \tag{15}$$

Finally, by inverting the term $E_{J_{\nu'}} - E_{J_\nu}$ in the first line and merging the summation in ν' we arrive at

$$R|0\rangle = \sum_{\nu=1}^{n}\left\{\left[1 - \sum_j \frac{G}{2\varepsilon_j - E_{J_\nu}} - \sum_{\nu'=1,\nu'\neq\nu}^{n}\frac{2G}{E_{J_\nu} - E_{J_{\nu'}}}\right] B_0^\dagger \prod_{\eta=1,\eta\neq\nu}^{n} B_{J_\eta}^\dagger\right\}|0\rangle.$$

Thus, the conditions which have to satisfy the parameters E_{J_ν} with $\nu = 1,\cdots,n$ (and n the number of pairs), is that all E_{J_ν} simultaneously satisfy the n non-linear equations, i.e. the Richardson equations [5]

$$1 - \sum_j \frac{G}{2\varepsilon_j - E_{J_\nu}} - \sum_{\nu'=1,\nu'\neq\nu}^{n}\frac{2G}{E_{J_\nu} - E_{J_{\nu'}}} = 0 \tag{16}$$

This complete the exact solution of the many-body pairing Hamiltonian with continuum in a box representation. In the next section we will make the limit to the continuum.

5. Real continuum energy spectrum

In this section we make the formal limit of the size R of the spherical box to infinity. In doing so, the negative energy spectrum stabilizes while the continuum spectrum became denser as the size of the box increases. For the infity limit we make the following substitution

$$\sum_{j\sigma} \xrightarrow{R\to\infty} \int_{-\infty}^{\infty} \tilde{g}(\varepsilon)\, d\varepsilon\ . \tag{17}$$

where the single particle density $\tilde{g}(\varepsilon)$ has two terms, (i) the first due to negative energy ε_{j_b} of bound states, and (ii) the second due to the continuum states ε with positive energy and density $g(\varepsilon)$,

$$\tilde{g}(\varepsilon) = \sum_{j_b\sigma} \delta(\varepsilon - \varepsilon_{j_b}) + g(\varepsilon) \tag{18}$$

The Richardson equations in a representation with real continuum energy (besides the bound states) reads

$$1 - \frac{G}{2}\sum_{j_b\sigma}\frac{1}{2\varepsilon_b - E_\nu} - \frac{G}{2}\int_0^\infty d\varepsilon\, \frac{g(\varepsilon)}{2\varepsilon - E_\nu} + 2G\sum_{\nu\neq\nu'}\frac{1}{E_\nu - E_{\nu'}} = 0\ . \tag{19}$$

note that unlike Eq. (16), here the degeneracy was included in the sum symbol. There are as many Eq. (19) as pairs in the system; all these equations have to be satisfied simultaneously in order to get all pair-energies. Then, summing up the n pair-energies we get eigen-energy for the $2n$-body system through Eq. (10). This result also implies that only the formulas which define the Richardson equations change in a continuum representation but not the formula which gives the eigen-energy.

6. Complex continuum energy spectrum

In this section we separate from the density, the resonant part from the non-resonant part. It is expected that if the system holds narrow resonances, them will be the most important correlations coming from the continuum configurations. For the single particle density we use the one given by Beth and Uhlenbeck [12] in terms of the phase shift,

$$g(\varepsilon) = \sum_{j_c\sigma} \frac{1}{\pi}\frac{d\delta_c}{d\varepsilon} = g_{\rm Res}(\varepsilon) + g_{\rm Bckg}(\varepsilon) \tag{20}$$

The presence of single particle resonances will show up in the density, and also in the cross section, as sharp structures. These sharp structures may be parametrized using the Lorentzian distribution in terms of the resonant energy ϵ_r and the resonant width Γ_r [13]

$$g_{\rm Res}(\varepsilon) \approx \sum_{j_r\sigma} \frac{1}{\pi}\frac{\Gamma_r/2}{(\varepsilon - \epsilon_r)^2 + (\Gamma_r/2)^2}\ . \tag{21}$$

The above approximation worsen as the width of the resonance increases.

The analytic extension of the resonant density $g_{\rm Res}(\varepsilon)$ to the lower complex energy plane, shows a pole at the complex energy $\varepsilon_r = \epsilon_r - i\,\Gamma_r/2$. By rotating the integration contour of the resonant part of the density in the Richardson equation (19) we get a contribution similar to the bound states but with complex energies instead. The following expression gives the Richardson equations in the complex energy representation [8]

$$1-\frac{G}{2}\sum_{j_b\sigma}\frac{1}{2\varepsilon_b-E_\nu}-\frac{G}{2}\sum_{j_r\sigma}\frac{1}{2\varepsilon_r-E_\nu}-\frac{G}{2}\int_0^\infty d\varepsilon\,\frac{g_{\text{Bckg}}(\varepsilon)}{2\varepsilon-E_\nu}-\frac{G}{2}\int_0^\infty d\varepsilon\,\frac{g_{\text{CxBckg}}(\varepsilon)}{2\varepsilon-iE_\nu}$$
$$+2G\sum_{\nu\neq\nu'}\frac{1}{E_\nu-E_{\nu'}}=0\,, \tag{22}$$

where $g_{\text{CxBckg}}(\varepsilon)$ is the density which remains after the poles contribution of the resonant part of the density was taken.

The solution of the Richardson equations (22) give the pair-energies to build the many-body eigenenergy from Eq. (10). If we ignore the densities g_{Bckg} and g_{CxBckg} in Eq. (22), we end up with the pole approximation of the Richardson equations given in Ref. [1]. Note that in this approximation, the eigen-energy given by Eq. (10) will no be any more real. The magnitude of the imaginary component of $\mathcal{E}_n$ is a measure of how good (or bad) the pole approximation is.

7. Conclusions and discussion

We have shown how to get the many-body eigenvalue for the constant pairing Hamiltonian for different model spaces including the continuum spectrum of energy. The procedure followed consisted of solving the many-body Schrödinger equation with the ansatz wave function (3) of Ref. [14]. The equations obtained are exact in the box representation, i.e., before the limit of the radius of the spherical box is taken to infinity. In the limit, a single particle density has to be introduced an some arbitrariness is given to the model solution depending of the density used. Since the total density diverges with the same divergence as the free density as the radius of the spherical box goes to infinity [15], we decided to take as density the difference between the mean-field and the free densities [12], i.e., in terms of the derivatives of the phase shift. Another advantage of this density is that it makes very natural the separation between resonant continuum and non-resonant continuum, which allow to use a simple parametrization for the resonant configurations. As a last step, in order to move from the real continuum to the complex continuum, we made the analytic continuation of the parametrized density and by using Complex Analysis we were able to changed the integral contribution of the resonant continuum by a single term (for each partial wave).

Acknowledgments

This work has been supported by the Consejo Nacional de Investigaciones Científicas y Técnicas PIP-625 (CONICET, Argentina).

Appendix A. Math tool box

This appendix collect three identities which are used in the process of derivation of the Richardson equations.

Isolation of one commutator: The following expression give the identity to factorize a single commutator from a commutator involving product of operators [9]

$$[A,\prod_{i=1}^n B_i]=\sum_{i=1}^n\left\{\left(\prod_{j=1}^{i-1}B_j\right)[A,B_i]\left(\prod_{j'=i+1}^{n}B_{j'}\right)\right\} \tag{A.1}$$

Changing indexes in double sums: By visualizing a double sum as the sum of the matrix elements of a matrix of order n, follows the identity below for the sum of all non diagonal terms

$$\sum_{i=1}^{n}\sum_{j=i+1}^{n} a_{ij} = \sum_{j=1}^{n}\sum_{i=1}^{j-1} a_{ij} \tag{A.2}$$

Partial fraction expansion: The partial fraction decomposition allows to write the quotient of two polynomials as a sum of terms with simpler numerator,

$$\frac{P(x)}{Q(x)} = \frac{P(x)}{\prod_i (x - x_i)} = \sum_i \frac{c_i}{x - x_i} \tag{A.3}$$

where x_i are the zeros of $Q(x)$. In order to calculate the coefficients c_i one has to evaluate some expression at the values of the roots x_i.

Appendix B. Calculation of $[H, B_J^\dagger]$

$$[H, B_J^\dagger] = \sum_j 2\varepsilon_j [b_j^\dagger b_j, B_J^\dagger] - G[B_0^\dagger B_0, B_J^\dagger] \tag{B.1}$$

The commutation in first term reduces to

$$[b_j^\dagger b_j, B_J^\dagger] = b_j^\dagger [b_j, B_J^\dagger] - [b_j^\dagger, B_J^\dagger] b_j \tag{B.2}$$

$$= b_j^\dagger \sum_{j'} \frac{1}{2\varepsilon_{j'} - E_J}[b_j, b_{j'}^\dagger] - \sum_{j'} \frac{1}{2\varepsilon_{j'} - E_J}[b_j^\dagger, b_{j'}^\dagger] b_j \tag{B.3}$$

$$= b_j^\dagger \sum_{j'} \delta_{jj'} \frac{1 - 2b_j^\dagger b_j}{2\varepsilon_{j'} - E_J} - \sum_{j'} \frac{1}{2\varepsilon_{j'} - E_J} 0 b_j = \frac{b_j^\dagger - 2(b_j^\dagger)^2 b_j}{2\varepsilon_{j'} - E_J} \tag{B.4}$$

$$= \frac{b_j^\dagger}{2\varepsilon_j - E_J} \tag{B.5}$$

While for the commutation of the second term we get

$$[B_0^\dagger B_0, B_J^\dagger] = B_0^\dagger [B_0, B_J^\dagger] - [B_0^\dagger, B_J^\dagger] B_0 \tag{B.6}$$

with

$$[B_0, B_J^\dagger] = [\sum_j b_j, \sum_{j'} \frac{b_{j'}^\dagger}{2\varepsilon_{j'} - E_J}] = \sum_{jj'} \frac{1}{2\varepsilon_{j'} - E_J}[b_j, b_{j'}^\dagger] \tag{B.7}$$

$$= \sum_j \frac{1 - 2b_j^\dagger b_j}{2\varepsilon_j - E_J} \tag{B.8}$$

and

$$[B_0^\dagger, B_J^\dagger] = [\sum_j b_j^\dagger, \sum_{j'} \frac{b_{j'}^\dagger}{2\varepsilon_{j'} - E_J}] = \sum_{jj'} \frac{1}{2\varepsilon_{j'} - E_J}[b_j^\dagger, b_{j'}^\dagger] \tag{B.9}$$

$$= 0 \tag{B.10}$$

Using the above identities, the commutator $[H, B_J^\dagger]$ reads

$$[H, B_J^\dagger] = \sum_j \frac{2\varepsilon_j b_j^\dagger}{2\varepsilon_j - E_J} - GB_0^\dagger \frac{1 - 2b_j^\dagger b_j}{2\varepsilon_j - E_J} \tag{B.11}$$

The term $\sum_j 2\varepsilon_j b_j^\dagger/(2\varepsilon_j - E_J)$ can be written in terms of $B_J^\dagger$and $B_0^\dagger$,

$$\sum_j \frac{2\varepsilon_j b_j^\dagger}{2\varepsilon_j - E_J} = \sum_j \frac{2\varepsilon_j b_j^\dagger}{2\varepsilon_j - E_J} - B_0^\dagger + B_0^\dagger \tag{B.12}$$

$$= \sum_j b_j^\dagger \left[\frac{2\varepsilon_j}{2\varepsilon_j - E_J} - 1\right] + B_0^\dagger \tag{B.13}$$

$$= E_J \sum_j b_j^\dagger \frac{1}{2\varepsilon_j - E_J} + B_0^\dagger \tag{B.14}$$

$$= E_J B_J^\dagger + B_0^\dagger \tag{B.15}$$

Gathering all together we get

$$[H, B_J^\dagger] = E_J B_J^\dagger + B_0^\dagger \left(1 - G\sum_j \frac{1 - 2b_j^\dagger b_j}{2\varepsilon_j - E_J}\right). \tag{B.16}$$

Appendix C. Calculation of the commutator $[\sum_j 2GB_0^\dagger b_j^\dagger b_j/(2\varepsilon_j - E_{J_\nu}), \prod_\mu B_{J_\mu}^\dagger]$

In order to invert the order of $b_j^\dagger b_j$ and $B_{J_\mu}^\dagger$ in eq. (11) we need to work out the following equation

$$[\sum_j \frac{2GB_0^\dagger b_j^\dagger b_j}{2\varepsilon_j - E_{J_\nu}}, \prod_{\mu=\nu+1}^{n} B_{J_\mu}^\dagger]|0\rangle = \sum_j \frac{2GB_0^\dagger b_j^\dagger b_j}{2\varepsilon_j - E_{J_\nu}} \left(\prod_{\mu=\nu+1}^{n} B_{J_\mu}^\dagger\right)|0\rangle. \tag{C.1}$$

By applying the indentity (A.1) we get

$$[\sum_j \frac{2GB_0^\dagger b_j^\dagger b_j}{2\varepsilon_j - E_{J_\nu}}, \prod_{\mu=\nu+1}^{n} B_{J_\mu}^\dagger]|0\rangle = \sum_{\nu'=\nu+1}^{n} \left\{\left(\prod_{\eta'=\nu+1}^{\nu'-1} B_{J_{\eta'}}^\dagger\right)[\sum_j \frac{2GB_0^\dagger b_j^\dagger b_j}{2\varepsilon_j - E_{J_\nu}}, B_{J_{\nu'}}^\dagger] \left(\prod_{\mu'=\nu'+1}^{n} B_{J_{\mu'}}^\dagger\right)\right\}|0\rangle, \tag{C.2}$$

where

$$[\sum_j \frac{2GB_0^\dagger b_j^\dagger b_j}{2\varepsilon_j - E_{J_\nu}}, B_{J_{\nu'}}^\dagger]|0\rangle = \sum_j \frac{2G}{2\varepsilon_j - E_{J_\nu}} \left\{B_0^\dagger[b_j^\dagger b_j, B_{J_{\nu'}}^\dagger] - [B_0^\dagger, B_{\nu'}^\dagger]b_j^\dagger b_j\right\}|0\rangle \tag{C.3}$$

$$= \sum_j \frac{2G}{2\varepsilon_j - E_{J_\nu}} B_0^\dagger \frac{b_j^\dagger}{2\varepsilon_j - E_{J_{\nu'}}}|0\rangle. \tag{C.4}$$

Then

$$\sum_j \frac{2GB_0^\dagger b_j^\dagger b_j}{2\varepsilon_j - E_{J_\nu}} \left(\prod_{\mu=\nu+1}^{n} B_{J_\mu}^\dagger \right) |0\rangle = \sum_{\nu'=\nu+1}^{n} \left\{ \left(\prod_{\eta'=\nu+1}^{\nu'-1} B_{J_{\eta'}}^\dagger \right) \sum_j \frac{2GB_0^\dagger b_j^\dagger}{(2\varepsilon_j - E_{J_\nu})(2\varepsilon_j - E_{J_{\nu'}})} \left(\prod_{\mu'=\nu'+1}^{n} B_{J_{\mu'}}^\dagger \right) \right\} |0\rangle. \quad \text{(C.5)}$$

Appendix D. Calculation of the term $\sum_j \frac{2gB_0^\dagger b_j^\dagger}{(2\varepsilon_j - E_{J_\nu})(2\varepsilon_j - E_{J_{\nu'}})}$

By using the simple fraction expansion of Eq. (A.3) we write,

$$\begin{aligned}
\sum_j \frac{2gB_0^\dagger b_j^\dagger}{(2\varepsilon_j - E_{J_\nu})(2\varepsilon_j - E_{J_\mu})} &= 2gB_0^\dagger \sum_j b_j^\dagger \frac{1}{(2\varepsilon_j - E_{J_\nu})(2\varepsilon_j - E_{J_\mu})} \\
&= 2gB_0^\dagger \sum_j b_j^\dagger \left[\frac{A}{2\varepsilon_j - E_{J_\nu}} + \frac{B}{2\varepsilon_j - E_{J_\mu}} \right] \\
&= 2gB_0^\dagger \sum_j b_j^\dagger \left[\frac{A(2\varepsilon_j - E_{J_\mu}) + B(2\varepsilon_j - E_{J_\nu})}{(2\varepsilon_j - E_{J_\nu})(2\varepsilon_j - E_{J_\nu})} \right],
\end{aligned}$$

with $1 = A(2\varepsilon_j - E_{J_\mu}) + B(2\varepsilon_j - E_{J_\nu})$. For $2\varepsilon_j = E_{J_\mu}$ we get $B = \frac{1}{E_{J_\mu} - E_{J_\nu}}$, while for $2\varepsilon_j = E_{J_\nu}$ we get $A = \frac{1}{E_{J_\nu} - E_{J_\mu}} = -B$. Then

$$\begin{aligned}
\sum_j \frac{2gB_0^\dagger b_j^\dagger}{(2\varepsilon_j - E_{J_\nu})(2\varepsilon_j - E_{J_\mu})} &= 2gB_0^\dagger \sum_j b_j^\dagger \frac{1}{E_{J_\nu} - E_{J_\mu}} \left(\frac{1}{2\varepsilon_j - E_{J_\nu}} - \frac{1}{2\varepsilon_j - E_{J_\mu}} \right) \\
&= \frac{2gB_0^\dagger}{E_{J_\nu} - E_{J_\mu}} \left[\sum_j \frac{b_j^\dagger}{2\varepsilon_j - E_{J_\nu}} - \sum_j \frac{b_j^\dagger}{2\varepsilon_j - E_{J_\mu}} \right] \\
&= \frac{2gB_0^\dagger}{E_{J_\nu} - E_{J_\mu}} \left(B_{J_\nu}^\dagger - B_{J_\mu}^\dagger \right). \qquad \text{(D.1)}
\end{aligned}$$

References

[1] Hasegawa M and Kaneko K 2003 *Phys Rev C* **67** 24304
[2] Suhonen J 2007 *From Nucleons to Nucleus, Concepts of Microscopic Nuclear Theory* (Berlin: Springer)
[3] Lane A M 1964 *Nuclear Theory. Pairing Force Correlations and Collective Motion* (New York, Benjamin)
[4] Dean D J and Hjorth-Jensen M 2003 *Rev Mod Phys* **75** 607
[5] Richardson R W 1963 *Phys Lett* **3** 277
[6] Bardeen J, Cooper L N and Schrieffer J R 1963 *Phys Rev* **108** 1175
[7] Id Betan R M 2012 *Nucl Phys A* **879** 14
[8] Id Betan R 2012 *Phys Rev C* **85** 064309
[9] Von Delft J and Braun F 1999 arXiv:cond-mat/9911058
[10] Richardson R W and Sherman N 1964 *Nucl Phys* **52** 221
[11] Schrieffer J R 1964 *Theory of superconductivity* (New York:. Benjamin)
[12] Beth E and Uhlenbeck G 1937 *Physica* **4** 915
[13] Kukulin V I, Krasnolpolsky V M and Horacek J 1988 *Theory of Resonances* (Dordrecht: Kluwer Academic Publishers)
[14] von Delft J, Zaikin A D, Golubev D S and Tichy W 1996 *Phys Rev Lett* **77** 3189
[15] Shlomo S, Kolomietz V M and Dejbakhsh H 1997 *Phys Rev C* **55** 1972

IOP Conf. Series: Journal of Physics: Conf. Series **839** (2017) 012004 doi:10.1088/1742-6596/839/1/012004

A Power Series Representation for the Characteristic Equation of Gamow-Siegert Eigenstates

V Barrera-Figueroa

Instituto Politécnico Nacional, Sección de Estudios de Posgrado e Investigación, UPIITA. Av. Inst. Politécnico Nal. 2580, Barrio la Laguna Ticomán, C.P. 07340, Ciudad de México

E-mail: `vbarreraf@ipn.mx`

Abstract. In this work we consider a quantum system described by a stationary Schrödinger equation and obtain an explicit representation for the characteristic equation of its Gamow-Siegert eigenstates in the form of a power series of the energy parameter. This representation is valid for both analytical or numerical calculations. We show that determining resonances reduces to calculating the roots of a polynomial equation. The effectiveness of this approach is demonstrated through several numerical examples.

1. Introduction

Solutions of the Schrödinger equation associated to complex energies and satisfying the purely outgoing wave conditions are called Gamow-Siegert functions, which describe the resonance states of quantum systems. These functions were first used in [1] for describing the spontaneous alpha-decay of nuclei, and later Siegert [2] obtained a dispersion formula where the *radioactive* states were represented by resonance terms. Let $|\Psi\rangle$ be a Gamow-Siegert function associated to the complex energy $\epsilon = E - \mathrm{i}\Gamma/2$ ($E > 0$, $\Gamma > 0$) and possessing the complex wave-number $k = \sqrt{2m\epsilon}/\hbar$, then $|\Psi\rangle$ describes states possessing an exponential decay behaviour

$$|\langle \Psi(q,0)|\ \Psi(q,t)\rangle|^2 = e^{-\Gamma t/\hbar}.$$

Several formalisms have been used for describing resonance states, including Darboux (supersymmetric) transformations [3–6], rigged Hilbert spaces [7–10], the S-matrix [11], the transfer matrix [5], the complex-scaling technique [12], and squeezing operators [13], among others (see, *e.g.*, [14–16] and references therein for a complete review of techniques used in quantum mechanics for analysing resonances). On the other hand, the theory of quantum resonances has been used for analysing leaky modes [17] in electromagnetic waveguides via suitable analogies between quantum mechanics and electromagnetism (see, *e.g.*, [18, 19]).

In the present work we apply the spectral parameter power series method [20] (SPPS for short) for obtaining explicitly the characteristic equation that describes the resonance states. This characteristic equation is given by the Taylor series of an analytic function of the complex variable λ, being λ the complex energy of a quantum particle. This series can be truncated for its numerical implementation. Then, analysing resonances reduces to determining zeros (or approximately, polynomial roots) of the characteristic equation. This approach provides an effective way for addressing resonance states in Schrödinger operators with almost arbitrary potential functions.

The outline of this work is as follows. In Section 1 we obtain a general solution to a one-dimensional Schrödinger equation equipped with an arbitrary short range potential. In Section 2 we obtain the characteristic equation that defines the resonance states. In Section 3 we obtain a power series representation of the characteristic equation, and use it in some numerical examples. Finally, in Section 4 we give some concluding remarks.

2. General solution of the Schrödinger equation with a short range potential

Let us consider the differential expression

$$\ell(u) := -\frac{d^2u(x)}{dx^2} + V(x)\,u(x), \qquad V(x) = \begin{cases} V_1, & -\infty < x < h_1, \\ V_2(x), & h_1 < x < h_2, \\ V_3, & h_2 < x < +\infty, \end{cases}$$

in $\mathbb{R} \setminus \{h_1, h_2\}$, corresponding to a stationary Schrödinger operator $(\hbar^2 = 1 = 2m)$ in its coordinate representation. The function V_2 represents a short range potential satisfying certain smoothness conditions. At the points of discontinuity the boundary conditions

$$[u(x)]_{x=h_j} = 0, \qquad [u'(x)]_{x=h_j} = 0, \qquad j = 1, 2, \tag{1}$$

define the behaviour of the functions on which ℓ acts. Here the notation $[f]_{x_0} := f(x_0+0) - f(x_0-0)$ represents the jump of f at the point x_0.

Let us search for a general solution of the equation $\ell(\psi) = \lambda\psi$ in $(-\infty, \infty)$, where λ is a parameter corresponding to the energy of a quantum particle. If ψ_1 and ψ_2 are two linearly independent solutions of the equation in (h_1, h_2), then the piecewise function

$$\psi(x;\lambda) = \begin{cases} a_1 e^{\mathrm{i}\kappa_1(x-h_1)} + a_2 e^{-\mathrm{i}\kappa_1(x-h_1)}, & -\infty < x < h_1, \\ b_1\psi_1(x;\lambda) + b_2\psi_2(x;\lambda), & h_1 \le x \le h_2, \\ c_1 e^{\mathrm{i}\kappa_3(x-h_2)} + c_2 e^{-\mathrm{i}\kappa_3(x-h_2)}, & h_2 < x < \infty \end{cases} \tag{2}$$

with $\kappa_1(\lambda) = (\lambda - V_1)^{1/2}$, $\kappa_3(\lambda) = (\lambda - V_3)^{1/2}$, satisfies the differential equation in $\mathbb{R} \setminus \{h_1, h_2\}$. By a_1, a_2, b_1, b_2, c_1, c_2 we denote arbitrary constants. Conditions (1) applied from right to left give the constants a_1, a_2, b_1, b_2 in terms of c_1, c_2 as follows

$$b_1 = \frac{1}{W}(c_1\beta_{22} + c_2\beta_{12}), \qquad b_2 = -\frac{1}{W}(c_1\beta_{21} + c_2\beta_{11}); \tag{3a}$$

$$a_1 = \frac{1}{2\mathrm{i}\kappa_1}(b_1\alpha_{11} + b_2\alpha_{12}), \qquad a_2 = -\frac{1}{2\mathrm{i}\kappa_1}(b_1\alpha_{21} + b_2\alpha_{22}), \tag{3b}$$

where the coefficients α_{ij}, β_{ij} $(i, j = 1, 2)$ are defined by

$$\alpha_{ij} := (-1)^{i+1}\mathrm{i}\kappa_1\psi_j(h_1;\lambda) + \psi_j'(h_1;\lambda), \quad \beta_{ij} := (-1)^{i+1}\mathrm{i}\kappa_3\psi_j(h_2;\lambda) + \psi_j'(h_2;\lambda),$$

and $W = W[\psi_1, \psi_2](x)$ is the Wronskian of ψ_1, ψ_2. Here the prime $'$ represents the derivative with respect to x.

Consider the Schrödinger equation $\ell(\psi) = \lambda\psi$ with $\lambda \in \mathbb{R}$ and suppose that $V_1, V_3 \in \mathbb{R}$. Scattering states are described by bounded solutions such that $\lambda > \max\{V_1, V_3\}$, while bound states are described by solutions belonging to an appropriate functional space (*e.g.*, the Hilbert space $L^2(\mathbb{R})$) such that $\lambda < \min\{V_1, V_3\}$. In this work, we are interested in the Gamow-Siegert

IOP Conf. Series: Journal of Physics: Conf. Series **839** (2017) 012004 doi:10.1088/1742-6596/839/1/012004

states possessing complex energies. For $\lambda \in \mathbb{C}$ the operation $(\lambda - V_j)^{1/2}$ $(j = 1, 3)$ leads to two branches defined by

$$(\lambda - V_j)^{1/2} = \sqrt{|\lambda - V_j|} e^{i \frac{\theta_j + 2\pi k}{2}}, \quad \theta_j = \arg(\lambda - V_j) \qquad (k = 0, 1;\ j = 1, 3).$$

The appropriate selection of a branch according to the constraints of the problem defines a proper Riemann sheet in the complex λ-plane in which $\kappa_j(\lambda)$ $(j = 1, 3)$ is unambiguously defined. Once a branch is chosen, general solution (2)-(3) is well-defined provided that $\kappa_1 \neq 0$.

3. Solutions satisfying purely outgoing wave conditions

Let $\mathcal{H}$ be an operator defined by the expression $\mathcal{H}\psi := \ell(\psi)$, with domain $\mathrm{Dom}(\mathcal{H})$ consisting of functions ψ satisfying continuity conditions (1), and the purely outgoing wave conditions

$$\lim_{x \to -\infty} \left(\psi'(x;\lambda) + i\kappa_1 \psi(x;\lambda)\right) = 0, \qquad \lim_{x \to +\infty} \left(\psi'(x;\lambda) - i\kappa_3 \psi(x;\lambda)\right) = 0. \tag{4}$$

Solutions in $\mathrm{Dom}(\mathcal{H})$ of the equation $\mathcal{H}\psi = \lambda\psi$ are constructed from general solution (2)-(3) as follows. Asymptotic condition as $x \to +\infty$ is satisfied if $c_2 = 0$. The resulting solution describes the dispersion of quantum waves generated by a source of particles to the left of the interaction potential V_2. The remaining constant c_1 is chosen so that the incoming wave $e^{i\kappa_1(x-h_1)}$ has unit amplitude, more precisely

$$\psi(x;\lambda) = \begin{cases} e^{i\kappa_1(x-h_1)} + (\widetilde{a}_2/\widetilde{a}_1)\, e^{-i\kappa_1(x-h_1)}, & -\infty < x < h_1, \\ (\beta_{22}/\widetilde{a}_1)\, \psi_1(x;\lambda) - (\beta_{21}/\widetilde{a}_1)\, \psi_2(x;\lambda), & h_1 \leq x \leq h_2, \\ (W/\widetilde{a}_1)\, e^{i\kappa_3(x-h_2)}, & h_2 < x < \infty, \end{cases}$$

$$\widetilde{a}_1 = \frac{1}{2i\kappa_1}(\beta_{22}\alpha_{11} - \beta_{21}\alpha_{12}), \quad \widetilde{a}_2 = -\frac{1}{2i\kappa_1}(\beta_{22}\alpha_{21} - \beta_{21}\alpha_{22}).$$

Let $R := \widetilde{a}_2/\widetilde{a}_1$ and $T := W/\widetilde{a}_1$ be the amplitudes of the reflected and transmitted waves, respectively, and let $\mathbb{R} := |R|^2$ and $\mathbb{T} := |T|^2$ be the corresponding reflection and transmission coefficients, which satisfy $\mathbb{T}(\lambda) + \mathbb{R}(\lambda) = 1$ for every $\lambda \in \mathbb{C}$. In order to fulfil asymptotic condition (4) as $x \to -\infty$, the term $R(\lambda)\, e^{-i\kappa_1(x-h_1)}$ must dominate over $e^{i\kappa_1(x-h_1)}$ for some complex energies λ_n, that is, $e^{i\kappa_1 x} = o\left(Re^{-i\kappa_1 x}\right)$ as $x \to -\infty$, $\lambda \to \lambda_n$, using Landau's notation. This is fulfilled if

$$\lim_{\lambda \to \lambda_n} e^{-2i\kappa_1 h_1} \left(\frac{\alpha_{11}\beta_{22} - \alpha_{12}\beta_{21}}{\alpha_{22}\beta_{21} - \alpha_{21}\beta_{22}}\right) = 0.$$

Since $e^{-2i\kappa_1 h_1} \neq 0$ for every $\lambda \in \mathbb{C}$, the limit is satisfied if $\lambda_n \in \mathbb{C}$ is a zero of the equation $\varrho(\lambda) = 0$, where $\varrho(\lambda) := \alpha_{11}(\lambda)\,\beta_{22}(\lambda) - \alpha_{12}(\lambda)\,\beta_{21}(\lambda)$. Let $\lambda_n = E_n - i\Gamma_n/2$ $(E_n > 0, \Gamma_n > 0)$ be a zero of $\varrho(\lambda) = 0$ such that $\mathrm{Re}[\kappa_1(\lambda_n)] > 0$ and $\mathrm{Im}[\kappa_1(\lambda_n)] < 0$, (these conditions define the *correct* direction in which an outgoing wave should propagate [3]) then λ_n represents the complex energy of a resonance state $\psi(x;\lambda_n) =: \psi_n(x) \in \mathrm{Dom}(\mathcal{H})$. The generalised eigenfunction ψ_n associated to the complex eigenvalue λ_n is a Gamow-Siegert eigenstates of the Hamiltonian $\mathcal{H}$. As usual, E_n is the energy of the resonance state and Γ_n is the inverse of its lifetime. Note that the zeros of $\varrho(\lambda) = 0$ coincide with the poles of the transmission amplitude

$$T(\lambda) = \frac{2i\kappa_1(\lambda)\, W}{\alpha_{11}(\lambda)\,\beta_{22}(\lambda) - \alpha_{12}(\lambda)\,\beta_{21}(\lambda)},$$

which is a meromorphic function of the complex variable λ. The function ϱ can be written in terms of ψ_1 and ψ_2 as follows

$$\begin{aligned} \varrho(\lambda) = {} & \left(\psi_1'(h_1;\lambda) + i\kappa_1(\lambda)\,\psi_1(h_1;\lambda)\right)\left(\psi_2'(h_2;\lambda) - i\kappa_3(\lambda)\,\psi_2(h_2;\lambda)\right) \\ & - \left(\psi_2'(h_1;\lambda) + i\kappa_1(\lambda)\,\psi_2(h_1;\lambda)\right)\left(\psi_1'(h_2;\lambda) - i\kappa_3(\lambda)\,\psi_1(h_2;\lambda)\right). \end{aligned}$$

In the next section we will obtain a series representation of this formula by means of the SPPS method [20], which is valid for a large class of interaction potentials V_2. It is worth noting that no general formula exists for obtaining ψ_1 and ψ_2 for, say, every continuous function V_2, thereby a numerical method should be employed (*e.g.*, the shooting method or the WKB method [21], [22, pp. 484–539], [23]). Numerical methods give approximations of ψ_1 and ψ_2 (see, *e.g.*, [24, Part III], and [25] for a review of the most commonly used numerical methods to solve ODEs), but the SPPS method gives exact solutions of Sturm-Liouville problems in the form of uniformly convergent power series of the spectral parameter, as is shown next.

4. SPPS form of the characteristic equation for the Gamow-Siegert eigenstates

The dispersion equation $\varrho(\lambda) = 0$ for the Gamow-Siegert eigenstates is constructed from two linearly independent solutions ψ_1, ψ_2 of Sturm-Liouville equation

$$-\psi''(x) + V_2(x)\psi(x) = \lambda\psi(x), \qquad h_1 < x < h_2 \tag{5}$$

Let ψ_0 be a non-vanishing particular solution of the associated homogeneous equation (*i.e.*, when $\lambda = 0$) satisfying the regularity conditions $\psi_0^2, {\psi_0}^{-2} \in \mathscr{C}[h_1, h_2]$, then the power series [20]

$$\psi_1(x;\lambda) = \psi_0(x)\sum_{k=0}^{\infty}(-\lambda)^k \widetilde{X}^{(2k)}(x), \qquad \psi_2(x;\lambda) = \psi_0(x)\sum_{k=0}^{\infty}(-\lambda)^k X^{(2k+1)}(x) \tag{6}$$

with $\widetilde{X}^{(n)}$, $X^{(n)}$ defined by the recursive integration procedure $\widetilde{X}^{(0)}(x) \equiv 1$, $X^{(0)}(x) \equiv 1$,

$$\widetilde{X}^{(n)}(x) = \begin{cases} \displaystyle\int_{x_0}^{x} \widetilde{X}^{(n-1)}(s)\,\psi_0^2(s)\,\mathrm{d}s, & \text{odd } n, \\ \displaystyle\int_{x_0}^{x} \widetilde{X}^{(n-1)}(s)\,\psi_0^{-2}(s)\,\mathrm{d}s, & \text{even } n; \end{cases} \qquad X^{(n)}(x) = \begin{cases} \displaystyle\int_{x_0}^{x} X^{(n-1)}(s)\,\psi_0^{-2}(s)\,\mathrm{d}s, & \text{odd } n, \\ \displaystyle\int_{x_0}^{x} X^{(n-1)}(s)\,\psi_0^{2}(s)\,\mathrm{d}s, & \text{even } n, \end{cases} \tag{7}$$

satisfy equation (5), being x_0 an arbitrary point in $[h_1, h_2]$. It is shown in [20] that series (6) converge uniformly in $[h_1, h_2]$. Hence, differentiation term by term leads to the series

$$\psi_1' = \frac{\psi_0'}{\psi_0}\psi_1 + \frac{1}{\psi_0}\sum_{k=1}^{\infty}(-\lambda)^k \widetilde{X}^{(2k-1)}, \qquad \psi_2' = \frac{\psi_0'}{\psi_0}\psi_2 + \frac{1}{\psi_0}\sum_{k=0}^{\infty}(-\lambda)^k X^{(2k)}.$$

Let $x_0 = h_1$ then, according to recursive procedure (7), ψ_1 and ψ_2 satisfy the initial conditions

$$\psi_1(h_1;\lambda) = \psi_0(h_1), \quad \psi_1'(h_1;\lambda) = \psi_0'(h_1); \qquad \psi_2(h_1;\lambda) = 0, \quad \psi_2'(h_1;\lambda) = 1/\psi_0(h_1).$$

By choosing ψ_0 such that $\psi_0(h_1) = 1$ and $\psi_0'(h_1) = \mathrm{i}$ (see Appendix A on the construction of such a particular solution), the characteristic equation takes the series representation

$$\begin{aligned} \mathrm{i}\,(1 + \kappa_1(\lambda)) &\left(\theta(\lambda)\sum_{k=0}^{\infty}(-\lambda)^k X^{(2k+1)}(h_2) + \sum_{k=0}^{\infty}(-\lambda)^k X^{(2k)}(h_2)\right) \\ &- \left(\theta(\lambda)\sum_{k=0}^{\infty}(-\lambda)^k \widetilde{X}^{(2k)}(h_2) + \sum_{k=1}^{\infty}(-\lambda)^k \widetilde{X}^{(2k-1)}(h_2)\right) = 0, \end{aligned} \tag{8}$$

where $\theta(\lambda) := (\psi_0'(h_2) - \mathrm{i}\kappa_3(\lambda)\psi_0(h_2))\,\psi_0(h_2)$.

Remark 1. *According to recursive integration procedure (7) one can see that the interaction potential V_2 is not explicitly invoked in the calculation of the formal powers $\widetilde{X}^{(n)}$ and $X^{(n)}$. However, this information is implicitly carried on in the particular solution ψ_0. If ψ_0 is obtained with the SPPS method (see Appendix A), then the condition $V_2 \in \mathscr{C}[h_1, h_2]$ should be fulfilled. This extends the applicability of our method to a large class of potentials.*

Without losing generality, let $V_1 = V_3 = 0$, so that $\kappa_1^2 = \kappa_3^2 = \lambda =: \kappa^2 \in \mathbb{C}$. In this case, the characteristic equation can be written in terms of the complex variable κ as $\widetilde{\varrho}(\kappa) = 0$, where

$$\widetilde{\varrho}(\kappa) := \mathrm{i}(1+\kappa)\left(\widetilde{\theta}(\kappa)\sum_{k=0}^{\infty}(-\kappa)^{2k}X^{(2k+1)}(h_2) + \sum_{k=0}^{\infty}(-\kappa)^{2k}X^{(2k)}(h_2)\right) - \left(\widetilde{\theta}(\kappa)\sum_{k=0}^{\infty}(-\kappa)^{2k}\widetilde{X}^{(2k)}(h_2) + \sum_{k=1}^{\infty}(-\kappa)^{2k}\widetilde{X}^{(2k-1)}(h_2)\right),$$

and $\widetilde{\theta}(\kappa) := (\psi_0'(h_2) - \mathrm{i}\kappa\psi_0(h_2))\,\psi_0(h_2)$. This is the Taylor series of an analytic function $\widetilde{\varrho}$ defined in the complex κ-plane. If κ_n is a zero of $\widetilde{\varrho}(\kappa) = 0$ then $\lambda_n = \kappa_n^2$ is a zero of the characteristic equation (8). Hence, in order for λ_n to be a complex energy of a resonance state it is necessary that $\mathrm{Re}(\kappa_n) > 0$ and $\mathrm{Im}(\kappa_n) < 0$. Numerically, the previous series should be truncated up to a *M-th* term. Let $\widetilde{\varrho}_M$ be the truncated version of $\widetilde{\varrho}$. Then, approximate zeros of the equation $\widetilde{\varrho}(\kappa) = 0$ can be found from the polynomial roots of $\widetilde{\varrho}_M(\kappa) = 0$. In other words, the calculation of resonances for a one-dimensional quantum system reduces to calculating polynomial roots.

Next we present some numerical examples using *Wolfram Mathematica*. For the truncated version $\widetilde{\varrho}_M$ of the characteristic equation we take $M = 200$. This implies calculating a finite set of formal powers according to recursive integration procedure (7). Each integration is carried out by segmenting the interval $[h_1, h_2]$ into 2000 segments on which the integrand is interpolated by cubic splines. Exact integration is then performed on the resulting interpolated integrands. For calculating polynomial roots we employ the instruction `FindRoot`, using machine precision.

Example 1. Consider a rectangular potential defined by $V_2(x) \equiv V_0$ on the symmetric support $[-h, h]$, see Figure 1.a. With the settings $V_0 = 10$ and $h = 1$ we calculate the transmission coefficient $\mathbb{T}(\epsilon)$ on the interval $\epsilon \in [0, 80]$ of real energies. In Figure 1.b the transmission coefficient shows a series of maxima corresponding to $|\mathbb{T}(\epsilon_n)| = 1$. The values ϵ_n shown in Table 1 are the transparency energies of the potential V.

Example 2. With the settings of Example 1 we calculate the transmission coefficient $\mathbb{T}(\lambda)$ in a region $\Omega = \{\lambda \in \mathbb{C} : 0 < \mathrm{Re}(\lambda) < 100,\ -60 < \mathrm{Im}(\lambda) < 10\}$ of the complex λ-plane, see Figure 2. Here we observe a series of peaks corresponding to the poles of the transmission coefficient $\mathbb{T}(\lambda)$. Some of the complex energies λ_n of this quantum system are shown in Table 1.

Remark 2. *Piecewise constant potentials are important since more complex potentials can be represented in terms of them. Their simplicity can lead to closed-form characteristic equations that can be solved graphically or using standard numerical methods* (e.g., *Newton method*). *Other potentials can be approached by performing Darboux transformations on the solutions corresponding to piecewise constant potentials (see,* e.g., *[4, 6]). In this work the analysis begins from a non-vanishing particular solution ψ_0 associated to the potential V_2, from which we calculate a finite set of formal powers $\widetilde{X}^{(n)}$, $X^{(n)}$ that compose the truncated characteristic equation $\widetilde{\varrho}_M(\kappa) = 0$, and calculate its polynomial roots. On the other hand, the transmission coefficient, which is usually approximated as a superposition of Breit-Wigner distributions (see,* e.g., *[3, 5]), here is totally characterized by its poles, which are the zeros of $\widetilde{\varrho}(\kappa) = 0$.*

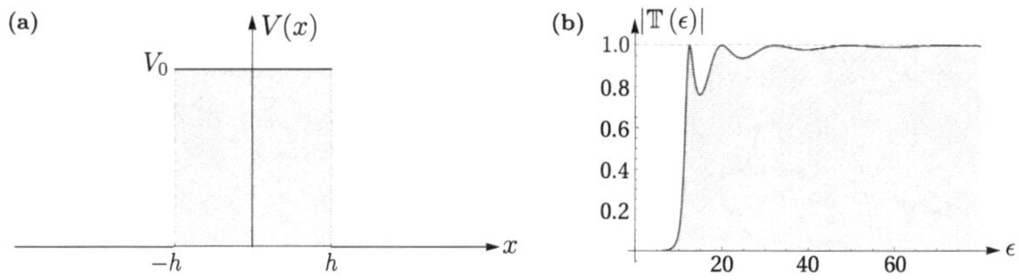

Figure 1. (a) Rectangular potential of Example 1. (b) Plot of the transmission coefficient.

Table 1. Some transparency energies and complex energies for the potential of Example 1.

n	Transparency energy ϵ_n	Complex energy $\lambda_n = E_n - \mathrm{i}\Gamma_n/2$
1	12.467401810269243	11.882874871230369 - i 1.2842071483079915
2	19.869815644509725	17.967206726724196 - i 4.9820599537268042
3	32.204502676194192	28.893263517977555 - i 10.532816267432139
4	49.436758521132234	44.892105206950130 - i 17.344759864814847
5	71.565365406214943	65.978437727543150 - i 25.052640345112376

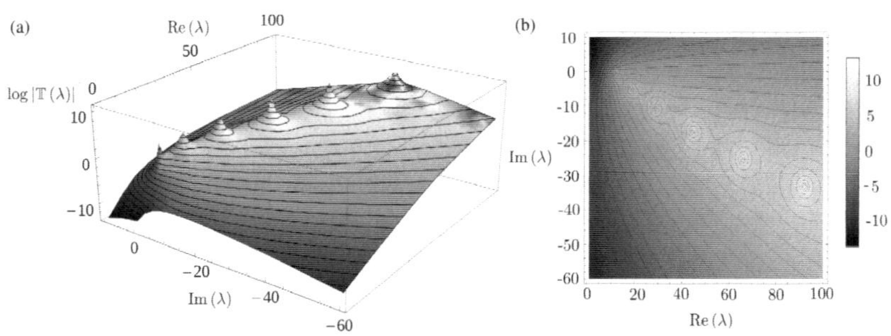

Figure 2. (a) Peaks representing the poles of the transmission coefficient $\mathbb{T}(\lambda)$ of Example 2. (b) Distribution of complex energies in the region Ω.

Example 3. In order to show the applicability of our method to a more interesting potential, let us consider a truncated version of the modified Pöschl-Teller potential, Figure 3.a,

$$V_2(x) = 10\frac{\mu(\mu-1)}{\cosh^2(x)}, \qquad -1 \le x \le 1.$$

By setting $\mu = 3$ and $h = 1$, the transmission coefficient $\mathbb{T}(\epsilon)$ evaluated on the interval $\epsilon \in [0, 160]$ of real energies is shown in Figure 3.b. Some transparency energies for this problem calculated from the polynomial roots of $\widetilde{\varrho}_M(\kappa) = 0$, are shown in Table 2.

Example 4. As a continuation of Example 4 we calculate the transmission coefficient $\mathbb{T}(\lambda)$ in a region $\Omega = \{\lambda \in \mathbb{C} : 0 < \mathrm{Re}(\lambda) < 180, -60 < \mathrm{Im}(\lambda) < 10\}$ of the complex λ-plane, see Figure 4. Here we observe that two complex energies (λ_1 and λ_2) are very close each other, forming a kind of cluster. In Table 2 we observe the complex energies λ_n calculated for this potential and the closeness of the two mentioned energies.

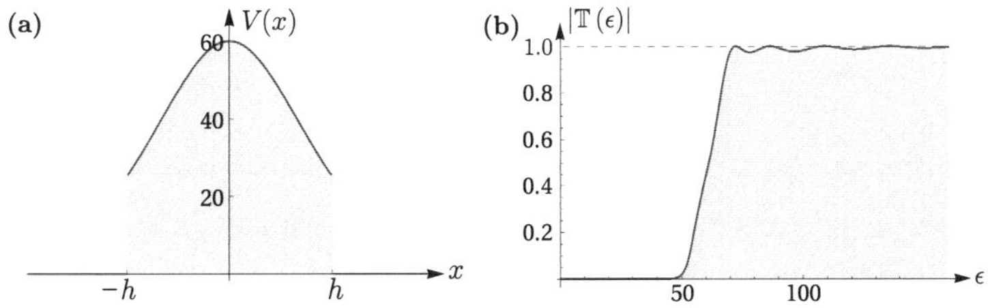

Figure 3. (a) Pöschl-Teller potential of Example 4. (b) Plot of the transmission coefficient.

Table 2. Some transparency energies and complex energies for the potential of Example 3.

n	Transparency energy ϵ_n	Complex energy $\lambda_n = E_n - \mathrm{i}\Gamma_n/2$
1	72.00775933091033	53.80659491247304 - i 5.796460655368762
2	85.01644361908951	54.57397005390005 - i 8.329055510961702
3	108.00016333744483	68.08250284745327 - i 8.590087191115057
4	135.00003458628376	105.04504696926546 - i 20.82541895927201
5	150.00060076698233	131.29281605305934 - i 27.47415336739937

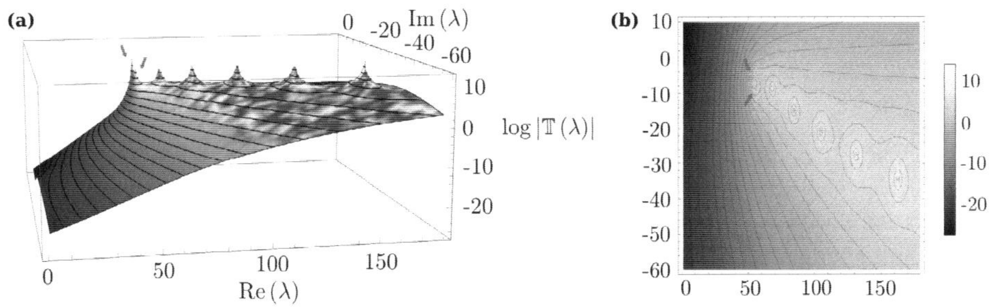

Figure 4. (a) Peaks representing the poles of the transmission coefficient $\mathbb{T}(\lambda)$ of Example 4. (b) Clustering of the complex energies λ_1 and λ_2 in the region Ω.

5. Concluding remarks

In this work we considered a one-dimensional Schrödinger equation with a short range potential V_2 defined in a region $[h_1, h_2]$, satisfying the smoothness condition $V_2 \in \mathscr{C}[h_1, h_2]$. We obtained an explicit representation of the characteristic equation for its Gamow-Siegert eigenstates in the form of a power series of the the complex energy λ. From a numerical point of view, such series should be truncated up to some M-th term. Thus, the problem of finding the complex energies of the Gamow-Siegert eigenstates of the Hamiltonian $\mathcal{H}$ reduces to calculating the zeros of equation (8) or equivalently to finding the polynomial roots of the equation $\widetilde{\varrho}_M(\kappa) = 0$. This constitutes an effective way of analysing the resonances of a quantum system. It is worth mentioning that no approximation nor simplification was used for obtaining the characteristic equation, in addition, the smoothness condition $V_2 \in \mathscr{C}[h_1, h_2]$ allows to apply the present theory to problems having quite arbitrary potentials. Finally, possessing such a series representation for the characteristic equation allows us to analyse other interesting phenomena such as the degeneration of resonances, which roughly consists in determining the multiplicity of the zeros of equation (8) by means of the standard techniques of complex analysis such as the argument principle or Rouché's theorem.

Appendix A. On the construction of the particular solution ψ_0

The SPPS method can also be used for constructing a non-vanishing particular solution ψ_0 of equation (5) for $\lambda = 0$. Let φ_1 and φ_2 be two linearly independent solutions, then the solution $\psi_0 = \varphi_1 + \mathrm{i}\varphi_2$ does not vanish in $[h_1, h_2]$ since φ_1 and φ_2 alternate their zeros according to Sturm's oscillation theorem. Let φ_0 be a non-vanishing particular solution of $-\varphi'' = 0$ in (h_1, h_2) satisfying the conditions $V_2\varphi_0^2$, $\varphi_0^{-2} \in \mathscr{C}[h_1, h_2]$, then the series

$$\varphi_1(x) := \varphi_0(x) \sum_{k=0}^{\infty} \widetilde{Y}^{(2k)}(x), \qquad \varphi_2(x) := \varphi_0(x) \sum_{k=0}^{\infty} Y^{(2k+1)}(x) \tag{A.1}$$

with $\widetilde{Y}^{(n)}$, $Y^{(n)}$ defined by the recursive integration procedure $\widetilde{Y}^{(0)}(x) \equiv 1$ and $Y^{(0)}(x) \equiv 1$,

$$\widetilde{Y}^{(n)}(x) := \begin{cases} \displaystyle\int_{x_0'}^{x} \widetilde{Y}^{(n-1)}(s) V_2(s) \varphi_0^2(s)\,\mathrm{d}s, & \text{odd } n, \\ \displaystyle\int_{x_0'}^{x} \widetilde{Y}^{(n-1)}(s) \frac{\mathrm{d}s}{\varphi_0^2(s)}, & \text{even } n, \end{cases} \qquad Y^{(n)}(x) := \begin{cases} \displaystyle\int_{x_0'}^{x} Y^{(n-1)}(s) \frac{\mathrm{d}s}{\varphi_0^2(s)}, & \text{odd } n, \\ \displaystyle\int_{x_0'}^{x} Y^{(n-1)}(s) V_2(s) \varphi_0^2(s)\,\mathrm{d}s, & \text{even } n, \end{cases}$$

satisfy equation (5) for $\lambda = 0$ in $[h_1, h_2]$. The point x_0' is an arbitrary point in $[h_1, h_2]$. In particular, taking $\varphi_0 \equiv 1$ and choosing $x_0' = h_1$ simplify significantly the work. According to the SPPS method [20], series (A.1) converge uniformly on $[h_1, h_2]$, and differentiation term by term gives an expression for the derivative ψ_0', that is

$$\psi_0'(x) = \sum_{k=0}^{\infty} \left(\widetilde{Y}^{(2k+1)}(x) + \mathrm{i} Y^{(2k)}(x) \right).$$

The solution ψ_0 thus constructed satisfies the initial conditions $\psi_0(h_1) = 1$, $\psi_0'(h_1) = \mathrm{i}$.

References

[1] Gamov G 1928 *Z. Phys.* **51** 204
[2] Siegert A J F 1939 *Phys. Rev.* **56** 750
[3] Fernández-García N and Rosas-Ortiz O 2008 *Ann. Phys.* **323** 1397
[4] Fernández-García N and Rosas-Ortiz O 2008 *J. Phys.: Conf. Series* **128** 012044
[5] Orozco-Cortés L F and Fernández-García N 2014 *J. Phys.: Conf. Series* **512** 012027
[6] Rosas-Ortiz O, Fernández-García N and Cruz y Cruz S G 2008 *AIP Conf. Proc.* **1077** 31
[7] Gadella M 1983 *J. Math. Phys.* **24** 1462
[8] Gadella M 1983 *J. Math. Phys.* **24** 2142
[9] Gadella, M 1984 *Physica A* **124A**,317
[10] De la Madrid R 2004 *J. Phys. A: Math. Gen.* **37** 8129
[11] Moiseyev N 2009 *J. Phys. Chem. A* **113** 7660
[12] Rosas-Ortiz O and Muñoz R 2003 *J. Phys. A: Math. Gen.* **36** 8497
[13] Fernández D J and Rosu H C 2001 *Phys. Scr.* **64** 177
[14] Brändas E and Elander N eds 1989 *Resonances. The Unifying Route Towards the Formulation of Dynamical Processes. Foundations and Applications in Nuclear, Atomic and Molecular Physics.* (Lecture Notes in Physics vol 325) eds Araki H *et al* (Berlin: Springer).
[15] Kukulin V I, Krasnopol'sky V M and Horáček J 1989 *Theory of Resonances. Principles and Applications* (Reidel Texts in the Mathematical Sciences Vol. 3) (Netherlands: Springer)
[16] Moiseyev N 2011 *Non-Hermitian Quantum Mechanics* (Cambridge: Cambridge University Press)
[17] Snyder A W 1974 *Appl. Phys.* **4** 273
[18] Cruz y Cruz S G and Razo R 2015 *J. Phys.: Conf. Series* **624** 012018

[19] Cruz y Cruz S G and Rosas-Ortiz O 2015 *Adv. Math. Phys.* **2015** 281472
[20] Kravchenko V V and Porter R M 2010 *Math. Method Appl. Sci.* **33** 459
[21] Bahar E 1967 *J. Math. Phys.* **8** 1735
[22] Bender C M and Orszag S A 1978 *Advanced Mathematical Methods for Scientists and Engineers* (New York: McGraw Hill)
[23] Cao Z, Zham L and Chen Y 1996 *Proc. SPIE* **2891** 289
[24] Ascher U M, Mattheij R M M and Russell R D 1995 *Numerical Solution of Boundary Value Problems for Ordinary Differential Equations* (Philadelphia: SIAM)
[25] Butcher J C 2003 *Numerical Methods for Ordinary Differential Equations* (West Sussex: John Wiley & Sons)

IOP Conf. Series: Journal of Physics: Conf. Series **839** (2017) 012005 doi:10.1088/1742-6596/839/1/012005

Resonances in Cut-Off Potentials Determined by the Exterior Complex-Scaling Method

A Jaimes-Nájera

Physics Department, Cinvestav, AP 14-740, 07000 México City, Mexico

E-mail: ajaimes@fis.cinvestav.mx

Abstract. The resonances of one-dimensional short-range potentials defined by a cutoff are analyzed in terms of a modification of the well known complex-scaling method. Such a modification consists in transforming the eigenfunctions only in the zone where the potential is identically zero. As usual, the complex-scaled Hamiltonian is non-Hermitian and the functions that represent resonances become square-integrable. The method is illustrated by calculating the resonances of a concrete cut-off potential.

1. Introduction

Resonances in quantum mechanics are special cases of scattering states for which the 'capture' of the incident wave (projectile) by the scatterer produces delays in the scattered wave. They are represented by solutions of the Schrödinger equation associated to complex energy eigenvalues $\epsilon = E - i\Gamma/2$ and satisfying purely outgoing conditions. Here E is interpreted as the binding energy of a decaying system which is composed by the scatterer and the projectile, and $\tau = \hbar/\Gamma$ as the corresponding life-time. In contrast with the conventional scattering wave-functions, such a kind of solutions to the Schrödinger equation are not finite at large distances. Therefore, some approaches have been introduced to extend the formalism of quantum theory so that the resonance states can be defined in precise form (for detailed information see e.g. [1]). Of special interest, the complex-scaling method is useful to transform the resonance eigenfunctions into square-integrable functions [1–4]. The latter at the price of transforming the initial Hamiltonian into a non-Hermitian operator.

In the present paper we analyze the resonances associated with cut-off potentials which are defined as $V(x) = v(x)\,\Theta\,(\xi - |x|)$, with Θ the step function, and v a real-valued smooth function. The interval $[-\xi, \xi] \in \mathbb{R}$ defines an *interaction zone* where the potential is not identically zero, so that our model represents a short-range interaction centered at the origin of coordinates and delimited by the *cutoff* $\xi \geq 0$. As these potentials are not analytic in general, the conventional complex-scaling method is not necessarily applicable. Therefore, we shall follow the *exterior complex scaling method* [5] to transform the cut-off potentials outside the interaction zone only. With this aim we revisit the main points of the conventional complex-scaling method in Sec. 2. The modifications of the method addressed to study the cut-off potentials are indicated in Sec. 3 and some examples are given. In Sec. 4 we give some final remarks.

IOP Conf. Series: Journal of Physics: Conf. Series **839** (2017) 012005 doi:10.1088/1742-6596/839/1/012005

2. Complex Scaling

Let $k = k_r + ik_i = |k|e^{-i\beta}$ be the wave number defining an scattering wave with $k_r, k_i, \beta \in \mathbb{R}$. If $0 < \beta < \pi/2$ (equivalently, $k_r > 0$ and $k_i < 0$), then k is in the fourth quadrant of the complex k-plane and defines a resonance with 'complex energy' $\epsilon = k^2$. Therefore, the corresponding eigenfunctions are such that $u_\epsilon(x) \sim e^{\pm i|k|x\cos\beta}e^{\pm|k|x\sin\beta}$ as $x \to \pm\infty$. Using the operator

$$S_0 = e^{i\,\theta\, x\, d/dx}, \qquad \theta \in \mathbb{R}, \tag{1}$$

and the Baker-Campbell-Hausdorff formula [6] $e^A B e^{-A} = \{e^A, B\} = \sum_{n=0}^{\infty} \frac{1}{n!}\{A^n, B\}$, one can show that the initial Hamiltonian H is transformed into the non-Hermitian operator

$$S_0 H S_0^{-1} \equiv H_\theta^0 = -e^{-2i\theta}\frac{d^2}{dx^2} + V(xe^{i\theta}), \tag{2}$$

and that the new eigenfunction is such that $\widetilde{u}_\epsilon(x \to \pm\infty) = e^{\pm i|k|x\cos(\theta-\beta)}e^{\mp|k|x\sin(\theta-\beta)}$. Therefore, $\widetilde{u}_\epsilon$ is square-integrable whenever $0 < \theta - \beta < \frac{\pi}{2}$ [1]. In particular, for $\beta = 0$ we realize that the scattering states are transformed as $e^{\pm i|k|x} \to e^{\pm i|k|x\cos\theta}e^{\mp|k|x\sin\theta}$. That is, plane waves are transformed into exponential decreasing or increasing functions for large values of $|x|$. To preserve the oscillating form of the scattering wave-functions we must make $k = |k| \to |k|e^{-i\theta}$. The latter induces a rotation of the positive real axis in the clockwise direction by the angle 2θ, so that the rotated energy is complex $\varepsilon = E(\cos 2\theta - i\sin 2\theta)$ [1]. Thus, the complex-rotation is such that bound energy states are preserved while the resonance states are now represented by square-integrable functions.

3. Exterior complex scaling

Following [5], we propose the transformation

$$\widetilde{x} = \begin{cases} e^{i\theta}(x+\xi) - \xi, & x \le -\xi \\ x, & |x| < \xi \\ e^{i\theta}(x-\xi) + \xi, & x \ge \xi \end{cases} \tag{3}$$

The rotated variable $\widetilde{x} = xe^{i\theta}$ is a complex-valued function of x which is continuous in $\mathbb{R}$ and has a derivative with discontinuities at $x = \pm\xi$. Let $\psi(x)$ be a solution to the Schrödinger equation associated with the cut-off potential $V(x) = v(x)\,\Theta(\xi - |x|)$. Outside the interaction zone, it can be denoted as ψ_I and ψ_{III} for $x \le -\xi$ and $x \ge \xi$ respectively. In the interaction zone $|x| < \xi$, we shall write ψ_{II}. The boundary conditions for the transformed solution $\psi(\widetilde{x}(x))$ are as follows

$$\psi_I(\widetilde{x})|_{x=-\xi} = \psi_{II}(\widetilde{x})|_{x=-\xi}, \qquad \psi_{II}(\widetilde{x})|_{x=\xi} = \psi_{III}(\widetilde{x})|_{x=\xi}, \tag{4}$$

$$\left.\frac{d\psi_I(\widetilde{x})}{d\widetilde{x}}\right|_{x=-\xi} = \left.\frac{d\psi_{II}(\widetilde{x})}{d\widetilde{x}}\right|_{x=-\xi}, \qquad \left.\frac{d\psi_{II}(\widetilde{x})}{d\widetilde{x}}\right|_{x=\xi} = \left.\frac{d\psi_{III}(\widetilde{x})}{d\widetilde{x}}\right|_{x=\xi}. \tag{5}$$

To avoid the non-differentiability of $\psi(\widetilde{x})$ at $x = \pm\xi$, let us introduce a twice-differentiable function of x in the form

$$w(x) = \begin{cases} e^{i\theta}(x+\xi) - \xi + w_1(x), & x \le -\xi \\ x, & -\xi \le x \le \xi \\ e^{i\theta}(x-\xi) + \xi + w_2(x), & x \ge \xi \end{cases} \tag{6}$$

where the functions w_1 and w_1 are as smooth as necessary and such that

$$\lim_{x\to-\infty} w_1(x) = 0, \qquad \lim_{x\to+\infty} w_2(x) = 0. \tag{7}$$

Therefore

$$w(x) \to e^{i\theta}x, \qquad \text{as } |x| \to \infty. \tag{8}$$

The differentiability requirements on w impose the following conditions on w_1 and w_2,

$$w_1(-\xi) = w_2(\xi) = 0, \qquad \left.\frac{dw_1}{dx}\right|_{x=-\xi} = \left.\frac{dw_2}{dx}\right|_{x=\xi} = 1 - e^{i\theta}, \tag{9}$$

$$\left.\frac{d^2w_1}{dx^2}\right|_{x=-\xi} = \left.\frac{d^2w_2}{dx^2}\right|_{x=\xi} = 0, \tag{10}$$

It can be shown that $w_1(x) = (1-e^{i\theta})(x+\xi)(1-(x+\xi))e^{x+\xi}$ and $w_2(x) = (1-e^{i\theta})(x-\xi)(1+x-\xi)e^{-(x-\xi)}$ satisfy the equations (7), (9) and (10). After some straightforward calculations one arrives at the differential operator

$$\frac{d^2}{dw^2} = (1 - V_2(x))\frac{d^2}{dx^2} - V_1(x)\frac{d}{dx}, \tag{11}$$

where the functions

$$V_1(x) = \begin{cases} \dfrac{w_1''}{(e^{i\theta}+w_1')^3}, & x \le -\xi, \\ 0, & |x| < \xi, \\ \dfrac{w_2''}{(e^{i\theta}+w_2')^3}, & x \ge \xi, \end{cases} \qquad V_2(x) = \begin{cases} 1 - \dfrac{1}{(e^{i\theta}+w_1')^2}, & x \le -\xi \\ 0, & |x| < \xi \\ 1 - \dfrac{1}{(e^{i\theta}+w_2')^2}, & x \ge \xi \end{cases} \tag{12}$$

are continuos in $\mathbb{R}$ since w is twice differentiable. Then, the complex-scaled Hamiltonian can be written as

$$H_\theta = -\frac{d^2}{dw^2} + V(w) = -\frac{d^2}{dx^2} + V_2(x)\frac{d^2}{dx^2} + V_1(x)\frac{d}{dx} + V(w(x)). \tag{13}$$

Notice the kinetic– and flux–like terms defined by the position-dependent coefficients $V_2(x)$ and $V_1(x)$, the latter respectively such that $\lim_{x\to\pm\infty} V_2(x) = 1 - e^{-i\theta}$ and $\lim_{x\to\pm\infty} V_1(x) = 0$. Besides, in the interaction zone, the Hamiltonians H_θ and H are identical because the transformation leaves invariant the function $v(x)$. In Fig 1 we show the behavior of the functions (12) outside the interaction zone.

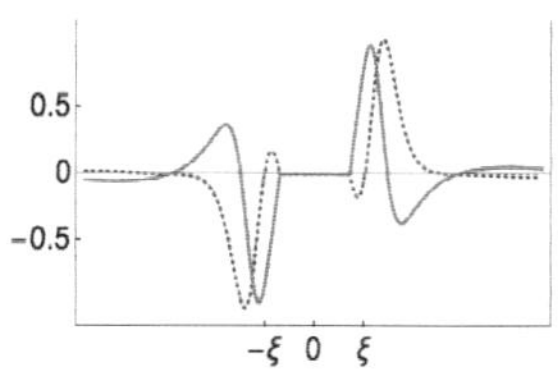

(a)

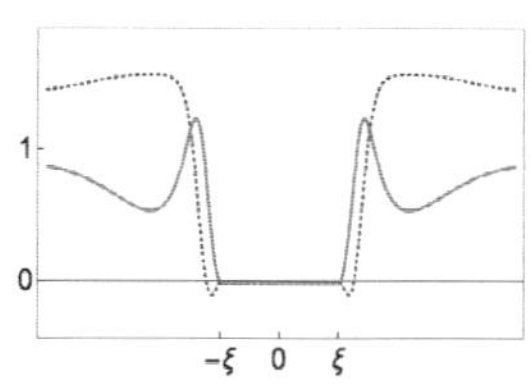

(b)

Figure 1. (Color online) Real (dotted-blue) and imaginary (continuous-red) parts of the functions V_1 (a) and V_2 (b) defined in (12) for $\theta = 1$. Notice that both functions are equal to zero in the interaction zone $[-\xi, \xi]$.

3.1. Examples

To obtain the transformed resonances of concrete problems, given a Hamiltonian, we only have to substitute the function (6) in the corresponding eigenfunction. For instance, consider the semi-harmonic barrier V_b defined by

$$V_b(x) = \begin{cases} v_2, & x \in (-\xi, -b), \\ x^2, & x \in (-b, b), \\ v_1, & x \in (b, \xi), \\ 0, & \text{otherwise}, \end{cases} \tag{14}$$

where $b, v_1, v_2 > 0$ [7]. The behavior of the new complex-valued potential is defined by the combination of (14) and (12), compare Figs. 1 and 2.

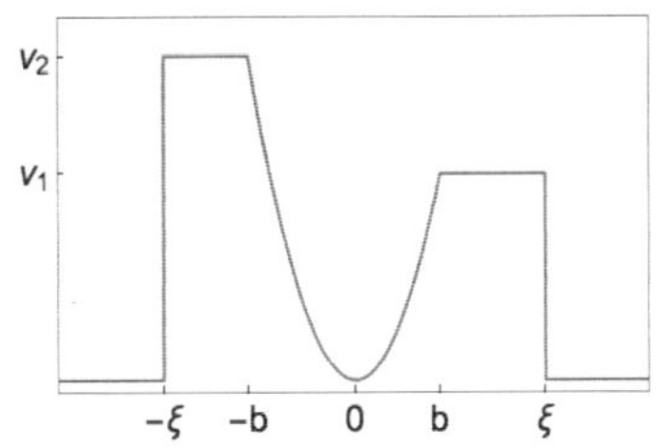

Figure 2. The semi-harmonic potential defined in (14).

In the panel of Fig. 3 we show the results of the exterior complex scaling for two different values of θ. In the upper row ($\theta = 0.05$), two poles in the fourth quadrant of complex k-plane have been rotated up to a position which is very close to the real axis, see Fig. 3(a). The wave-function of the resonance which is closest to the real axis is plotted in Fig. 3(b). Such a function is still not square-integrable, see the corresponding squared modulus in Fig. 3(c). In turn, the kinetic– and flux–like terms defined by V_1 and V_2 in (13) are shown in figures 3(d) and 3(e). For a greater value of θ, namely $\theta = 0.77$, in Fig. 3(f) we appreciate that both resonances have been rotated to the first quadrant of the k-plane. As a result, the above described wave-function is now square-integrable, see Figs. 3(g) and 3(h). Besides, as indicated in Figs. 3(i) and 3(j), the kinetic– and flux–like terms are not oscillating any more and show extremal behavior in the vicinities of $x = \pm\xi$ only.

4. Concluding remarks

We have studied the complex scaling method for cut-off potentials. We have shown that performing the transformation outside the interval where the discontinuities of the potential lie, the complex scaling can be done if the transformed variable is well behaved at the cut-off. We have shown that additional terms are present in the scaled Hamiltonian which can be interpreted as diffusion and flux terms, and that in the case of cut-off potentials they are the unique change in the scaled Hamiltonian, so that they carry all the physical information about the complex scaling. Our results are in agreement with those reported in e.g. [8]. The applicability of our approach is shown with a concrete cut-off potential. The study of the physical information of the diffusion and flux terms is part of work under progress.

Acknowledgments

The author acknowledges the funding received through a CONACyT scholarship. Fruitful discussions with Prof. Oscar Rosas-Ortiz are appreciated.

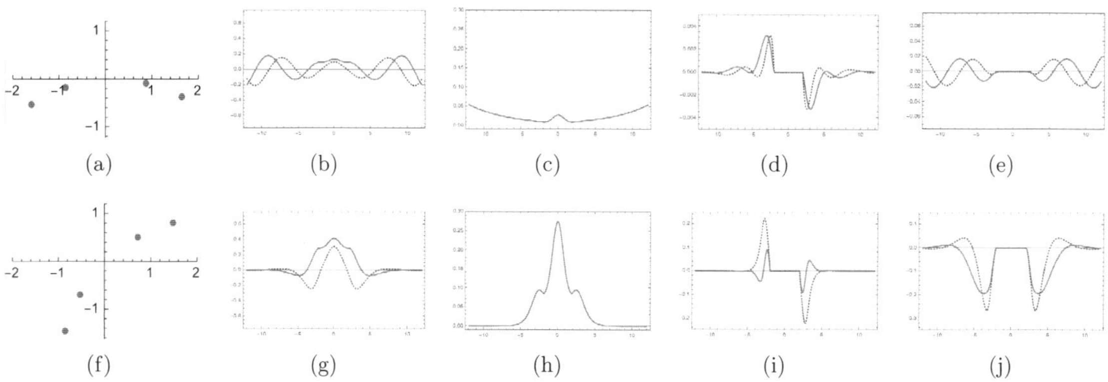

Figure 3. Results of the exterior complex-scaling method applied to the semi-harmonic barrier potential (14) with $v_1 = v_2 = b^2$, $b = 1$, $\xi = 2$, for $\theta = 0.05$ (upper row) and $\theta = 0.77$ (lower row). In both cases, from left to right, the columns correspond to the poles in the complex k-plane that define the resonances, the real (dotted-blue) and imaginary (continuous-red) parts of the wave-function belonging to the resonance which is closest to the real axis, the corresponding squared modulus, the real (dotted-blue) and imaginary (continuous-red) parts of the kinetic– and the flux–like terms defined by the functions V_1 and V_2 in (13).

References

[1] Rosas-Ortiz O, Fernández-García N and Cruz y Cruz S 2008, A Primer on Resonances in Quantum Mechanics, *AIP Conf. Proc.* **1077** 31

[2] Moiseyev N 1998 Quantum theory of resonances: calculating energies, widths and cross-sections by complex scaling, *Phys. Rep.* **302** 211-293

[3] Aguilar J and Combes J M 1971 A class of analytic perturbations for one-body Schrdinger Hamiltonians. *Comm. Math. Phys.* 22(4), 269-279

[4] Balslev E and Combes J M 1971 Spectral properties of many-body Schrdinger operators with dilatation-analytic interactions. *Comm. Math. Phys.*, 22(4), 280-294

[5] Simon B 1979 The definition of molecular resonance curves by the method of exterior complex scaling, *Phys. Lett. A* **71** 211

[6] Mielnik B and Plebanski J 1970 Combinatorial approach to Baker-Campbell- Hausdorff exponents, *Ann Inst Henry Poincaré* XII 215

[7] Jaimes-Nájera A 2016 Transmission and Reflection of Wave Packets by Asymmetric Semi-Harmonic Potential Barriers, *J. Phys.: Conf. Ser.* **698** 012030

[8] Rom N, Engdahl E and Moiseyev N 1990 Tunneling rates in bound systems using smooth exterior complex scaling within the framework of the finite basis set approximation, *J. Chem. Phys.* **93** 3413

IOP Conf. Series: Journal of Physics: Conf. Series **839** (2017) 012006 doi:10.1088/1742-6596/839/1/012006

Empty Bottle: The Revenge of Schrödinger's Cat?

B Mielnik
Physics Department, Cinvestav, A.P. 14-740, 07000, México City, Mexico
E-mail: bogdan@fis.cinvestav.mx

Abstract. We discuss some measurement problems which in spite of all progress of the statistical interpretation seem not well understood in the present day quantum theory.

1. Introduction

Since the basic idea of E. Schrödinger [1, 2] about the linear evolution of the wave packets, the quantum theories show an impressive mathematical progress, but rather little fundamental changes. The pure states of a microsystem are always represented by vectors in the Hilbert spaces in which they always navigate linearly, except if the process is interrupted by a measurement. The results of a measurement are statistical and can be predicted with certainty only for some special states. The probabilities correspond to the quadratic forms of the state vectors and their statistical averages are $\langle\psi|A|\psi\rangle$, where A are certain self-adjoint operators. This form of probabilities was predicted already by Schrödinger who, however, was against their present day interpretation.

A key was the *stability problem*, represented most naively on Fig. 1. In the first instants or in the fast repetitions of any measurement, the apparatus readings (e.g., the indication of the needle) should not change, or must be almost the same. Otherwise, the measurement could not be performed, because of the instability of the results (i.e. the dancing of the needle on the apparatus scale). So, the initial micro-object state must suddenly be *reduced* to one of states for which the measurement result is *certain* (and therefore stable). Almost all physicists accepted the argument...

But not everybody! While he himself guessed the quadratic form of the probabilities[1], Schrödinger was in strong disagreement with the sudden jumps ('reductions'). He exclaimed: *Diese Verdammte Quantenspringerei!*, cited by Sudbery [5]. In fact, is there any fundamental barrier between the measuring devices and the rest of physical objects? Is there any fundamental difference between the microscopic and macroscopic bodies? If not, then why certain macroscopic objects should cause some unpredictable jumps? Moreover, why the quantum 'indeterminacy', originally restricted to micro-objects, could not be shared by macroscopic systems? In his ample and fundamental paper of 1935 [6] (see also [2]), Schrödinger presents his short but famous tale about a *half alive* and a *half dead* animal:

[1] It is also worth noticing that if the interference patterns on our screens were non-quadratic, it could mean some different quantum theories [3, 4].

IOP Conf. Series: Journal of Physics: Conf. Series **839** (2017) 012006 doi:10.1088/1742-6596/839/1/012006

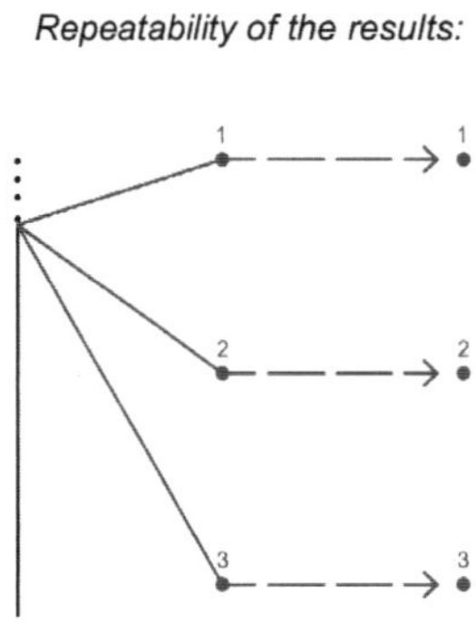

Figure 1. The stability of the observed result needs the assumption that after each measurement the system state jumps to one of special states for which the statistical distribution is reduced to certainty. This state is no longer changed by fast repetitions of the measurement.

> "One can even set up quite ridiculous cases. A cat is penned up in a steel chamber, along with the following device (it must be secured against direct interference by the cat): in a Geiger counter there is a tiny bit of radioactive substance, *so* small that *perhaps* in the course of the hour one of the atoms decays, but also, with equal probability, perhaps none; if it happens, the counter tube discharges and through a relay releases a hammer which shatters a small flask of hydrocyanic acid. if one has left this entire system to itself for an hour, one would say that the cat still lives *if* meanwhile no atom has decayed. The psi-function of the entire system would express this by having in it the living and the dead cat (pardon the expression) mixed or smeared in equal parts."

The fragment occupies less than 1/300 part of an ample 22 page article. Yet, almost nobody remembers the profound ideas of E. Schrödinger, but almost everybody heard about his humorous story about the cat in a superposed state of being *dead* and being *alive*. However, almost everybody thinks, it is just a dark anecdote of the past. But is it?

• **The Hawking's gun.** The situation antagonized strongly Stephen Hawking: *When I hear of the Schrödinger's cat, I reach for my gun!* However, the Schrödinger's cat will probably survive the Hawking's gun. The point is that until now we cannot construct a truly consistent picture of the quantum state reduction.

2. The decoherence?

One of widely announced answers is that the mechanism of wave packet reductions, "these damned quantum jumps", during the measurements are indeed the *decoherence* effects of the dynamical evolution, leading always to an entropy increase [7]. In fact, some increase of disorder in our prediction of the future states is rather obvious. The data about the present state of either classical or quantum particle system, even if good, are never infinitely exact. So even the little errors of our "initial knowledge" might cause an increasing blurring in our predictions of the future. The problem, however, is whether this lost of information is in any way responsible for the sudden state jumps (interpretable rather as an escape from the disorder?)

By trying to understand the phenomenon in an arbitrarily complex but still micro-reversible theory, one inevitably faces the seldom remembered *Loschmidt paradox.* It states that while the prediction of the future might be increasingly blurred (entropy increase!) then if only the theory is micro-reversible, i.e. the change of t into $-t$ is irrelevant for the dynamical laws, the disorder (blurring) has the tendency to grow either if we try to predict the future or to reconstruct the past. So the 'entropy' (or at least the 'entropy of our personal information') is rather bad defined: the smallest is always at the 'present moment', growing both in the future and in the past! This, however is not limited to our subjective feeling, since in micro-reversible physical theories the system evolution in the direction of the past is governed by an identical law as its evolution toward the future [8, 9] (c.f. Fig 2).

Figure 2. A naive image of Lozschmidt paradox: an extremely regular particle state shows an increasing disorder in the future, but also in the past.

The paradox makes senseless the persistent belief of growing disorder (entropy) in a large class of physical theories. It includes all known cases of classical (multi-particle) dynamics without friction. No different in quantum theories. No matter, how many subsystems describes a quantum state, in all customarily considered cases of quantum dynamics, the evolution 'toward the past', after a simple CPT conjugation, is identical to the 'motion to the future.

An independent argument in favor of the decoherence is frequently presented by considering a macroscopic object with two macro-states A and B (for instance $A =$ cat alive, $B =$ cat dead). Now, if the statistical measurements reduce any initial superposition into a random mixture of either A or B, this means the decoherence; but it does not yet mean an inverse statement, that the decoherence reduces the wave packet. Indeed, a different measurement, with the eigenstates $C = \frac{1}{\sqrt{2}}(A + B)$ and $D = \frac{1}{\sqrt{2}}(A - B)$ might create the same mixture (decoherence) without 'reducing' the state to either A or B. The question then remains, why an increment in he entropy must 'reduce' the system state to either A or B (alive or dead) but not to e.g. $C + D$ (either alive+dead or alive−dead), or to a continuous mixture of all qubit states representing the same state of complete chaos [10]?

Hence, if the unitary evolution is the only element of the story, there is no such miracle, that one of these directions will introduce more disorder than the other (c.f. also [8, 9]). The situation can be different if the unitary evolution is not the only law, but is completed by some additional mechanism. Only then, it might explain the growing decoherence in the future. Yet, the question still remains, whether apart of the growing disorder (entropy) it will indeed describe the sudden 'reduction jumps' of quantum states?

• **The computer experiment of Gisin and Percival.** An intriguing computer study of Gisin and Percival published in 1992 [11] shows a distinguished role of the energy eigenstates for the harmonic oscillator perturbed by a dissipative term consistent with Lindblad formalism [12]. It turns out, that the average number of photons emitted stabilizes asymptotically around the energy levels (see e.g.Fig. 10 in Gisin and Percival paper [11]). The results awoke an enormous interest (more than 600 citations, though curiously, it is not even mentioned in the latest Gisin article [13]).The authors of [11] choose the Lindblad dissipative term which privileges the direction of time and is determined by the oscillator algebra. Would it represent the missing element of the theory?

3. The relativistic problems

The equally serious problems with the *Verdammte Quantenspringerei* become obvious in the relativistic theories. As it seems, they must appear even in the special relativity, if only the evolving (and reduced) states are defined on the infinite space-like hyperplanes. The simplest case is presented on Fig. 3.

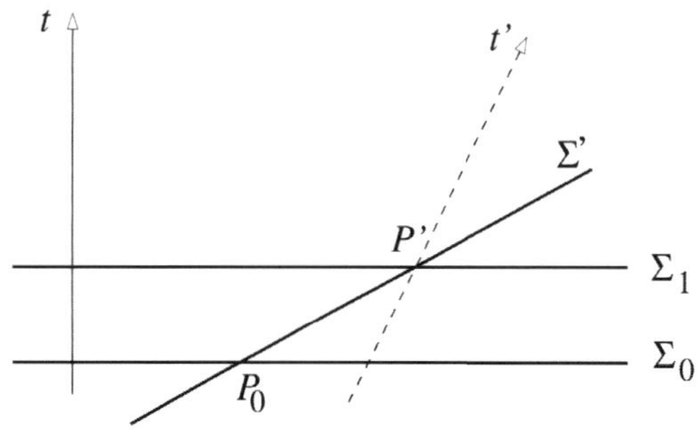

Figure 3. The measurement performed in point P' of the plane Σ_1 by a moving observer with different simultaneity hyperplane Σ' can affect the probability distribution around the point P_0 of the hyperplane Σ_0.

The more general symptoms of the trouble are reported e.g. as the *Paradox of two bottles* [14], see Fig. 4. Suppose, a wave packet of one particle is trapped in a closed bottle which is then split carefully into two parts (without affecting coherence), tentatively with 1/2 and 1/2 probabilities of containing the particle. The bottles then travel into widely separated places of our galaxy, both still containing 50% of the initial packet. Now, the problem is about the wave packet reduction on some extended space-like hyperplane Σ_0.

Suppose an observer whose simultaneity hyperplane is Σ_0 checks the interior of bottle 1 and finds the particle. Hence, the probability of finding the particle in bottle2 becomes 0. But not only on Σ_0. If now another observer, who moves differently and has the slightly distinct simultaneity hyperplane (Σ_1) checks the interior of the bottle 1 immediately after, by continuity, he must find the particle in bottle 1 as well. Hence, the bottle 2 was empty already on Σ_1, etc. By induction, the bottle 2 was already empty on the hyperplanes Σ_1, Σ_3, $\Sigma_6, \ldots$, and the bottle 1 certainly contained the particle on Σ_0, Σ_2, $\Sigma_4, \ldots$ By an obvious limiting transition, one must conclude that the bottle 2 was empty and the particle was from the very beginning in the bottle 1; the wave packet reduction affected the whole past, from the very moment when the original bottle was split.

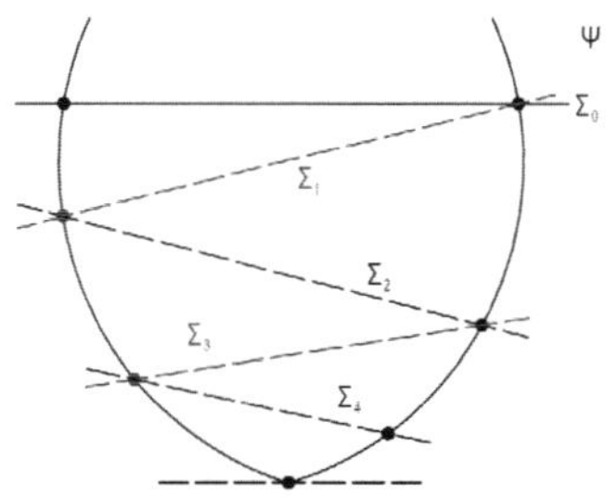

Figure 4. A case of the relativistic reduction for two traveling bottles containing two parts of the wave packet of one particle.

About one year later I suddenly noticed that exactly the same problem was already considered much earlier by Y. Aharonov and D. Albert [15, 16], though in less simplified way and with

IOP Conf. Series: Journal of Physics: Conf. Series **839** (2017) 012006 doi:10.1088/1742-6596/839/1/012006

different conclusions. In some panic I have published a corrigendum [17], recognizing their contribution, though maintaining my point of view. Soon, some new contributions appeared.

In 1992 J. Finkelstein criticize the very idea of an observer dealing with (and reducing) the quantum state defined on a space-like hyperplane in Minkowski space [18]. The measurements are always local, he argued, so the wave packet reduction can affect only the future causality cone of the measuring apparatus, never an entire space-like hyperplane! A polemic answer was published by Mosley in 1993 [19], with an opposite argument, following Hellwig and Krauss [20]. Any local observer can construct the wave packet of a micro-object only on the basis of some signals arriving from the past causal cone. If some experiment brings a new information, it means that the state description must be corrected, implying the state reduction on the past cone. Both arguments show some logical merits. However, if the wave packet reduction affected only the future light cones (Finkelstein) then it could not explain the (benign) teleportation, cryptography, etc of the entangled states. On the other hand, the quantum states limited to the past cones (Mosley), could not offer unitary evolution laws which require more ample dynamical informations (on the simultaneity hyperplanes). What we need are the quantum evolution equations for states defined on the families of Cauchy hyper-surfaces (Fig. 3): the information which suffers a difficulty even for the simplest state reductions represented on Fig. 5. The situation certainly is not easier in the "interaction free experiment" of Elitzur and Vaidman.

Figure 5. The reduction doctrines of Finkelstein (left) and Mosley (right).

4. Interaction free interaction?

In their inspired article [21, 22], Elitzur and Vaidman consider a photon in a system of optical fibers, with beam splitters and mirrors of Mach-Zehnder interferometer, see Fig. 6. The photon wave function is divided into two coherent parts by the first beam splitter, then reflected by two mirrors toward the second splitter, where they unify again, recovering their original state of motion. So, if there is no obstacle, the photon recovers its original propagation momentum, falling into the detector D. However, if one of the branches is blocked e.g. by a perfectly absorbing obstacle, then what occurs is the first state reduction. Either the obstacle detects (by absorbing) the photon which therefore arrives neither to D nor to E. Or it reduces the entire propagation scheme, cancelling the blocked trajectory. The second splitter will then receive the photon which propagated only along the unblocked path (as if the first splitter acted as a mirror). Its state is then decomposed by the second splitter into the superposition of two parts, tentatively reaching D or E. The choice of one of them will be the second state reduction. The peculiar effect is that in $1/2$ cases the first reduction eliminates completely one of the trajectories, while the second one makes an equilibrated choice between D and E, so in $1/4$ of cases the photon will appear in E, thus revealing the existence of an obstacle, to which it has never approached. Elitzur and Vaidman choose still a more challenging version of the experiment, assuming that the obstacle is a supersensitive bomb, which would explode immediately under any contact with the photon. Hence, if the detector E clicks, it would mean that the bomb could be detected (without exploding) by a photon which could pass many kilometers away!

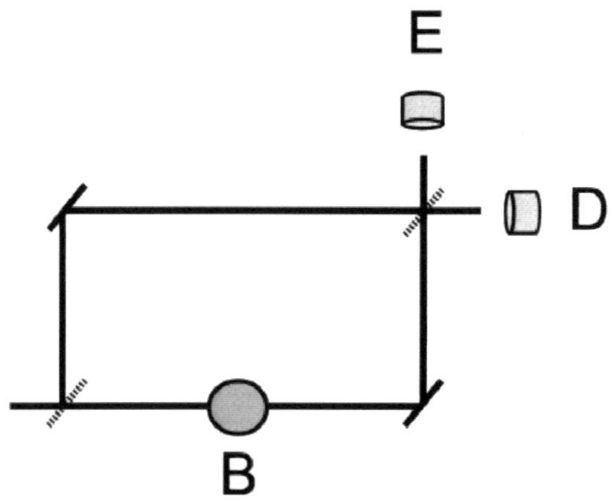

Figure 6. Elitzur and Vaidman 'interaction free experiment'. An absorbing obstacle B (e.g. a bomb) eliminates one of possible photon trajectories. The photon, moving only along the obstacle-free way, has the probability 1/4 to fall to the detector E, thus revealing the existence of an obstacle with which it never collided.

All this can seem natural if the photon is just an instantaneous pulse. However, what is precisely the *single photon*? Must it propagate always as an infinitely short pulse? Or perhaps, it can also form a very long, narrow wave divided by the first splitter into a pair of still weaker but as long components which laboriously reconstruct their initial form at the second splitter, falling then (gradually) into the detector D? If so, the problem is, at which moment precisely the detector responds to the *single photon*? At the beginning or at the end of the process? In fact, whatever we try, it seems completely impossible to imagine the single photon wave in presence of an obstacle. Can it *know* from the very beginning that one of the trajectories is blocked? If not, then what can it do after arriving to the obstacle. Does it announce: "Oh excuse me, I was wrong, I have to return!". But then, too late, since it would have no time to reappear in the free fiber![2].

Worse, because if one of the (EV) trajectories is blocked by the bomb, then after what time the bomb explodes? If it does not, then after what time the (large but incomplete) photon trajectory which *tried to cross the bomb* is mysteriously annihilated and contributes (again mysteriously), to the other weak component creating finally the (entire) one-photon state, which arrives to second beam splitter, but now with the probability to activate the second detector E? We can only conclude that the story is incomplete: indeed, it is impossible to form any mental picture of the obligatory *linear propagation* if it means the extinction of the whole propagation arm, detecting an obstacle which exists precisely in the place where the photon, finally, newer was! Here, it is worth to remind the point made by Sudbery [23]:

> "It is often stated that however puzzling some of its features may be, quantum mechanics does constitute a well defined algorithm for calculating physical quantities. Unless some form of continuous projection postulate is included as a part of the algorithm, this is not true."

All this, except some simple cases, does not yet solve entirely the photon behavior in the topological nets of optical fibers. The tensor products of Hilbert spaces were introduced to assure the *linear propagation of the entangled states* between the multiple detectors. An enormous

[2] According to our colleagues from Stony Brooks, the single photon wave can be arbitrarily weak but arbitrarily long and still obeys the measurement axioms. (The discussion on the annual meeting of DICU, Monterrey Oct. 2016).

success of the picture was the prediction of the *benign* teleportation acts, (which did not permit to read immediately the teleported messages) [24]. It started the ample works on *quantum cryptography* with hopes to program the efficient quantum computers [13, 25, 26]. Despite all interpretational questions, the (EV) idea [21] opened also unsuspected perspectives of "seeing in the dark" [27], with the perspectives of new age in quantum information permitting the use of powerful quantum computing [28]. On the other hand, sequence of studies of the *imperfect* cases of (EV) bombs was undertaken [29–32].

Apart all these achievements, some ambitious trends in Quantum Field Theories (QFT) started already to talk about the "Theory of Everything" as if the end of fundamental research was not too far away. Yet, some signs indicate that the satisfaction might be premature. Besides all known relativistic problems, some simple evaluations indicate that the experiments with linearly propagating entangled states can affect the past [29]. The similar paradoxical conclusion on *quantum steering into the past* seems to reemerge from the 2012 study of Vienna-Innsbruck group [33]. In spite of the 'benign' teleportation mechanism one can wonder whether our quantum theories do not to cross some common sense limits neglected also in the past.

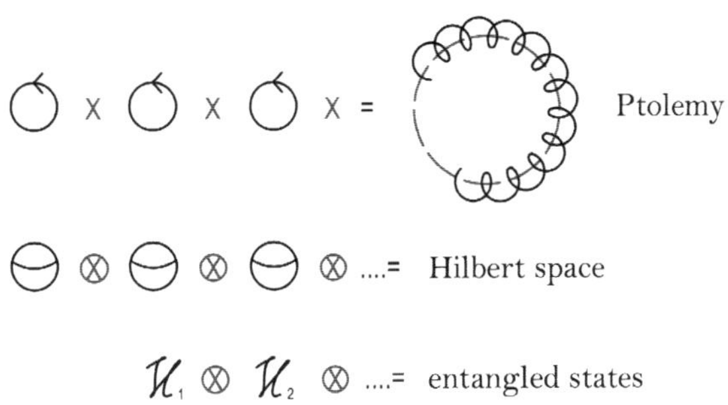

Figure 7. Ptolemean aspects of the present day quantum theories, created by the universal idea of linear navigation.

Indeed, the persistent image of the entangled states navigating in the tensor product spaces has some similarity to the thousands of years of the ancient astronomy. Just remember the Ptolemean idea to describe all trajectories by the *holy shape* of the circle. The present day "tensor products" use another *holy shape* of the 2D surface of the 3D euclidean sphere (the quantum *qubit*, see Fig. 7). The Ptolemean astronomy was, in fact, not wrong, even if complicated. However, what about our theory of microscopic phenomena?

In order to recover some independent view, let us postpone for a moment the locality problems, returning to the traditional quantum paradoxes still waiting for some better understanding.

5. Half full, half empty

Our story will concern a quantum system in a superposed energy state, which will be reduced–though not when the experimentalist decides, but when the system itself decides by emitting a photon (compare with the 'time of arrival' [34–36]). As a simplified model, we consider a bottle containing an atom in a state of equitable *superposition* of two lowest energy levels, ground state ϕ_0 and an excited state ϕ_1.

Since our theory until today obeys a persistent intuition of quantum (pure) state which must always linearly navigate until suffering some 'reduction', let me try to adopt it in the most naive way and examine consequences.

In some distant past, the experimentalists examining the spectral lines imagined an atom always in one of the energy eigenstates. Today the picture changed. The existence of the

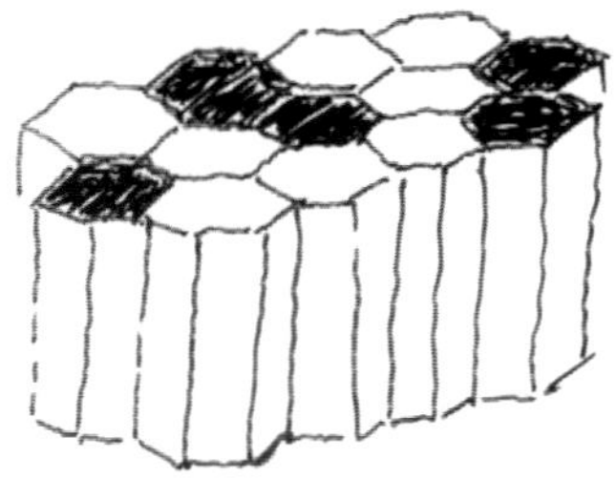

Figure 8. The idealized bottles.

superposed (but pure) energy states is (or it seems?) unavoidable if one takes seriously the quantum mechanical formalism. Now, if the atom is in the excited state, we shall say that "the bottle is full", but if in the ground state, "the bottle is empty". The bottle is just to assure that the atom is left in peace, isolated from the external perturbations. It should be wide enough to neglect the influence of its surface on the atom behavior, but it should be also able to detect the events of radiation. So, at the top of each bottle, at some safe distance, there is a sensitive screen, prepared to detect the photon, should the atom radiate. If it does, the top of the cell turns black (it is burned!). For purely illustrative reasons, the bottles on Fig. 8, are painted hexagonal, resembling the bee hive (bees today are also in troubles!). By observing the detectors which turned black, we can see, how many photons already 'incubated'.

Now, in almost all studies of the atom radiation one can find the description of the process starting from the excited state ϕ_1 but not from the superposed one. This allows the suggestive representation of the excited states as some narrow superpositions of slightly different energy eigenstates, forming an unstable composition, with the average lifetime τ inverse to the (little) energy width δE, in agreement with the time-energy uncertainty (even though, all this awakes a lot of unfinished discussions [34–36]). Anyhow, by reading the literature you can always find the considerations in which the beginning of the decay process is an *excited state*, with a slightly diffused spectral line, which seems to confirm the validity of the idea. However, what about the study of a decay starting from the superposition of quite distant levels? Perhaps, the difference is superfluous, but it may be worth to examine.

To fix attention, let us thus assume that our initial state ϕ is an equitable superposition $\phi = a_0\phi_0 + a_1\phi_1$, where $|a_0|^2 = |a_1|^2 = 1/2$ (bottle half full, half empty). From a credible phenomenology we know the behavior of an atom in its ground state ϕ_0. If unperturbed, it just remains in ϕ_0 forever, $\phi_0(t) = exp(-itE_0)\phi_0$. We also know something about the behavior of the excited state ϕ_1. In purely quantum mechanical approximation, this state is as stationary

$$\phi_1(t) = e^{-itE_1}\phi_1 \tag{1}$$

In reality though, the stationary evolution might be interrupted by an unpredictable photon emission with a sudden jump to the ground state ϕ_0. The number of bottles $N(t)$ with undecayed atoms thus goes down; assuming an almost exponential decay, it behaves approximately as $N(t) = N_0 e^{-\Gamma t}$. This is sometimes represented in the literature by assigning to the unstable state a complex energy $E = E_1 + i\frac{\Gamma}{2}$ so that (1) can be replaced by $\phi_1(t) = e^{-itE_1 - \frac{\Gamma t}{2}}\phi_1$. Yet, this is just a verbal allegory, pretending to attribute to the single atoms the properties of their ensemble. Suppose, however, that the transitions $\phi_1 \to \phi_0$ are incomparably faster than the average lifetime of the whole ensemble in the excited state ϕ_1. Then we are again in the old idea of the excited states in their *linear navigation* interrupted by sudden jumps (indeed, the state reductions), caused by the photon detectors. This does not yet affect essentially our model of the excited state radiation.

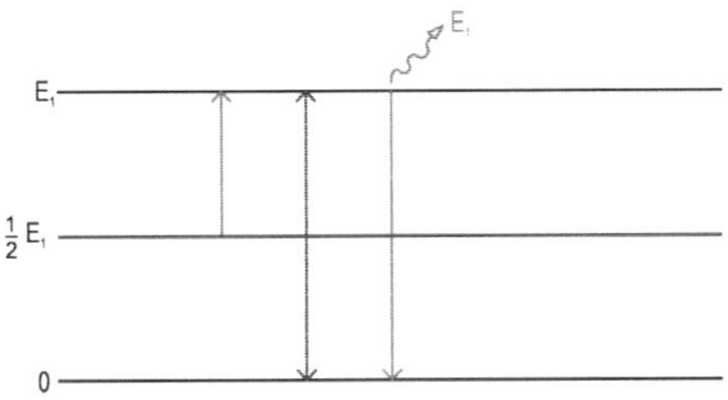

Figure 9. An image of spontaneous reduction: before radiating each 'undecided' atom must first choose whether it is in the excited or in the ground state.

However, what happens with the atoms in the initial *superposed energy state* $\phi = \frac{1}{\sqrt{2}}(\phi_0 + \phi_1)$ (the bottle *neither full, nor empty*)? At the first sight, it may seem that there is hardly any problem here: the evolution of the atom must simply obey the standard law

$$\phi(t) = \frac{1}{\sqrt{2}}(e^{-iE_0 t}\phi_0 + e^{-iE_1 t}\phi_1), \tag{2}$$

granted by the superposition principle, except if (bam!), it suddenly radiates emitting a photon of energy $E_1 - E_0$, colocating itself on the ground state ϕ_0. Yet, this plausible picture contains certain disquieting gap. We are accustomed to think, that the quantum system in presence of a measuring device performs first of all a unitary evolution process in a certain Hilbert space (an extremely continuous picture?), until interrupted by the sudden reduction (an extremely discontinuous picture?).

On the other hand, in our initial state the average energy is $\frac{1}{2}(E_1 - E_0)$ and nobody observed the photons emitted with (incomplete) energies smaller than $\Delta E = E_1 - E_0$. Does it means that before radiating, some atoms must perform first of all a spontaneous (*introspective?*) state reduction (see Fig. 9), deciding whether they are in ϕ_0 or ϕ_1? The question then is, whether they must ask some energy credit from their detector? If so, is the detector kindness due to its very existence, even if the photon was not yet emitted [37,38], or is it a kind of *shadowing* [39], or generous grant from the polarized vacuum? (Though notice the criticism of Penrose about the very concept of the 'polarized vacuum' [40]) Furthermore, if the *spontaneous reduction* failed to bring some more energy, locating the semi-excited atom on the ground state ϕ_0 then it will stay there without never emitting anything. Even if the total energy balance is not violated, the single atom behavior hides still some mystery.

6. The experiment of vanishing hope?

It may be therefore curious to imagine a population of N atoms in the initial state $\phi = \frac{1}{\sqrt{2}}(\phi_0 + \phi_1)$, each closed in its own bottle, in form of a little, mesoscopic cell. The top surface of each cell is simultaneously a detector, sensitive to the photons of particular energy $\hbar\omega = E_1 - E_0$. By calculating the (increasing) number of the black cells, we know how many atoms have already radiated, see Fig. 9. If all atoms are initially in an identical superposed state ϕ, then for $t \to +\infty$ all must end up in the ground state ϕ_0, though for different reasons: 1/2 of them, since they have radiated and settled down on ϕ_0; the remaining 1/2, just because the long waiting was a kind of yes-no measurement (i.e., asking whether the atom was able to radiate at all), the negative answer reducing the ϕ_1 component to nonexistence, even though no photon was emitted. As the matter of fact, what has caused the state collapse in this last case, was not any active intervention, but just the vanishing hope (that the atom could have been in the excited state ϕ_1). Thus e.g., supposing that the average lifetime of the excited state ϕ_1, was 1 nsec., but the atom in the initial state ϕ did not radiate for 1 year, then, we can be practically certain that it will radiate never: the atom state can no longer correspond to ϕ, but it is indistinguishable from ϕ_0.

Even if the global energy balance is not affected, the situation seems a bit strange. While the atoms which have radiated cause already some problem, the ones which did not contain a true puzzle! Their superposed energy state vanished, giving place to ϕ_0. The only external factor was our vanishing hope (take it as a rhetoric figure if you dislike!), or perhaps shadowing [39], instability or some kind of non-linearity? Anyhow, the bottle was *half full, half empty*, nothing escaped, yet the bottle is *empty!*

The way to avoid the trouble would be to answer that we have *post-selected* our ensemble. The principle of "not post-select" (equivalently, "not retrospect") is reasonable if some micro-objects were submitted to a measurement. Their initial states (in general) were reduced and it makes no sense to look for their past. The principle is not so clear if the micro-objects *escaped* the detection. In their fascinating paper the Swiss group considers the non-orthogonal transformations of (pure into pure) states for an ensemble whose particles escaped absorption [37, 38]. The (EV) story of interaction free experiment too contain an element of retrospection. The same, the famous "delayed choice measurements" of J.A. Wheeler [2]. Similarly, as all paradoxes implying the reduction of the past states. So have we to restrict our theory only to strictly pragmatic rules, like *ensemble and only ensemble, correlation and only correlations*, resigning from considering the pure states of single particles? Limited by "do not retrospect" and other "do not think" principles? However the naive ideas were the true source of our sophisticated theories!

On the other hand, what if our idea of "spontaneous reductions" was overdone? Note that all difficulties of our strange story vanish if we simply assume that no coherent superposition of two bound atomic states can be indeed created. Then a variant of Bohr-Sommerfeld formula would turn sufficient to define atomic energy levels (back in the beginning of XX-century?). The quantum state of undecided energy could be only a mixtures and our over-sophisticated problems would not exist. Is it better?

7. Einstein boxes

About a year ago, I had an occasion to discuss with D. Albert in Tepoztlan, and I learned about something unexpected. "Oh!", said Albert after hearing my arguments on the measurement problems, "then you are the follower of Einstein boxes!". This turned out an old subject, unknown to participants of many discussions. After participating in the EPR paper [41], (truly written by Podolsky), deeply frustrated Einstein exchanged some angry opinions with Luis de Broglie, not officially published but reconstructed from some materials, see Norsen [42, 43]. The discussion was about a single particle state closed in two boxes (like in the paradox of 2 bottles [14], though without any reference to the relativistic aspects. The boxes were simply traveling to two distinct cities, Fig. 10). Let me just quote the categoric opinion of de Broglie:

> "Suppose, a particle is enclosed in a box B with impermeable walls. The associated wave ψ is confined in the box and cannot leave it. (...) Let us suppose that by some process or other, for example, by inserting a partition into the box, the box B is divided into two separate parts B_1 and B_2 and that B_1 and B_2 are then transported to two very distant places, for example to Paris and Tokyo. (...) The particle which has not yet appeared thus remains potentially in the assembly of the two boxes and its wave function ψ consists of two parts, one of which ψ_1 is located in B_1 and the other ψ_2 in B_2. (...) According to the usual interpretation (...) the particle (...) would be immediately localized in box B_1 in the case of a positive result in Paris. This does not seem to me to be acceptable. If we show that the particle is in box B_1 it implies that it was already there prior to localization", see Fig. 10.

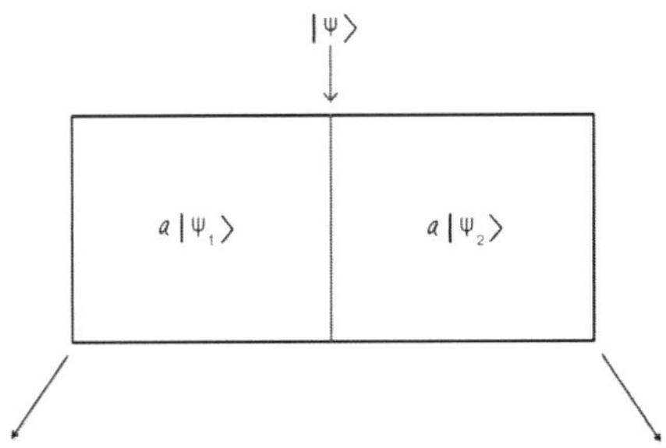

Figure 10. Einstein boxes.

I must admit that if not the discussion with Albert, I would still knew nothing about the Einstein boxes (the same as many other colleagues discussing the conceptual problems of teleportation, quantum computing etc.). The subject proves that the quantum paradoxes were quite well understood already by L. de Broglie, perhaps already before the 'official' doctrine of the quantum state reduction? Now, however, if the argument of de Broglie was right and if the linear superposition cannot exist for two perfectly isolated quantum states, then what about the ground and the first excited state of any hydrogen or hydrogen-like atom? True, they can never be completely isolated, but what if our present day theories are not exactly true, and so, in some circumstances (e.g. absence of Rabi rotations), ψ_0 and ψ_1 tend to behave like the Einstein boxes, with the linear combination inexistent? –so that in spite of illusions of creating such state we would always end up with the mixture instead of the superposition. In this case, the evolution problem of the linear combination of $\frac{1}{\sqrt{2}}(\psi_0 + \psi_1)$ would disappear, the number of emitted photons after any time τ would correspond exactly to the decaying ψ_1 i.e, $N(\tau) = N_0(1 - e^{-\Gamma\tau})$? This form of $N(\tau)$ would mean the decay a bit faster than the decay slightly delayed if the decayed atoms should wait for the spontaneous reductions of their $\phi = \frac{1}{\sqrt{2}}(\phi_0 + \phi_1)$ to ϕ_1 states. Can it be experimentally distinguished? If so, some information about the universal linearity (and the fate of the Shrödinger's cats) could be obtained.

Acknowledgments

The author is indebted to P. Kielanowski and D. Fernández C. for illuminating discussions; to M. Zelikin for a valuable information about the cyclone and magnetic vortices, to A. Odzijewicz for his interest in the subject, and to M.J. Rivero-Monrroy (ESIME-IPN, Mexico) for editing the manuscript. The support of the CONACyT project 49253-F is acknowledged.

References

[1] E. Schrödinger, *Collected papers on wave mechanics*, Chelsea, New York, 1978
[2] J.A. Wheeler and W.H. Żurek, *Quantum theory and measurement*, Princeton Univ. Press, Princeton, 1983
[3] J. Bell and M. Hallett, *Phil. Sci.* **49** (1981) 355
[4] D.C. Brody, L.F. Houghston, *J. Geom. Phys.*, **38** (2001) 19
[5] A. Sudbery, *Stud. Hist. Phil. Sci. B* **33** (2002) 387
[6] E. Schrödinger, *Naturwiss* **23** (1935) 807
[7] H.D. Zeh, *Found. Phys.* **1** (1970) 69
[8] S. Dolev, A.C. Elitzur, M. Hemmo, arXiv:quant-ph/0101088v1
[9] H. Price, *Time's Arrow and Archimedes Point*, Oxford University Press, Oxford, 1996
[10] E. Okon and D. Sudarsky, *Found. Phys.* **47** (2017) 120
[11] N. Gisin and J.C. Percival, *J.Phys. A* **25** (1992) 5677
[12] G. Lindblad, *Commun. Math. Phys.* **48** (1976) 119
[13] N. Gisin, *Can Relativity be Considered Complete? From Newtonian Nonlocality to Quantum Nonlocality and*

Beyond, in P. Blanchard and J. Fröhlich (Eds), The Message of Quantum Science Attempts, Towards a Synthesis, Springer, New York, 2015, Ch. 8

[14] B. Mielnik, *Found.Phys.* **20** (1990) 745

[15] Y. Aharonov and D.Z. Albert, *Phys.Rev. D* **24** (1981) 359

[16] Y. Aharonov and D.Z. Albert, *Phys.Rev. D* **29** (1984) 228

[17] B. Mielnik, *Found.Phys.* **21** (1991) 869

[18] J. Finkelstein, *Found.Phys.Lett.* **5** (1992) 383

[19] S.N. Mosley, *Phys.Lett.A* **182** (1993) 1

[20] K.E. Hellwig and K. Krauss, *Phys. Rev. D* **1** (1970) 566

[21] A.C. Elitzur and L. Vaidman, *Found. Phys.* **23** (1993) 987

[22] L. Vaidman, arXiv:quant-ph/0006077v1

[23] A. Sudbery, *Continuous state reduction*, in Quantum Concepts in Space and Time, Ed. R. Penrose, C.J. Isham, Oxford University Press, Oxford, 1986

[24] C.H. Bennett, G. Brassard and C. Crepeau, *Phys. Rev.Lett.* **70**, 1895 (1993)

[25] J.W. Pan, D. Bouwmeester, M. Damell, H. Weinfurter, A. Zeilinger, *Nature* **403** 515 (2000)

[26] T. Jennewein, C. Simon, G. Weights, H. Weinfurter, A. Zeilinger, *Phys.Rev.Lett.* **84** 4729 (2000)

[27] P. Kwiat, H. Weinfurter, and A. Zeilinger, *Sci. Am.* **278** (1996) 52

[28] S. Lloyd, M. Mosheni, P. Rebentrost, *Nature Phys.* **10** (2014) 631

[29] S. Lloyd, *Science* **338** (2012) 621

[30] A. Peruzzo, P. Shadbold, and N. Brunner *Science* **338** (2012) 634

[31] F. Kaiser, T. Coudreau, and P. Milman *Science* **338** (2012) 637

[32] Z. Blanco-Garcia and O. Rosas-Ortiz, *J.Phys.Conf.* **698** (2016) 012013

[33] X. S. Ma, S. Zotter, J. Kofler, R. Ursin, T. Jennewein, C. Brukner, A. Zeilinger, *Nature Phys.* **8** (2012) 479

[34] B. Mielnik, *Found. Phys.* **24** (1994) 1113

[35] C.R. Leavens, *Phys.Lett.A* **303** (2002) 154

[36] C.R. Leavens, *Phys. Lett. A* **338** (2005) 19

[37] B. Huttner, A. Muller, J.D. Gautier, H. Zbinden, and N. Gisin, *Phys. Rev. A* **54** (1996) 3783

[38] H. Bechmann-Pasquinucci, B. Huttner and N. Gisin, *Phys. Lett. A* **242** (1998) 198

[39] P.Ph. Blanchard and A. Jadczyk, *Helv. Phys. Acta* **69** (1996) 613

[40] R. Penrose, *The Road to Reality*, Publ. Jonatan Cape G.B., 2004, Ch.26.5 'Alternative vacua', p. 664

[41] A. Einstein, B. Podolsky, N. Rosen, *Phys. Rev.* **47** (1935) 777

[42] T. Norsen, *Am. J. Phys.* **70** (2002) 664

[43] T. Norsen, *Am. J. Phys.* **73** (2005) 164

IOP Conf. Series: Journal of Physics: Conf. Series **839** (2017) 012007 doi:10.1088/1742-6596/839/1/012007

Towards Modelling QFT in Real Metamaterials: Singular Potentials and Self-Adjoint Extensions

L M Nieto[1], M Gadella[1], J Mateos Guilarte[2], J M Muñoz-Castañeda[3], and C Romaniega[1,4]

[1] Departamento de Física Teórica, Atómica y Óptica and IMUVA, Universidad de Valladolid, Spain

[2] Departamento de Física Fundamental, Universidad de Salamanca, Spain

[3] Departamento de Física ETSIAE, Universidad de Politécnica de Madrid, Spain

[4] Departamento de Física Teórica II, Universidad de Complutense de Madrid, Spain

E-mail: luismiguel.nieto.calzada@uva.es, gadella@fta.uva.es, guilarte@usal.es, jose.munoz.castaneda@upm.es, csromaniega@gmail.com

Abstract. Solutions of the one-dimensional Schrödinger equation are found when point interactions of the type $a\delta(x-q)+b\delta'(x-q)$ are placed either in a couple of points or in a regular lattice. The results obtained in the present study are a first step toward a rigorous mathematical model of real metamaterials is Solid State Physics.

1. Introduction

One dimensional (1D) models with point interactions have received much attention in the recent literature (see Ref. [1] and references therein). They can model small and abrupt defects in materials allowing to perform analytic calculations such as the tunnel effect and the Casimir force. They are also interesting in the analysis of heterostructures in connection to an abrupt effective mass change. The general inquiry of point interactions of the free Hamiltonian $H_0=-\frac{\hbar^2}{2m}\frac{d^2}{dx^2}$ is mainly due to Kurasov [2] and it is based on the construction of self-adjoint extensions of symmetric operators with identical deficiency indices. Muñoz-Castañeda and coworkers [3] reformulated the theory of self-adjoint extensions of symmetric operators over bounded domains in terms of meaningful quantities from a Quantum Field Theory (QFT) point of view, allowing a rigorous study of the theory of quantum fields over bounded domains and the quantum boundary effects. In this context, it is well known that set of self-adjoint extensions of H_0 defined over the space $(-\infty,q)\cup(q,\infty)$ is given by the unitary group $U(2)$ and therefore is a 4-parameter family of operators. Some of some of these self-adjoint extensions are associated to the Hamiltonians H_0 plus point-supported potential $a\delta(x-q)+b\delta'(x-q)$, see [4]. This perturbation is of great interest in Quantum Mechanics from mathematical and physical points of view, and has been largely discussed in [4, 5]. Although it is totally clear which self-adjoint extension corresponds to the 1D-Dirac delta, there is no consensus on which one should be assigned to its derivative.

In the framework of QFT the spectra and eigenstates of non-relativistic Schrödinger operators provide one-particle states in scalar (1+1)-QFT systems. Point supported potentials can be used in scalar QFT on a line to model impurities and classical singular backgrounds that interact with quantum vacuum fluctuations. The most successful results in the frame of QFT are obtained in Casimir-type set-ups where the scattering data allow to compute the distance dependence finite part of the quantum vacuum energy for the quantum vacuum fluctuations that live between two plane parallel plates. In particular, configurations of two pure delta potentials added to the free Schrödinger Hamiltonian allowed to compute the quantum vacuum interaction energy between two parallel semitransparent Dirac δ-plates in terms of the scattering data of the pure double delta potential on the real line, see [5]. The same configuration of delta interactions is addressed as a perturbation of the Salpeter Hamiltonian. Additionally, generalized Dirichlet boundary conditions were discussed, showing that the use of δ-δ' potentials provides a much larger set of admissible boundary conditions, and the quantum vacuum energies between two plane parallel plates represented by a δ-δ' potential in arbitrary space-time dimension were computed [5].

Since the zero range potentials mimic the plates in a Casimir effect setup, it is interesting to consider a three-plate configuration (Casimir pistons) and investigate the Casimir forces acting of the plate in the middle that is supposed to be mobile in a piston configuration. In particular, it is meaningful to displace the central plate towards one of the other two placed in the boundary; thus, we find the main physical motivation to study the particular situation where two $\delta - \delta'$ interactions are superimposed. This is the first problem that will be addressed in Section 2, and the outcome is surprising: a non-abelian addition law emerges and it corresponds to the group law of the Borel subgroup of the $SL_2(\mathbb{R})$ group. The other focus of interest is the case where the δ'-couplings are exceptional, i.e., those couplings for which the transmission coefficients are zero and the plates become completely opaque: the left-right decoupling limit. It happens that the distinguished Dirichlet/Robin boundary conditions are implemented in this case by the $\delta - \delta'$-potentials but the superposition law just mentioned becomes singular [6].

An ideal model of electric conductivity in solids is provided by a δ-Dirac comb where δ-point interactions sit at the ions sites. This model can be enriched by adding a δ' potentials at every site of the lattice. This is the objective of Sec. 3, which is a work presently in progress [7].

2. One and two singular perturbations on a line

First of all, let us consider first the one-dimensional Schrödinger equation with singular potential

$$-\frac{\hbar^2}{2m}\frac{d^2}{dx^2}\psi(x) + a\delta(x-d)\psi(x) + b\delta'(x-d)\psi(x) = E\,\psi(x). \tag{1}$$

Introducing the dimensionless quantities $x = \frac{\hbar}{mc}y$, $d = \frac{\hbar}{mc}p$, $w_0 = \frac{2a}{\hbar c}$, $w_1 = \frac{mb}{\hbar^2}$, $\varepsilon = \frac{2E}{mc^2}$, and $\varphi(y) = \psi(x)$, the Hamiltonian is simply

$$-\frac{d^2}{dy^2}\varphi(y) + w_0\delta(y-p)\varphi(y) + 2w_1\delta'(y-p)\varphi(y) = \varepsilon\,\varphi(y). \tag{2}$$

The point potential we are interested in is usually defined via the theory of self-adjoint extensions of symmetric operators of equal deficiency indices, so that the total Hamiltonian in (2) is self-adjoint. The crucial point is to find the domain of wave functions $\varphi(y)$ that makes H_0 self-adjoint over the domain $\mathbb{R}/\{p\}$ and characterizes the potential $V(y)$. As these functions and their derivatives should have a discontinuity at $y = p$, we have to define the products of the form $\delta(y-p)\varphi(y)$ and $\delta'(y-p)\varphi(y)$. These can be done in several ways, and we choose the following:

$$\delta(y-p)\varphi(y) = \frac{\varphi(p^+) + \varphi(p^-)}{2}\,\delta(y-p)\,, \tag{3}$$

$$\delta'(y-p)\varphi(y) = \frac{\varphi(p^+) + \varphi(p^-)}{2}\,\delta'(y-p) - \frac{\varphi'(p^+) + \varphi'(p^-)}{2}\,\delta(y-p)\,. \tag{4}$$

To obtain a self-adjoint determination of $H = H_0 + V(y) = H_0 + w_0\delta(y-p) + 2w_1\delta'(y-p)$, we have to find a self-adjoint extension of H_0. To do it, we have to find a domain on which this extension acts, which is given by the Sobolev space $W_2^2(\mathbb{R}\setminus\{p\})$ of absolutely continuous functions $f(y)$ with absolutely continuous derivative $f'(y)$, both having arbitrary discontinuities at p, such that the Lebesgue integral given by

$$\int_{-\infty}^{\infty} \{|f(y)|^2 + |f''(y)|^2\}\, dy$$

converges, and such that at p satisfy the following matching conditions:

$$\begin{pmatrix} \varphi(p^+) \\ \varphi'(p^+) \end{pmatrix} = \begin{pmatrix} \dfrac{1+w_1}{1-w_1} & 0 \\ \dfrac{w_0}{1-w_1^2} & \dfrac{1-w_1}{1+w_1} \end{pmatrix} \begin{pmatrix} \varphi(p^-) \\ \varphi'(p^-) \end{pmatrix}. \tag{5}$$

These results could be extended to interactions of the type $\sum\big(a_i\delta(y-p_i) + 2b_i\delta'(y-p_i)\big)$, where the sum could be either finite or infinite.

2.1. Two singular perturbations on a line

We assume now that the total Hamiltonian is $H = H_0 + V + W$, with

$$H_0 = -\frac{d^2}{dy^2}, \qquad V = v_0\delta(y) + 2v_1\delta'(y), \qquad W = w_0\delta(y-p) + 2w_1\delta(y-p). \tag{6}$$

V is supported at the origin $y = 0$ and W is supported at $p > 0$. The potential $V + W$ is related to the Casimir effect. One of the motivations of this work is the study of the effect of the superposition of the two point potentials V and W at the same point (taking the limit as $p \to 0$). We shall see that we obtain a point potential of the type $u_0\delta(y) + 2u_1\delta'(y)$, where u_0 and u_1 are not the sums $v_0 + w_0$ and $v_1 + w_1$.

2.1.1. Matching conditions and scattering coefficients. In order to analyze $\varphi(y)$, the solution of (6), let us split the real line into the regions (1) $y < 0$, (2) $0 < y < p$, and (3) $p < y$, where

$$\varphi_i(y) = A_i\, e^{-iky} + B_i\, e^{iky}\,, \quad \varphi_i'(y) = -ik(A_i\, e^{-iky} - B_i\, e^{iky})\,, \quad i = 1,2,3\,, \quad k^2 = \varepsilon > 0.$$

Since the point potential V is defined by matching conditions like those in (5), we have

$$\begin{pmatrix} A_2 \\ B_2 \end{pmatrix} = K^{-1}\, M_v\, K \begin{pmatrix} A_1 \\ B_1 \end{pmatrix}, \quad M_v = \begin{pmatrix} \dfrac{1+v_1}{1-v_1} & 0 \\ \dfrac{v_0}{1-v_1^2} & \dfrac{1-v_1}{1+v_1} \end{pmatrix}, \quad K = \begin{pmatrix} 1 & 1 \\ -ik & ik \end{pmatrix}. \tag{7}$$

This expression gives the matching conditions at $y = 0$. At the point $y = p$ we get

$$\begin{pmatrix} A_3 \\ B_3 \end{pmatrix} = Q^{-1}\, K^{-1}\, M_w\, K\, Q\, K^{-1}\, M_v\, K \begin{pmatrix} A_1 \\ B_1 \end{pmatrix} = T_p \begin{pmatrix} A_1 \\ B_1 \end{pmatrix}, \tag{8}$$

where the definition of T_p is obvious, while Q and M_w are respectively:

$$Q = \begin{pmatrix} e^{-ipk} & 0 \\ 0 & e^{ipk} \end{pmatrix}, \qquad M_w = \begin{pmatrix} \dfrac{1+w_1}{1-w_1} & 0 \\ \dfrac{w_0}{1-w_1^2} & \dfrac{1-w_1}{1+w_1} \end{pmatrix}. \tag{9}$$

The T_p-matrix in (8) relates the asymptotic behavior of the two linearly independent Jost scattering solutions at $y \ll 0$ with their counterparts at $y \gg 0$. Remark that

$$\det T_p = T_p^{11} T_p^{22} - T_p^{12} T_p^{12} = 1 \ , \ T_p^{11} = \bar{T}_p^{22} \ , \ T_p^{12} = \bar{T}_p^{21} \quad \Rightarrow \quad T_p \in SL_2(\mathbb{C}).$$

From the T_p-matrix one obtains the 2×2 unitary scattering matrix S_p:

$$S_p = \frac{1}{T_p^{11}} \begin{pmatrix} 1 & -T_p^{12} \\ T_p^{21} & \det T_p \end{pmatrix} \ , \ S_p^\dagger = \frac{1}{\bar{T}_p^{11}} \begin{pmatrix} 1 & \bar{T}_p^{21} \\ -\bar{T}_p^{12} & \det T_p \end{pmatrix} \ , \ S_p^\dagger S_p = \begin{pmatrix} 1 & 0 \\ 0 & 1 \end{pmatrix} . \tag{10}$$

The transmission and reflection coefficients provide the usual form of writing the S_p matrix:

$$S_p = \begin{pmatrix} t(k) & r_R(k) \\ r_L(k) & t(k) \end{pmatrix}, \tag{11}$$

where $t(k)$ is the amplitude of the transmitted waves coming from the far left or from the far right. Time-reversal invariance implies $t_R(k;p) = t_L(k;p) = t(k;p)$. However, as the potential is not parity invariant, the reflection amplitudes are different for incoming waves either from the far right or the far left: $r_R(k;p) \neq r_L(k;p)$. We explicitly find:

$$r_L(k;p) = -[e^{-2ipk}\left(2k\left(v_1^2+1\right)+iv_0\right)(4kw_1 - iw_0) + (4kv_1 - iv_0)\left(2k\left(w_1^2+1\right) - iw_0\right)]/\Delta(k),$$
$$r_R(k;p) = [e^{2ipk}\left(2k\left(v_1^2+1\right)-iv_0\right)(4kw_1 + iw_0) + (4kv_1 + iv_0)\left(2k\left(w_1^2+1\right) + iw_0\right)]/\Delta(k),$$
$$t(k;p) = 4k^2(1-v_1^2)(1-w_1^2)/\Delta(k),$$

where $\Delta(k) = e^{2ikp}(v_0 + 4ikv_1)(w_0 - 4ikw_1) + (2kv_1^2 + 2k + iv_0)(2kw_1^2 + 2k + iw_0)$ is a function of k and the other parameters of the problem.

2.2. Bound/antibound states and resonances

Complex zeroes of $T_p^{11}(k) = \Delta(k)/[4k^2(1-v_1^2)(1-w_1^2)]$ (poles of the S_p-matrix) give rise to bound or antibound states if are located on the purely imaginary, respectively positive or negative half-axis. Complex zeroes coming in pairs having opposite real part and identical imaginary part correspond to resonances. Then the relevant physical information is encoded in the analysis of complex zeroes of $\Delta(k)$. If we make the changes

$$2kp = z, \quad \sigma = \frac{pv_0}{1+v_1^2}, \quad \tau = \frac{pw_0}{1+w_1^2}, \quad v = \frac{v_1}{1+v_1^2}, \quad w = \frac{w_1}{1+w_1^2}, \tag{12}$$

then, $\Delta(k) = 0$ is equivalent to

$$e^{iz} = -\left(\frac{z+i\sigma}{\sigma+2ivz}\right)\left(\frac{z+i\tau}{\tau-2iwz}\right), \quad z = z_r + iz_i, \ z_r, z_i \in \mathbb{R}. \tag{13}$$

Equation (13) is a generalization of the so-called *generalized Lambert equation*, from where we can obtain two real equations, corresponding to its real and imaginary parts

$$\left[4vwz_r^2 - (2vz_i - \sigma)(2wz_i + \tau)\right]\cos z_r - 2z_r(\tau v + 4vwz_i - \sigma w)\sin z_r = \left[(z_i+\sigma)(z_i+\tau) - z_r^2\right]e^{z_i}.$$

$$2z_r(\tau v + 4vwz_i - \sigma w)\cos z_r + \left[4vwz_r^2 - (2vz_i - \sigma)(2wz_i + \tau)\right]\sin z_r = -z_r(2z_i + \sigma + \tau)e^{z_i}.$$

A simpler compatibility condition is this Lambert equation in two real variables

$$e^{2z_i} = \frac{\left(4v^2z_r^2 + (2vz_i - \sigma)^2\right)\left(4w^2z_r^2 + (2wz_i + \tau)^2\right)}{\left(z_r^2 + (z_i+\sigma)^2\right)\left(z_r^2 + (z_i+\tau)^2\right)}. \tag{14}$$

We can easily check that a simple solution of the system is $z_r = 0$, $z_i = 0$, but this corresponds to $k = 0$, which is a pole of $T_p^{11}(k)$. The typical behavior of the solutions of these equations is shown in Fig. 1, where the blue curves are the real part, the yellow ones the imaginary part and the green curve is the compatibility condition (14).

IOP Conf. Series: Journal of Physics: Conf. Series **839** (2017) 012007 doi:10.1088/1742-6596/839/1/012007

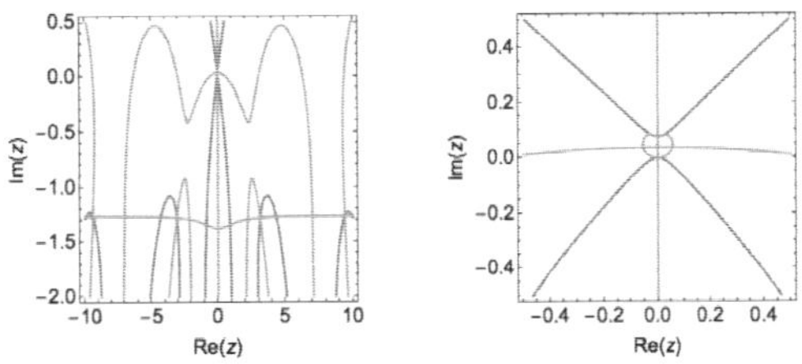

Figure 1.

2.3. *The addition law at the $p = 0$ limit*

We can summarize what happens in the limit $p \to 0$ as follows: the T_p matrix for two-pairs of $\delta - \delta'$ interactions displaced from each other a distance p becomes:

$$\begin{pmatrix} A_3 \\ B_3 \end{pmatrix} = K^{-1} M_w \cdot M_v K \begin{pmatrix} A_1 \\ B_1 \end{pmatrix} = K^{-1} M_u K \begin{pmatrix} A_1 \\ B_1 \end{pmatrix} . \tag{15}$$

Then, the superposition of these two $\delta - \delta'$ is equivalent to a single $\delta - \delta'$ where the matching conditions are characterized by the Kurasov matrix

$$M_u = \begin{pmatrix} \dfrac{1+u_1}{1-u_1} & 0 \\ \dfrac{u_0}{1-u_1^2} & \dfrac{1-u_1}{1+u_1} \end{pmatrix}, \qquad u_1 = \frac{v_1 + w_1}{1 + v_1 w_1}, \quad u_0 = \frac{v_0(1-w_1)^2 + w_0(1+v_1)^2}{(1+v_1 w_1)^2}. \tag{16}$$

This addition law is quite interesting. It can be proved [6] that the Kurasov matrices is the Borel subgroup of $SL_2(\mathbb{R})$, with the product law $M_v \cdot M_w = M_u$ given by (16). It can be also shown that the $p \to 0$ limit of the scattering coefficients of two a priori separated pairs of $\delta - \delta'$ interactions is

$$\lim_{p\to 0} S_p(M_v, M_w) = S(M_v \cdot M_w) = S(M_u) \tag{17}$$

Hence, the scattering matrices produced by two superimposed pairs of $\delta - \delta'$ interactions give rise to a representation of the non-abelian composition law of Kurasov matrices. In other words, the non-abelian addition law (16) also works for the scattering coefficients.

2.4. *The left-right decoupling values of the δ' couplings*

It is interesting to stress that there are singularities of the Kurasov matrix M_v, M_w in (7) and (9) at the critical points $v_1 = \pm 1$ and $w_1 = \pm 1$. One may expect that these critical cases, where the transmission coefficients are zero, do not contribute to the group structure that we have previously explained. The completely opaque potentials have one of the following forms:

$$V = v_0 \delta(y) \pm 2\delta'(y) \quad \text{and/or} \quad W = w_0 \delta(y-p) \pm 2\delta'(y-p) . \tag{18}$$

In order to define these potentials, we cannot use the same matching conditions as before. Instead, following the work of Kurasov, we shall impose:

(i) For $v_1 = 1$: $\varphi(0^-) = 0\,, \quad \varphi'(0^+) = v_0\, \varphi(0^+)/4.$

(ii) For $w_1 = 1$: $\varphi(p^-) = 0\,, \quad \varphi'(p^+) = w_0\, \varphi(p^+)/4\,.$

(iii) For $v_1 = -1$: $\varphi(0^+) = 0\,, \quad \varphi'(0^-) = -v_0\, \varphi(0^-)/4.$

(iv) For $w_1 = -1$: $\varphi(p^+) = 0\,, \quad \varphi'(p^-) = -w_0\, \varphi(p^-)/4\,.$

Thus, we set Dirichlet boundary conditions on one side of the two points $y = 0$ and $y = p$, whereas Robin boundary conditions are chosen on the other. There are eight possible configurations involving at least one decoupling configuration of the couplings: either the two $\delta - \delta'$ interactions build opaque walls both at $y = 0$ and $y = p$, or there is no transmission only at one point. We discuss first the four cases when the two δ' couplings take the decoupling limit.

2.4.1. Two δ' couplings in the decoupling limit. There are four cases in which we have decoupling situations both at $y = 0$ and $y = p$. Let us consider them separately.

Case 1: $v_1 = 1, w_1 = 1$. The boundary conditions are given by $\varphi(0^-) = 0$, $\varphi'(0^+) = v_0\,\varphi(0^+)/4$, $\varphi(p^-) = 0$, $\varphi'(p^+) = w_0\,\varphi(p^+)/4$. For $y = 0^+$, the condition is written as:

$$-\,ik(A_2 - B_2) = \frac{v_0}{4}\,(A_2 + B_2) \Longrightarrow \frac{A_2}{B_2} = -\frac{v_0 - 4ik}{v_0 + 4ik} = -\exp\left(-2i\arctan\frac{4k}{v_0}\right). \tag{19}$$

For $y = p^-$, we obtain $A_2\,e^{-ikp} + B_2\,e^{ikp} = 0$. Then, we have a transcendental equation, which can be written either in the form of a generalized Lambert equation $e^{2ikp} = (v_0 - 4ik)/(v_0 + 4ik)$, or $\tan(kp) = -4k/v_0$. This equation has a countably infinite number of solutions, k_n, that give the energy levels corresponding to this case. In the limit $p = 0$, we have only one solution, $k = 0$. Outside the interval $[0, p]$, we have the equations

$$\varphi(0^-) = 0 \Longrightarrow A_1 + B_1 = 0 \qquad \varphi'(p^+) = \frac{w_0}{4}\,\varphi(p^+) \Longrightarrow \frac{A_3}{B_3} = -e^{2ikp}\,\frac{w_0 - 4ik}{w_0 + 4ik}. \tag{20}$$

These are relations between coefficients, which do not provide of any further information, so that we shall not refer for similar situations which will appear for all other cases.

Case 2: $v_1 = 1$, $w_1 = -1$. The boundary conditions are $\varphi(0^-) = 0$, $\varphi'(0^+) = v_0\,\varphi(0^+)/4$, $\varphi(p^+) = 0$, $\varphi'(p^-) = -w_0\,\varphi(p^-)/4$. The condition at $y = 0$ has already been studied. From the new boundary condition at $y = p$:

$$-\,kp = \arctan\frac{4k}{w_0} + \arctan\frac{4k}{v_0} \quad\Longrightarrow\quad \tan(kp) = -\frac{4k(w_0 + v_0)}{w_0 v_0 - 16k^2}. \tag{21}$$

This new transcendental equation gives another set of energy eigenvalues, seen in the left hand side of next Figure. In the limit $p \to 0$, we have only one solution: $k = 0$.

Case 3: $v_1 = -1$, $w_1 = 1$. The boundary conditions are such that $A_2 + B_2 = 0$ and $\varphi(p^-) = 0$, which is $A_2/B_2 = -e^{2ikp}$, so that

$$\frac{A_2}{B_2} = -1 = -e^{-2ikp} \Longrightarrow k = \frac{\pi n}{p}. \tag{22}$$

Again, in the limit $p = 0$, the only solution is $k = 0$.

Case 4: $v_1 = -1$, $w_1 = -1$. This correspond to $\varphi(0^+) = 0$, $\varphi'(0^-) = -v_0\,\varphi(0^-)/4$, $\varphi(p^+) = 0$, $\varphi'(p^-) = -w_0\,\varphi(p^-)/4$. Then, the transcendental equation giving the energy levels is given by

$$e^{2ikp} = \frac{w_0 - 4ik}{w_0 + 4ik} \quad\Longleftrightarrow\quad \tan(kp) = -(4k)/(w_0). \tag{23}$$

This transcendental equation also has a countably infinite number of solutions, k_n, that give the energy levels corresponding to this situation, seen in the right hand side of Fig. 2. Once more, in the limit $p \to 0$, only the solution $k = 0$ remains.

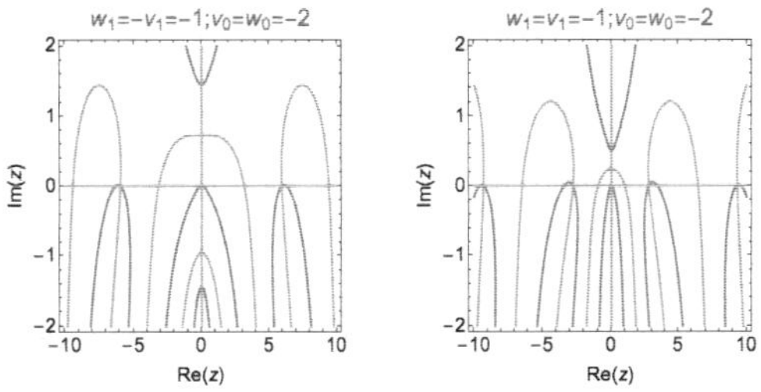

Figure 2.

2.4.2. Only one δ′ coupling in the decoupling limit. Let us consider the following four cases:

Case 1: $v_1 = 1, w_1 \neq \pm 1$. Skipping the technical details, the system behaves as if there is an impenetrable barrier at $y = 0$. The coefficients A_3, B_3 are obtained from A_2, B_2:

$$\begin{pmatrix} A_3 \\ B_3 \end{pmatrix} = \begin{pmatrix} \dfrac{2k(1+w_1^2)+iw_0}{2k(1-w_1^2)} & e^{2ikp}\dfrac{4kw_1+iw_0}{2k(1-w_1^2)} \\ e^{-2ikp}\dfrac{4kw_1-iw_0}{2k(1-w_1^2)} & \dfrac{2k(1+w_1^2)-iw_0}{2k(1-w_1^2)} \end{pmatrix} \begin{pmatrix} A_2 \\ B_2 \end{pmatrix}. \tag{24}$$

All the relevant information about bound states, anti-bound states and resonances is obtained imposing the *purely outgoing boundary condition*, which in our case is $A_3 = 0$. Then, we get

$$e^{2ikp} = \frac{v_0 - 4ik}{v_0 + 4ik}\,\frac{2k(1+w_1^2)+iw_0}{4kw_1 + iw_0}. \tag{25}$$

This equation is the equivalent of $\Delta(k) = 0$, which provides the relevant information in the non-decoupling case. Indeed, (25) is obtained making $v_1 = 1$ in $\Delta(k) = 0$. The analysis in this case is similar to the one carried out previously. We are very interested in the limit $p \to 0$. Then, from (25) we get $(v_0 + 4ik)(4kw_1 + iw_0) = (v_0 - 4ik)(2k(1 + w_1^2) + iw_0)$. The complex solutions of this equation are $k_0 = 0$, $k_1 = -i(4w_0 + v_0(1 - w_1)^2)/(4(1 + w_1)^2)$. The solution $k = 0$ is not relevant, but the other produces interesting results:

- If $4w_0 + v_0(1 - w_1)^2 < 0$, k_1 is on the positive imaginary axis and it is a bound state.
- If $4w_0 + v_0(1 - w_1)^2 > 0$, k_1 is on the negative imaginary axis and it is an anti-bound state.

The curves solving the equations $\mathrm{Re}(\Delta) = 0$ (blue) and $\mathrm{Im}(\Delta) = 0$ (yellow) are represented on the left hand side of the next Figure (the green line is the real exis).

Case 2: $v_1 = -1$, $w_1 \neq \pm 1$. The system behaves also as if there were an impenetrable barrier at $y = 0$. Now, the *purely outgoing boundary condition* $A_3 = 0$ implies that $e^{2ikp} = (2k(1 + w_1^2) + iw_0)/(4kw_1 + iw_0)$. See the right hand side of Fig. 3. In the limit $p \to 0$ we get the unique solution $k = 0$.

Case 3: $v_1 \neq \pm 1$, $w_1 = 1$. It is similar to the previous one.

Case 4: $v_1 \neq \pm 1$, $w_1 = -1$. This is similar to the first case studied before. Imposing the *purely outgoing boundary condition* $B_1 = 0$, we get

$$e^{2ikp} = -\frac{w_0 - 4ik}{w_0 + 4ik}\,\frac{2k(1+v_1^2)+iv_0}{4kv_1 - iv_0}. \tag{26}$$

In the limit $p \to 0$, we have the complex solutions $k_0 = 0$, $k_1 = -i(4v_0 + w_0(1+v_1)^2)/(4(1-v_1)^2)$. The solution $k = 0$ is not relevant, but the other one, k_1, corresponds either to a bound state if $4v_0 + w_0(1 + v_1)^2 < 0$, or to an anti-bound state if $4v_0 + w_0(1 + v_1)^2 > 0$.

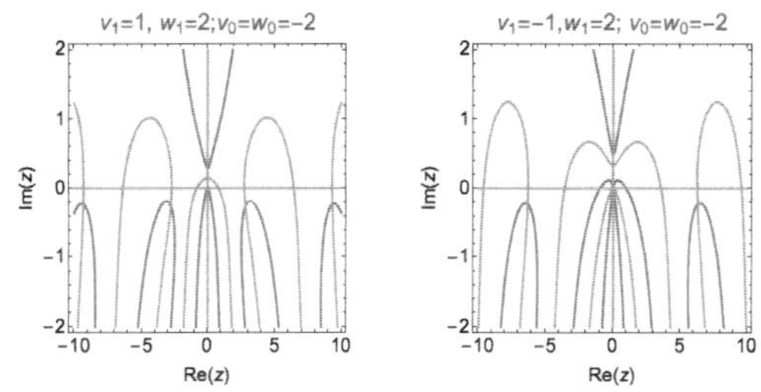

Figure 3.

3. A hybrid $\delta - \delta'$ Dirac comb

A very well known example of exactly solvable periodic potential used in Solid State Physics is the Kronig-Penney model, which describes electron motion in a period array of rectangular barriers. The Dirac-Kronig-Penney model can be considered as a special case of the Kronig-Penney model obtained by taking the appropriate limit in such a way that the rectangular barriers become Dirac delta distributions. Hence, very naturally comes the idea of considering a modified Dirac comb adding a δ' in every singular point. The dimensionless Schrödinger equation for such an hybrid Dirac comb becomes:

$$-\frac{d^2\psi(y)}{dy^2} + \left(\sum_{n\in\mathbb{Z}} \left[w_0\delta(y-na) + 2w_1\delta'(y-na)\right]\right)\psi(y) = \varepsilon\,\psi(y). \tag{27}$$

To solve it, first, we will solve the equation in $(-a,0)$ (region I) and $(0,a)$ (region II):

$$\psi_I(y) = A_I\,e^{iky} + B_I\,e^{-iky}, \qquad \psi_{II}(y) = A_{II}\,e^{iky} + B_{II}\,e^{-iky}\ , \quad \varepsilon = k^2. \tag{28}$$

Hence, $\psi_I'(y) = ikA_I\,e^{iky} - ikB_I\,e^{-iky}$, $\psi_{II}'(y) = ikA_{II}\,e^{iky} - ikB_{II}\,e^{-iky}$. In compact matrix form, these solutions can be written as follows ($J = I, II$):

$$\begin{pmatrix} \psi_J(y) \\ \psi_J'(y) \end{pmatrix} = K\,Q_y \begin{pmatrix} A_J \\ B_J \end{pmatrix}, \qquad K = \begin{pmatrix} 1 & 1 \\ ik & -ik \end{pmatrix}, \quad Q_y = \begin{pmatrix} e^{iky} & 0 \\ 0 & e^{-iky} \end{pmatrix}. \tag{29}$$

Now, we impose Kurasov's matching conditions at $y = 0$, as in the previous Section:

$$\begin{pmatrix} A_{II} \\ B_{II} \end{pmatrix} = K^{-1}T_wK \begin{pmatrix} A_I \\ B_I \end{pmatrix}. \tag{30}$$

Finally, as the potential is periodic, Floquet-Bloch theory establishes that for a periodic potential of period a, the wave functions are quasi-periodic, a plane wave times a periodic function:

$$\psi(y) = e^{iqy}\phi(y), \quad q \in \mathbb{R}, \ q \simeq q + \frac{2\pi}{a}n, \ n \in \mathbb{Z}, \qquad \phi(y+a) = \phi(y),$$

which implies that $\psi(y+a) = e^{iqa}\psi(y)$, henceforth $\psi'(y+a) = e^{iqa}\psi'(y)$. The Floquet index q is conventionally referred to as *quasi-momentum*. Thus, we write for $y \in (-a,0)$,

$$\begin{pmatrix} \psi_{II}(y+a) \\ \psi_{II}'(y+a) \end{pmatrix} = e^{iqa} \begin{pmatrix} \psi_I(y) \\ \psi_I'(y) \end{pmatrix}. \tag{31}$$

Using the explicit form of the solutions in that equation we get

$$KQ_yQ_a \begin{pmatrix} A_{II} \\ B_{II} \end{pmatrix} = e^{iqa}KQ_y \begin{pmatrix} A_I \\ B_I \end{pmatrix}. \tag{32}$$

In order to find non-trivial solution, we must have det $\left[e^{iqa}\,\mathbb{I} - Q_a K^{-1} T_w K\right] = 0$. After some algebra, we obtain the following *dispersion relation* guaranteeing that the determinant is null:

$$\cos qa = f(w_1)\left[\cos ka + \frac{a}{2} w_0\, g(w_1)\,\frac{\sin ka}{ka}\right], \qquad f(w_1) = \frac{1+w_1^2}{1-w_1^2}, \quad g(w_1) = \frac{1}{1+w_1^2}. \tag{33}$$

If the δ' interactions are switched off, $w_1 = 0$, the band spectral equation for the standard Dirac comb reappears. On the other hand, if we define the 2×2-matrix $\Omega^{(1)} = \mathbb{M}_a \mathbb{K}^{-1} \mathbb{T}_w \mathbb{K}$, the band spectrum condition or dispersion relation can re-written as $\cos(qa) = \mathrm{tr}(\Omega^{(1)})/2$. Since $\cos(qa) \in (-1, 1)$ only those momenta/energies that satisfy $\left|\mathrm{tr}(\Omega^{(1)})\right| \leq 2$ comply with the spectral condition and therefore characterize the *allowed band spectrum* of the ideal 1D crystal described by the $\delta - \delta'$ comb.

3.1. Analysis of the structure of the band spectrum

The analysis of the band structure of the energy spectrum of a periodic but not even potential with period a demands the solution of the transcendent equation:

$$\cos qa = \frac{\left[t^2(k) - r_R(k) r_L(k) + 1\right]\cos ka + i\left[t^2(k) - r_R(k) r_L(k) - 1\right]\sin ka}{2t(k)} \tag{34}$$
$$= \frac{\cos\left[ka + \delta(k)\right]}{|t(k)|},$$

which depends only on the scattering data for a single point-potential. In the case of a single $\delta - \delta'$ interaction in the $y \in [0, a)$ interval these magnitudes are:

$$t(k) = \frac{(1-w_1^2)k}{(1+w_1^2)k + iw_0/2}, \qquad |t(k)| = \frac{2|k||1-w_1^2|}{\sqrt{4k^2(1+w_1^2)^2 + w_0^2}},$$

$$r_R(k) = -\frac{2kw_1 + iw_0/2}{(1+w_1^2)k + iw_0/2}, \qquad r_L(k) = \frac{2kw_1 - iw_0/2}{(1+w_1^2)k + iw_0/2},$$

$$\delta(k) = \frac{1}{2i}\log\left(t^2(k) - r_R(k) r_L(k)\right) = \frac{1}{2i}\log\frac{2ik(1+w_1^2) + w_0}{2ik(1+w_1^2) - w_0} + \theta(w_1^2 - 1)\pi,$$

where $\theta(x)$ is the Heaviside step function. Notice that the reflection amplitudes for incoming particles from either the left or the right are different since the δ' coupling breaks parity invariance. Time reversal invariance, however, is preserved and the transition amplitude is the same whatever the incoming direction of the particle is.

It follows that a given value of k belongs to an allowed band if

$$\frac{1}{|t(k)|}\cos\left(ka + \delta(k)\right) \leq 1.$$

The right hand side of (34) is an oscillating function of k that tends asymptotically to a periodic function of amplitude $1/|t(k \to \pm\infty)| \geq 1$. Thus, there are pairs of points k_j and k_{j+1} where

$$\cos\left(k_j a + \delta(k_j)\right)/|t(k_j)| = \pm 1 \quad \text{and} \quad \cos\left(k_{j+1} a + \delta(k_{j+1})\right)/|t(k_{j+1})| = \mp 1,$$

which are the edges of allowed bands: in the intervals $[k_j, k_{j+1}]$ the spectral function varies between ± 1 and ∓ 1, and there are intersections with the k-independent functions $g = \cos qa \in [-1, 1]$, giving solutions of the spectral equation. In the interval (k_{j+1}, k_{j+2}), where k_{j+2} is the

IOP Conf. Series: Journal of Physics: Conf. Series **839** (2017) 012007 doi:10.1088/1742-6596/839/1/012007

next point for which $\cos\left(k_{j+2}a+\delta(k_{j+2})\right)/|t(k_{j+2})| = \mp 1$, the modulus of the spectral function is greater than one and there are no solutions: we find a forbidden band or gap.

The distribution of allowed and forbidden bands in the hybrid Dirac comb depends thus of the w_0 and w_1 couplings through the dependence in these parameters of the scattering data. There are several points and/or lines in the parameter space (w_0, w_1) that lead to critical behavior:

(i) The origin: $(w_0 = 0, w_1 = 0)$. In this case $|t(k)| = 1$, $\forall k$, and there is only one allowed band: the whole real line. Trivially, we end with the continuous spectrum of the free particle. All the forbidden bands disappear.

(ii) The infinite lines: $(w_0 = \pm\infty, w_1)$. When the point δ-interactions are either infinitely repulsive walls or attractive wells the transition amplitudes are null, $|t(k)|_{w_0=\pm\infty} = 0$, for all k. There are only solutions to the secular equation (or dispersion relation) (34) if $\cos\left(ka+\delta(k)\right)|_{w_0=\pm\infty}$ is also zero. Because $\delta(k)|_{w_0=\pm\infty} = i\pi$ this happens only if

$$k_n = \frac{\pi}{a}n, \quad \text{where} \quad n \in \mathbb{Z}.$$

The spectrum is purely discrete and all the allowed bands collapse to a point. The reflection amplitudes are constant, $r(k, \pm\infty, w_1) = -1$, and the eigenfunctions are standing waves with n nodes between two adjacent impenetrable walls/wells. Notice that, given a fixed n, there is an infinite number of eigenfunctions with energy proportional to $\varepsilon_n = k_n^2$. All the other possibilities interpolate between these two extreme cases. There are however other exceptional cases.

(iii) The singular lines: $(w_0, w_1 = \pm 1)$. Over thes two lines in the space of couplings the $\delta - \delta'$ potentials are also opaque: $t[k, w_0, \pm 1] = 0$. There are solutions of the equation (34) if and only if:

$$k_n a + \delta(k_n) = k_n a + \frac{1}{2}\text{Arg}\left[\frac{w_0 i - 4k_n}{w_0 i + 4k_n}\right] + \frac{\pi}{2} = (n + \frac{1}{2})\pi \ , \ n \in \mathbb{Z} \,.$$

All the allowed bands collapse to points and the spectrum is discrete. The dependence on n of the physical momenta k_n is not linear like in the previous case because the reflection amplitudes are not trivial and phase shifts vary with k, specially at low momenta:

$$r_R[k, w_0, -1] = r_L[k, w_0, 1] = \frac{4k - iw_0}{4k + iw_0} \quad , \quad r_R[k, w_0, 1] = r_L[k, w_0, -1] = -1\,.$$

The standing wave eigenfunctions incorporate in this case phase shifts after rebounding at the walls.

(iv) The ordinate axis: $(w_0 = 0, w_1)$. In this limit where only the δ'-coupling w_1 is non null we find a pathological behaviour. All the scattering amplitudes are independent of the energy:

$$t[k, 0, w_1] = \frac{1 - w_1^2}{1 + w_2^2}\,, \qquad r_R[k, 0, w_1] = -\frac{2w_1}{1 + w_1^2}\,, \qquad r_L[k, 0, w_1] = \frac{2w_1.}{1 + w_1^2}\,.$$

Moreover, the phase shifts in the even and odd channels are also k-independent

$$\delta_+(k, 0, w_1) = \frac{1}{2}\text{Arg}\left[\frac{2|w_1|}{1 - w_1^2}\right], \qquad \delta_-(k, 0, w_1) = -\frac{1}{2}\text{Arg}\left[\frac{2|w_1|}{1 - w_1^2}\right],$$

but the total phase shift is $\delta(k, 0, w_1) = \delta_+(k, 0, w_1) + \delta_-(k, 0, w_1) = 0$. Thus, the secular equation simplify to

$$\cos qa = \frac{1 + w_1^2}{|1 - w_1^2|} \cos ka. \tag{35}$$

3.2. More details on the distribution of allowed and forbidden bands in the hybrid Dirac comb

Another interesting line in the plane of couplings is the axis $w_1 = 0$. There is no δ' interaction which means that $f(0) = g(0) = 1$ and the spectral equation reduces to the well known secular equation for the Dirac comb potential:

$$\cos qa = \cos ka + aw_0(2ka)^{-1}w_0 \sin(ka)\,, \tag{36}$$

which identifies the band edge points as the discrete solutions of the transcendent equations:

$$\sqrt{\frac{4k_j^2 + w_0^2}{4k_j^2}} \cos\left[k_j a + \frac{1}{2}\text{Arg}(4k_j^2 - w_0^2 + 4ik_j w_0)\right] = \pm 1\,. \tag{37}$$

Implicitly we have focused until now on solutions to the secular equations for *real momenta*, $k \in \mathbb{R}$, corresponding to propagating (conducting) band states.

Another important band state solutions respond to purely imaginary momenta, $k = i\kappa$, $\kappa \in \mathbb{R}$, which give rise to non-propagating states filling the so called valence bands. For the Dirac comb arrangement with attractive couplings the secular equation for purely imaginary momenta is well known:

$$\cos qa = v(\kappa, a, |w_0|)\,, \qquad v(\kappa, a, |w_0|) = \cosh \kappa a - \frac{a}{2}|w_0|\frac{\sinh \kappa a}{\kappa a}\,, \tag{38}$$

One checks easily that $v(0, a, |w_0|) = 1 - \frac{a}{2}|w_0|$ and that $\frac{\partial v}{\partial \kappa}(\kappa, a, |w_0|)\Big|_{\kappa=0} = 0$. Thus, $\kappa = 0$ is a critical point of the even function $v(\kappa, a, |w_0|)$. In the range $0 < |w_0| < \frac{4}{a}$ the spectral function v intersects with the straight line $v_0 = 1$ at two points $\pm\kappa_1$. Moreover, in this range the function v is always less than 1 but greater than -1 in the interval $\kappa \in [-\kappa_1, \kappa_1]$. Thus, there is one valence band with upper and lower edge points respectively of zero and $-\kappa_1^2$ energy.

Note that $\kappa = 0$ is a minimum of v, whereas $1 > v > -1$. Observe also that $\lim_{\kappa\to\pm\infty}(\kappa, a, |w_0|) = +\infty$. If $\frac{4}{a} < |w_0| < \frac{6}{a}$ the spectral function v intersect also with the straight line $v_0 = -1$ at the points $\pm\kappa_2$. $\kappa = 0$ is still the minimum of v but it does not belong to the valence band and the upper edge energy $-\kappa_2^2$ is lower than zero. For $\frac{6}{a} < |w_0|$ the band structure of negative energy states is similar but a little more subtle because $\kappa = 0$ becomes a local maximum of v prompting also the appearance of two minima of v at $\kappa = \pm\kappa_0$. For attractive coupling stronger than this the valence band width becomes more and more thin.

3.3. Differences between the band structures of the hybrid and standard Dirac combs

The strategy to analyze the differences between the band structures will be to fix a value of w_0, the strength of the δ-coupling, and to study, mainly graphically, how the solutions of the secular equation vary with w_1. In particular, variations in the width of the forbidden and allowed bands as function of the δ'-coupling w_1 will inform us about the modulation of the band distribution of the hybrid Dirac comb.

To start with we recall again that, since $|\cos(qa)| \leq 1$, the allowed bands are characterized by the inequality:

$$\left| f(w_1)\left[\cos(a\sqrt{\varepsilon}) + \frac{a}{2}w_0\, g(w_1)\,\frac{\sin(a\sqrt{\varepsilon})}{a\sqrt{\varepsilon}}\right]\right| \leq 1\,. \tag{39}$$

Here, we have written the dispersion relation in terms of $\varepsilon = k^2$ instead of using k as independent variable. This election of independent variable allows us to encode as a function of a single real variable the spectral function determining both the valence and conduction bands, without the need of going to the complex k-plane. The use of k as the independent variable in the spectral function that characterize the bands requires that the conducting bands lie in the real k-axis while the valence bands are found in the pure imaginary $k = i\kappa$-axis. Thus, we shall plot both the valence bands, if they exists, the values of the energy where the inequality (39) is satisfied with $\varepsilon < 0$, and the conduction bands, energy intervals compatible with (39) for $\varepsilon > 0$, in a single $2D$ graphic.

This work is presently in progress [7] and we present here just some plots. For example, in Fig. 4 we show a plot of the allowed (yellow) and forbidden (blue) energy bands inferred from the spectral condition for $w_0 = -0.5$ and $a = 1$. In the graphic on the left it is seen how the width of several energy bands complying with the inequality (39) varies with w_1. Forbidden energy bands missing this inequality are also plotted in blue. The graphic on the right shows only the lowest allowed and forbidden energy bands for identical choice of w_0. This zoom of the graphic allows to recognize the merging of the valence with the lower conduction band at values of w_1 slightly greater than 0. Note also that the secular equation is invariant under the exchange $w_1 \leftrightarrow -w_1$. Thus, there are mirror images of the Figures shown here on the negative w_1-axis.

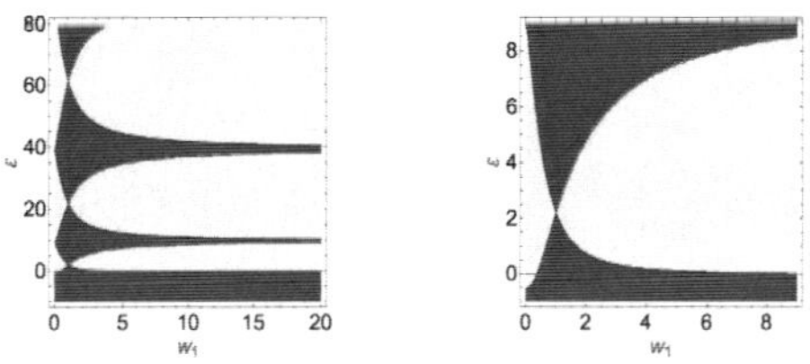

Figure 4.

Acknowledgments

LMN thanks the organizers of Quantum Fest 2016 (IPN-México City) for their kind invitation to give three talks. Partial financial support of Spanish MINECO (MTM2014-57129-C2-1-P) and Junta de Castilla y León (VA057U16) is acknowledged.

References

[1] Albeverio S, Gesztesy F, Hoegh-Krohn R and Holden H 2005 *Solvable models in quantum mechanics* (Providence, RI: AMS Chelsea Publishing)

[2] Kurasov P 1996 *J. Math. Anal. Appl.* **201** 297; S Albeverio, P Kurasov 2000 *Singular perturbations of differential operators. Solvable Schrödinger type operators*, London Mathematical Society Lecture Notes 271, (Cambridge: Cambridge UP).

[3] Asorey M, García-Alvarez D and Muñoz-Castañeda J M 2006 *J. Phys A: Math. Theor* **39** 6127-6136; Asorey M and Muñoz-Castañeda J M 2013 *Nucl. Phys. B* **874** 852876; Muñoz-Castañeda J M, Kirsten K and Bordag M 2015 *Lett. Math. Phys.* **105** 523549.

[4] Muñoz-Castañeda J M, Mateos Guilarte J and Moreno Mosquera A 2013 *Phys. Rev. D* **87** 105020; Muñoz-Castañeda J M and Mateos Guilarte J 2015 *Phys. Rev. D* **91** 025028.

[5] Díaz J I, Negro J, Nieto L M and Rosas-Ortiz O 1999 *J. Phys. A: Math. Gen.* **32** 8447; Gadella M, Heras F J H, Negro J and Nieto L M 2009 *J. Phys. A: Math. Theor.* **42** 465207; Gadella M, Negro J, Nieto L M 2009 *Phys. Lett. A* **373** 1310; Gadella M, Glasser M L and Nieto L M 2011 *Int. J. Theor. Phys.* **50** 2144; Álvarez-Sánchez J J, Gadella M and Nieto L M 2011 *Int. J. Theor. Phys.* **50** 2161; Gadella M, Glasser M L and Nieto L M 2011 *Int. J. Theor. Phys.* **50** 2191.

[6] Gadella M, Mateos-Guilarte J, Muñoz-Castañeda J M and Nieto L M 2016 *J. Phys. A: Math. Theor.* **49** 015204.

[7] Gadella M, Mateos-Guilarte J, Muñoz-Castañeda J M and Nieto L M, *Adding δ'-point interactions to the theeth of the Dirac δ-comb potential* (in progress).

IOP Conf. Series: Journal of Physics: Conf. Series **839** (2017) 012008 doi:10.1088/1742-6596/839/1/012008

Quantum Theory From a Non-Conventional Perspective

D Schuch

Institut für Theoretische Physik, J.W. Goethe-Universität Frankfurt am Main, Max-von-Laue-Str. 1, D-60438 Frankfurt am Main, Germany

E-mail: `Schuch@em.uni-frankfurt.de`

Abstract. Quantum mechanics is essentially described in terms of complex quantities like wave functions. The interesting point is that phase and amplitude of the complex quantities are not independent of each other, but coupled by some kind of conservation law. This coupling exists in time-independent quantum mechanics as well as in in its time-dependent form. It can be traced back to a reformulation of quantum mechanics in terms of complex nonlinear Riccati equations, where the quadratic term in the latter equation explains the origin of the phase-amplitude coupling. Since realistic physical systems are always in contact with some kind of environment this aspect is also taken into account. It turns out that this dissipative effect can either affect the amplitude or the phase of the complex quantity describing the open quantum system. This suggests a relation between non-unitary transformations and gauge-transformations for these systems. A change of the amplitude also seems to be connected with a second quantum of action for "radial" changes , compared to $\hbar$ for "angular" changes, leading to an interpretation of Sommerfeld's constant.

1. Introduction

Looking at quantum theory in the conventional way one would say that it is a linear theory. Particularly in the Schrödinger picture, which is the one adopted in the following, the Schrödinger equation (SE) is a linear differential equation; therefore the superposition principle is valid which is in agreement with wave-particle duality. In the most common version it is also a Hamiltonian theory – although other formulations are possible like the one based on a Lagrangian, as in Feynman's path integral formulation. The operator corresponding to the Hamiltonian function determines the *energetics* of the system. It not only represents the energy of the system that can be determined via its mean value and corresponding eigenvalues but, as the Hamiltonian operator is usually hermitian, this also guarantees real eigenvalues. The energy of the quantum system, as in the classical case, is a conserved quantity (for time-independent (TI) potentials) or, at least, is not dissipated into heat. For certain systems the energy is also quantized, providing the spectrum of the system.

Moreover, the Hamiltonian also determines the *dynamics* of the system. The Hamiltonian enters the (imaginary) exponent of a time-evolution operator that transforms an initial state into a state at another time. As this transformation is unitary (only a "rotation" in Hilbert space with constant norm), this time-evolution is reversible, i.e., there is no direction of time.

While quantum theory is therefore a linear theory with reversible time-evolution and conserved energy, the macroscopic world we are observing daily is characterized by nonlinear

(NL) evolution processes. These are usually irreversible, thus having a direction of time, and (mechanical) energy is not conserved but transferred as heat into the environment of the system being considered.

A theory well-suited for describing the latter phenomena is *Nonlinear Dynamics*. Furthermore, this theory is usually scale invariant, i.e., independent of the size of the system; only relative changes matter. In this context, the Mandelbrot set and the aesthetic pictures of fractals [1] are also familiar. Interestingly enough, the customary spiral-type patterns of these fractals arise from angular changes that are accompanied by radial ones during the evolution process (in time or space). Manifestations of such processes are ubiquitous in nature; just look, for example, at the shell of a nautilus or the horn of a ram. The radial changes in particular determine the definition of a direction of time; as for an expanding system, a larger radius indicates a later state of the system than one with a smaller radius and vice versa for a contracting system. Since the time-evolution of quantum systems is usually described by a unitary transformation with constant norm, i.e., constant "length" of the state vector, this radial component is missing a priori in quantum theory. Or is it just hidden? Could this possibly be revealed using a different, maybe even NL, formulation of quantum theory? The following discussion is an attempt at providing some answers to these questions.

Such a formulation of quantum theory must be able to take into account aspects that are present in the theories mentioned previously, like nonlinearity, scale-invariance, radial changes, etc. In order to resolve this, one must first determine which properties of the established form of quantum theory are indispensable and cannot therefore be relinquished. As the name "quantum" theory already suggests, there are so-called quanta, i.e., fixed units of a physical quantity. These quantities can be expressed in terms of multiples (or sometimes fractions) of integers of these quanta. As Planck had shown, the essential quantity in nature that is quantized is *action*, quantized in terms of Plancks quantum h (or $\hbar = \frac{h}{2\pi}$).

Another essential ingredient of quantum theory is the occurrence of complex quantities –and not just as a mathematical tool to simplify calculations, but with intrinsic physical meaning. This was stressed by C.N. Yang in his talk "Square root of minus one, complex phase and Erwin Schrödinger" [2] presented 30 years ago at a conference in London celebrating the centenary of the birth of Schrödinger and also E. Wigner points this out in his paper "The unreasonable effectiveness of mathematics in the natural sciences" [3]. Remarkably, Schrödinger introduced his wave function $\Psi(\mathbf{r},\mathbf{t})$ via the action S as $S = K \ln \Psi$ [4] where, later on, it turned out that $K = \frac{\hbar}{\mathrm{i}}$ with $\mathrm{i} = \sqrt{-1}$, so Schrödinger's action and also Ψ are complex quantities.

Which other essential property of quantum theory remains? The superposition principle was mentioned above as it seems to be necessary for the description of the wave-like properties of material systems. This is definitely in agreement with the linearity of the SE; but there are also NL differential equations that possess a kind of superposition principle. One of them is the NL Riccati equation. Can quantum theory therefore be rewritten in terms of NL Riccati equations? Or are these equations already somehow hidden in the conventional formulation of quantum theory? An initial answer is given in the next section where it is shown that the dynamics of a time-dependent (TD) quantum system can equally well be described by a complex NL Riccati equation. Besides other treatments, this equation can be linearized to a complex Newtonian equation. During their time-evolution in the complex plane, the amplitude and phase of the corresponding variable can display the simultaneous change of angle and radius mentioned previously. A similar treatment in terms of complex Riccati equations, or equations that can be derived directly thereof, is also possible in the TI case. In both cases this is connected with a conservation law that couples phase and amplitude of the complex quantity.

In Sec. 3 it is shown that the same systems considered in Sec. 2 can also be treated, including a linear velocity-dependent friction force. For this purpose, in the TD case, a NLSE with complex logarithmic nonlinearity is used that can be uniquely linked to other approaches (including conventional ones). This leads to similar Riccati equations that can again be linearized to complex Newtonian equations. Integrating the dissipative friction force into the NL formalism developed in the non-dissipative case requires only minimal changes. A similar treatment is also possible in the TI case.

Comparison of the complex quadratic nonlinearity appearing in the Riccati equation and the problem of finding the so-called Pythagorean triples shows in Sec. 4 that the angular as well as the radial changes of the complex quantities in the Riccati version of quantum theory can always be quantized in principle. Searching for an example where quantization of the radial changes can be observed in nature leads to the quantum Hall effect (QHE) and the fractional quantum Hall effect (FQHE). The latter gives rise to the introduction of a second quantum of action that enables an interpretation of Sommerfeld's fine-structure constant. Sec. 5 summarizes the results and highlights future perspectives.

2. Complex nonlinear Riccati equations in time-dependent and time-independent quantum mechanics

In order to start the reformulation of quantum theory from save grounds, in the following systems are considered whose SEs have exact, even analytic solutions. In the TD case, these are Hamiltonians that are at most quadratic (or bilinear) in the variables of position and momentum, particularly in one dimension (although generalization to two or three dimensions usually does not cause a problem), in the TI case, in general three-dimensional problems are permitted.

Beginning with the TD case, the discussion focuses on the harmonic oscillator (HO) with potential $V = \frac{m}{2}\omega^2 x^2$ with constant frequency $\omega = \omega_0$, the parametric oscillator with TD frequency $\omega = \omega(t)$ and, in the limit $\omega \to 0$, the free motion, $V = 0$. For all three potentials the TDSE

$$\mathrm{i}\hbar\frac{\partial}{\partial t}\Psi(x,t) = \left\{-\frac{\hbar^2}{2m}\frac{\partial^2}{\partial x^2} + V(x)\right\}\Psi(x,t) \tag{1}$$

has to be solved, what is possible using Gaussian wave packets (WPs) of the form

$$\Psi(x,t) = N(t)\exp\left\{\mathrm{i}\left[y(t)\tilde{x}^2 + \frac{\langle p\rangle}{\hbar}\tilde{x} + K(t)\right]\right\} \tag{2}$$

as solutions.

The variable $\tilde{x}$ in WP (2) is a shifted coordinate $\tilde{x} = x - \langle x\rangle = x - \eta(t)$, where the mean value $\langle x\rangle = \int_{-\infty}^{+\infty}\Psi^* x\Psi dx = \eta(t)$ corresponds to the classical trajectory, $\langle p\rangle = m\dot{\eta}$ represents the classical momentum and the coefficient of the quadratic term in the exponent, $y(t) = y_{\mathrm{R}}(t) + \mathrm{i}y_{\mathrm{I}}(t)$, is a complex function of time and related to the WP width. The (possibly) time-dependent normalization factor $N(t)$ and the purely time-dependent function $K(t)$ in the exponent are not relevant for the dynamics of the WP maximum and width and therefore ignored in the folowing.

As a Gaussian function is totally determined by its maximum and width, the equations of motion that determine the dynamics of these properties are considered now. They are obtained by inserting WP (2) into the TDSE (1) and have the form

$$\ddot{\eta} + \omega^2\eta = 0, \tag{3}$$

and, multiplying $y(t)$ by the constant factor $\frac{2\hbar}{m}$ to obtain, just for formal reasons, the complex variable $\mathcal{C}(t) = \frac{2\hbar}{m}y(t)$,

$$\frac{d}{dt}\mathcal{C} + \mathcal{C}^2 + \omega^2 = 0, \tag{4}$$

the latter being the desired complex Riccati equation.

The Newtonian equation (3) simply conveys that the maximum of the WP, located at $x = \langle x \rangle = \eta(t)$, follows the classical trajectory. The equation for $\mathcal{C}(t)$ determines the time-dependence of the WP width that is related with the position uncertainty via $y_I = \frac{1}{4\langle \tilde{x}^2 \rangle}$ with $\langle \tilde{x}^2 \rangle = \langle x^2 \rangle - \langle x \rangle^2$ being the mean square deviation of position.

The connection with the WP width and the corresponding equation of motion can be made even more transparent by introducing a new (real) variable α via $\mathcal{C}_{\rm I} = \frac{1}{\alpha^2}$, where α can be expressed as $\alpha = \left(\frac{2m\langle \tilde{x}^2(t) \rangle}{\hbar} \right)^{1/2}$ and is directly proportional to the WP width. Inserting this definition of $\mathcal{C}_{\rm I}$ into the Riccati equation (4), the real part $\mathcal{C}_{\rm R}$ can be determined as $\mathcal{C}_{\rm R} = \frac{\dot{\alpha}}{\alpha}$ and, inserting $\mathcal{C}_{\rm I}$ and $\mathcal{C}_{\rm R}$ in this form into the comple NL Riccati equation turns it into the NL real Ermakov equation (more details about this equation and a corresponding invariant can be found, e.g., in [5–7] and references cited therein, but are not relevant for the following)

$$\ddot{\alpha} + \omega^2 \alpha = \frac{1}{\alpha^3} \, . \tag{5}$$

As indicated in the Introduction, the Riccati equation can be linearized, in this case using the logarithmic derivative of a complex quantity $\lambda(t)$, $\mathcal{C} = \frac{\dot{\lambda}}{\lambda}$ (note, only the relative change of λ matters), to yield the complex Newtonian equation

$$\ddot{\lambda} + \omega^2 \lambda = 0, \tag{6}$$

First, a kind of geometric interpretation of the motion of $\lambda(t)$ in the complex plane shall be given. Expressed in Cartesian coordinates, λ can be written as $\lambda = u + {\rm i}z$, turning Eq. (6) into two equivalent equations for u and z, respectively, without showing any connections between them. Writing λ in polar coordinates as $\lambda = \alpha e^{{\rm i}\varphi}$ turns $\mathcal{C} = \frac{\dot{\lambda}}{\lambda}$ into

$$\mathcal{C} \; = \; \frac{\dot{\alpha}}{\alpha} + {\rm i}\, \dot{\varphi} \, , \tag{7}$$

where the real part is already identical to $\mathcal{C}_R$, as defined above.

The quantity $\alpha(t)$ defined above in $\mathcal{C}_I$ as being proportional to the position uncertainty is identical to the absolute value of λ, as it can be shown that

$$\dot{\varphi} = \frac{1}{\alpha^2} \, . \tag{8}$$

This relation corresponds to the "conservation of angular momentum", but here for the motion in the complex plane (for further details, see [5–7])!

Relation (8) also shows that real and imaginary parts, or phase and amplitude, respectively, of the complex quantity $\lambda(t)$ are not independent of each other but uniquely coupled. This coupling is due to the quadratic nonlinearity in the Riccati equation.

An analogous situation can be found in the TI case, as shown by Reinisch [8]. For this purpose, Madelung's hydrodynamic formulation of quantum mechanics [9] is taken as starting point. Using the polar ansatz

$$\Psi(\mathbf{r},t) = a(\mathbf{r},t) \; \exp \left(\frac{\rm i}{\hbar} S(\mathbf{r},t) \right) \tag{9}$$

for the (complex) wave function $\Psi(\mathbf{r},t)$ (where $a^2(\mathbf{r},t) = \varrho(\mathbf{r},t) = \Psi^* \Psi$), turns the linear SE (1)[1] into two coupled equations for the amplitude $a^2(\mathbf{r},t)$ and the phase $S(\mathbf{r},t)$; i.e., the continuity equation

$$\frac{\partial}{\partial t} a^2 \; + \; \frac{1}{m} \nabla (a^2 \, \nabla S) = 0, \tag{10}$$

[1] In the following $\frac{\partial^2}{\partial x2}$ is replaced by the three-dimensional Laplace operator $\Delta = \nabla^2$ with ∇ = nabla operator.

and the modified Hamilton–Jacobi-type equation

$$\frac{\partial}{\partial t}S + \frac{1}{2m}(\nabla S)^2 + V - \frac{\hbar^2}{2m}\frac{\Delta a}{a} = 0. \tag{11}$$

Already here, the coupling of phase and amplitude can be seen clearly because the Hamilton–Jacobi equation for the phase[2] S contains a term (misleadingly called "*quantum potential*", $V_{qu} = -\frac{\hbar^2}{2m}\frac{\Delta a}{a}$) depending on the amplitude $a(\mathbf{r},t)$ and the continuity equation for the probability density $\varrho = a^2$ contains ∇S. In the following, it is shown that, also in the TI case, this coupling is not arbitrary but related to a conservation law originating from the quadratic nonlinearity of a complex Riccati equation.

For stationary states, the energy of the system is related to the action S via $\frac{\partial}{\partial t}S = -E = const$, and the density is TI, i.e., $\frac{\partial}{\partial t}a^2 = 0$. The continuity equation (10) then turns into

$$\nabla(a^2\nabla S) = 0 \tag{12}$$

and the modified Hamilton–Jacobi equation (11) into

$$-\frac{\hbar^2}{2m}\Delta a + (V - E)\,a = -\frac{1}{2m}(\nabla S)^2 a. \tag{13}$$

Equation (12) is definitely fulfilled for $\nabla S = 0$, turning Eq. (13) into the usual TISE for the real wave function $a = |\Psi|$ with position-independent phase S.

However, Eq. (12) can also be fulfilled for $\nabla S \neq 0$ if only the conservation law

$$\nabla S = \frac{C}{a^2} \tag{14}$$

is fulfilled with constant (or, at least, position-independent) C.

This expression now shows explicitly the coupling between phase and amplitude of the wave function a'nd is equivalent to Eq. (8) in the TD case. Inserting (4) into the rhs of Eq. (13) changes this to the Ermakov equation

$$\Delta a + \frac{2m}{\hbar^2}(E - V)\,a = (\frac{1}{\hbar}\nabla S)^2\,a = \left(\frac{C}{\hbar}\right)^2\frac{1}{a^3} \tag{15}$$

equivalent to Eq. (5) in the TD case. The corresponding *complex Riccati* equation equivalent to Eq. (4) in the TD case is given here by

$$\nabla\left(\frac{\nabla\Psi}{\Psi}\right) + \left(\frac{\nabla\Psi}{\Psi}\right)^2 + \frac{2m}{\hbar^2}(E - V) = 0 \tag{16}$$

where the following substitutions must be made

$$\frac{\partial}{\partial t} \leftrightarrow \nabla\,,\ \left(\frac{2\hbar}{m}y\right) = \mathcal{C} = \frac{\dot{\lambda}}{\lambda} \leftrightarrow \frac{\nabla\Psi}{\Psi}\,,\ \lambda = \alpha\,e^{i\varphi} \leftrightarrow \Psi = a e^{i\frac{S}{\hbar}}. \tag{17}$$

Note that whereas in the TD case the NL formulation is essentially possible for quadratic potentials, in the TI case the formulation in terms of a complex Riccati equation is possible for *any* potential $V(\mathbf{r})$.

Again, in linearizing the complex Riccati equation (16), only the logarithmic derivative $\frac{\nabla\Psi}{\Psi} = \nabla\ln\Psi$ is of importance; so the absolute value of Ψ is irrelevant. This might be a reason why the value is arbitrary and can also just be defined to be "one", thus leading to the probabilistic interpretation of the wave function[3].

[2] In the classical case, S is identical to the (real) action function of the system.

[3] The change of amplitude in the TI case is somehow "hidden" via normalization in order to fulfil this probabilistic interpretation. In the TD case, the change of the amplitude, e.g., caused by the spreading of a WP, is compensated by a TD normalization coefficient to guarantee normalizability at any time t, not only at the beginning $t = t_0$.

IOP Conf. Series: Journal of Physics: Conf. Series **839** (2017) 012008 doi:10.1088/1742-6596/839/1/012008

3. Dissipative systems with irreversible time-evolution

Realistic physical systems are always in contact with some kind of environment, leading to phenomena like irreversible time-evolution and dissipation of energy. So the question is how can this be taken into account in classical, particularly Hamiltonian (or Lagrangian) mechanics and then be transferred to a quantum mechanical description of those systems?

Without the boundary condition that necessitates a Hamiltonian theory in classical mechanics, a so-called phenomenological description is possible. Using the trajectory picture, this leads to the Langevin equation

$$m\ddot{x} = m\dot{v} = -m\gamma v - \frac{\partial}{\partial x}V \tag{18}$$

with the linear velocity dependent friction force $-m\gamma v$ (where γ is a friction coefficient), written here without a purely random TD stochastic force that generally vanishes on average. The same Brownian motion-type scenario can also be described in the picture of a (one-particle) distribution function ϱ by adding a diffusion term breaking the time-reversal symmetry. Particularly for a distribution function $\varrho(x,t)$ in position space, this leads to the Smoluchowski equation

$$\frac{\partial}{\partial t}\varrho_{\rm cl} + \frac{\partial}{\partial x}\left(\frac{F(x)}{m\gamma}\varrho_{\rm cl}\right) - \frac{k_{\rm B}{\rm T}}{m\gamma}\frac{\partial^2}{\partial x^2}\varrho_{\rm cl} = 0, \tag{19}$$

a Fokker–Planck-type equation where the diffusion coefficient obeys the Einstein relation $D = \frac{k_{\rm B}{\rm T}}{m\gamma}$ with $k_{\rm B}$ being Boltzmanns constant and T, temperature; $F(x)$ is an external force.

However, Eqs. (18) and (19) do not fit consistently into the Lagrange/Hamilton formalism that is invariant under canonical transformations and provides a basis for quantization.

There are numerous approaches in the literature for finding a Hamiltonian description of dissipative systems that can roughly be divided into three groups of methods:

1. the system-plus-reservoir approaches;
2. modifications of the classical Lagrange/Hamilton formalism with subsequent canonical qauntization;
3. modifications of the Hamiltonian operator, usually leading to NLSEs.

It has been shown [10,11] that most of the approaches belonging to any of these three groups can be uniquely be linked to a particular NLSE with complex logarithmic nonlinearity that is free of the shortcomings of other related NLSEs and can uniquely be connected with the approaches of group two via non-unitary transformations on the quantum mechanical level (corresponding to non-canonical transformations on the classical level). As a connection between the canonical approaches of group two and the conventional system-plus-reservoir approach of group one is possible [12,13], this NLSE can also be related to the conventional approaches.

Starting point for the derivation of the logarithmic NLSE is the irreversible Smoluchowski equation

$$\frac{\partial}{\partial t}\varrho + \frac{\partial}{\partial x}(\varrho\, v_-) - D\frac{\partial^2}{\partial x^2}\varrho = 0, \tag{20}$$

but now for the quantum mechanical probability density $\varrho(x,t) = \Psi^*(x,t)\Psi(x,t)$ and with the velocity field $v_-(x,t) = \frac{\hbar}{2mi}\left(\frac{\frac{\partial}{\partial x}\Psi}{\Psi} - \frac{\frac{\partial}{\partial x}\Psi^*}{\Psi^*}\right)$. For the continuity equation, Madelung [14] and Mrowka [15] had shown that with a bilinear ansatz for the probability current density $j = \varrho\left(\frac{1}{m}\frac{\partial}{\partial x}S\right) = \varrho v_-$, this can be separated into the TDSE and its complex conjugate where the separation "constant" is proportional to the potential V. However, due to the diffusion term in (20) this is no longer possible in general. In order to achieve separability, an additional condition must be imposed. One possibility is the choice

$$-D\frac{\frac{\partial^2}{\partial x^2}\varrho}{\varrho} = \gamma\left(\ln\varrho - \langle\ln\varrho\rangle\right), \tag{21}$$

because of $\ln\varrho = \ln\Psi + \ln\Psi^* = F_1(\Psi) + F_2(\Psi^*)$ the separation of terms containing either Ψ or Ψ^{ast} is possible. This leads to an additional *complex* logarithmic term in the SE [16],

$$i\hbar\frac{\partial}{\partial t}\Psi_{\mathrm{NL}}(x,t) = \{H_{\mathrm{L}} + \gamma\frac{\hbar}{i}(\ln\Psi_{\mathrm{NL}} - \langle\ln\Psi_{\mathrm{NL}}\rangle)\}\Psi_{\mathrm{NL}}(x,t) = \{H_{\mathrm{L}} + W_{\mathrm{lgNL}}\}\Psi_{\mathrm{NL}}(x,t), \tag{22}$$

where H_{L} is the usual linear Hamiltonian.

Also the logarithmic NLSE (22) possesses Gaussian WP solutions in the cases that were discussed for the TDSE without dissipation. Maximum and width of this Gaussian function are now determined by the modified equations of motion

$$\ddot{\eta} + \gamma\dot{\eta} + \omega^2\eta = 0 \tag{23}$$

for the maximum and

$$\frac{d}{dt}\mathcal{C}_{\mathrm{NL}} + \mathcal{C}_{\mathrm{NL}}^2 + \gamma\mathcal{C}_{\mathrm{NL}} + \omega^2 = 0 \tag{24}$$

for the width with an additional linear term depending on the friction coefficient γ.

The relation between the imaginary part of C_{NL} and the position uncertainty or Ermakov variable $\alpha(t)$ remains unchanged as in the non-dissipative case, $\mathcal{C}_{\mathrm{NL,I}} = \frac{\hbar}{2m\langle\tilde{x}^2(t)\rangle} = \frac{1}{\alpha_{\mathrm{NL}}^2}$, but the real part differs by a contribution from the friction coefficient, $\mathcal{C}_{\mathrm{NL,R}} = \frac{\dot{\alpha}_{\mathrm{NL}}}{\alpha_{\mathrm{NL}}} - \frac{\gamma}{2}$.

Inserting $C_{\mathrm{NL,I}}$ and $C_{\mathrm{NL,R}}$ into the Riccati equation (24) turns this into the Ermakov equation

$$\ddot{\alpha}_{\mathrm{NL}} + \left(\omega^2 - \frac{\gamma^2}{4}\right)\alpha_{\mathrm{NL}} = \frac{1}{\alpha_{\mathrm{NL}}^3}, \tag{25}$$

i.e., in comparison with the non-dissipative case, only ω^2 has been replaced by $\Omega = \left(\omega^2 - \frac{\gamma^2}{4}\right)$.

Riccati equation (24) can also be linearized, now using the ansatz $C_{\mathrm{NL}} = \frac{\dot{\tilde{\lambda}}}{\tilde{\lambda}} = \frac{\dot{\lambda}}{\lambda} - \frac{\gamma}{2}$ to yield the Newtonian equation with linear friction term

$$\ddot{\tilde{\lambda}} + \gamma\dot{\tilde{\lambda}} + \omega^2\tilde{\lambda} = 0\,, \tag{26}$$

for the complex variable $\tilde{\lambda}(t) = \lambda e^{-\gamma t/2} = \alpha_{\mathrm{NL}}e^{-\gamma t/2 + i\varphi}$. Inserting the polar form into Eq. (24) leads to the unchanged conservation law $\dot{\varphi} = \frac{1}{\alpha_{\mathrm{NL}}^2}$ as in the non-dissipative case and to the modified Ermakov equation (25).

Comparison of Eqs. (23), (24) and (26) with the corresponding Eqs. (3), (4) and (6) in the non-dissipative case shows the following modifications. In the linear second-order differential equations for η and λ, a linear term with first derivative has been added while in the first-order Riccati equation an additional term linear in C_{NL} appears. All the additional terms depend on the coefficient of the friction force.

Now the question is: is it possible to modify the corresponding equations in a similar way to include the effect of a dissipative environment into the TISE? The answer has been given in [9]. Here only a few details necessary for the following discussion will be cited.

The above-mentioned modifications in the TD case would mean the addition of a term $\Gamma\left(\frac{\nabla\Psi}{\Psi}\right)$ to Eq, (16) or a term $\Gamma\nabla\Psi$ to the TISE, respectively, where the coefficient Γ should somehow be related to the friction force but, in general, could be a complex function of $\mathbf{r}$ and t, $\Gamma(\mathbf{r},t)$.

As Ψ is complex, a complex contribution from the imaginary part of the additional term $\Gamma\nabla\Psi$ to the continuity equation for $\varrho = \Psi^*\Psi$ of the form $\Gamma\ \nabla\varrho\ =\ \Gamma\left(\Psi^*\nabla\Psi\ +\ \Psi\nabla\Psi^*\right)$ should arise in the evolution equation for ϱ. The contribution from this term would enter the SE with an *imaginary* coefficient.

The additional term in the equation for ϱ, at least for cases with Gaussian WP solutions (i.e., $V = 0$ and HO) discussed in this paper, should have the same effect as the diffusion term or the $\ln\varrho$-term, i.e.,

$$-D\Delta_x\varrho\ =\ \gamma(\ln\varrho - \langle\ln\varrho\rangle)\varrho\ =\ \frac{\gamma}{2}\left(1\ -\ \frac{\tilde{x}^2}{\langle\tilde{x}^2\rangle}\right)\varrho. \tag{27}$$

Comparison shows that

$$\nabla_x\left(\frac{\gamma}{2}\tilde{x}\varrho\right)\ =\ \frac{\gamma}{2}\left(1\ -\ \frac{\tilde{x}^2}{\langle\tilde{x}^2\rangle}\right)\varrho \tag{28}$$

yields the desired result![4]

Taking into account this additional term, the modified continuity equation requires a modified action function $S'\ =\ S\ +\ m\frac{\gamma}{4}\tilde{x}^2\ +\ f(t)$ in order to provide a conservation law of the form $\nabla S'\ =\ \frac{C}{a^2}$.

Separation of the continuity equation including the additional term (28) and achieving consistency of the modified Hamilton–Jacobi equation with the modified action function S' leads to a form of the dissipative term that is essentially identical to the one Hasse [17] and provides the full dissipative friction force. Further details can be found in [10].

4. Quantization of angular and radial aspects

We have seen so far that the changes of radial and angular components of a complex quantity are not independent of each other, if this complex quantity is somehow related to a complex Riccati equation due to the quadratic nonlinearity of this equation. This applies also, when the complex quantity is linearized via a logarithmic derivative. In the case of TD quantum mechanics, the complex quantity fulfilling the Riccati equation is $\mathcal{C} = \mathcal{R} + \mathrm{i}\mathcal{I} = \frac{\dot{\alpha}}{\alpha} + \mathrm{i}\dot{\varphi}$ where α is the amplitude and φ the phase angle of another complex quantity, $\lambda = \alpha\mathrm{e}^{\mathrm{i}\varphi}$ that was used to linearize the Riccati equation (4) to the linear complex Newtonian equation (6). How can this be connected with a quantization of radial and angular components?

Unlike modern physics that is looking for smallest building blocks of our material world (with largest accessible energy - and money), Plato's view of the world, expressed in his work "Timaios", was that it is build up in terms of right triangles that could be used form cube, tetrahedron, octahedron and icosahedron representing the four elements earth, fire, air and water of our material world - there is no word about the size of the triangles, i.e., scale invariance!

Certainly, in connection with right triangles one also thinks about Pythagoras and his law $a^2 + b^2 = c^2$. But there are special right triangles that can in a certain way be quantized (after Platon's "quantization", a kind of "second quantization").

These so-called Pythagorean triples are right triangles where all three sides can be expressed in terms of integers, e.g., 3, 4, 5 leading to 9 + 16 = 25. Most people have difficulties to find even a second example for those triples, although infinitely many of them exist. The problem of finding them can easily be solved considering the square of a complex quantity as it occurs in our Riccati equations.

Assuming a complex quantity like $\mathcal{C}\ =\ \mathcal{R} + \mathrm{i}\mathcal{I}\ =\ \frac{\dot{\alpha}}{\alpha} + \mathrm{i}\dot{\varphi}$ represents a right triangle in the complex plane, made out of its real and imaginary parts and its absolute value. If $\mathcal{C}$ is squared, the result is again a complex quantity with real part $\mathcal{R}\{\mathcal{C}^2\} = \mathcal{R}^2 - \mathcal{I}^2$, imaginary part $\mathcal{I}\{\mathcal{C}^2\} = 2\ \mathcal{R}\mathcal{I}$ and absolute value $|\mathcal{C}^2| = \mathcal{R}^2 + \mathcal{I}^2$. It is straightforward to show that for

[4] In our one-dimensional case ∇_x means $\frac{\partial}{\partial x}$.

IOP Conf. Series: Journal of Physics: Conf. Series **839** (2017) 012008 doi:10.1088/1742-6596/839/1/012008

"quantized" $\mathcal{C}$, i.e., integer values of $\mathcal{R}$ and $\mathcal{I}$ (with $\mathcal{R} > \mathcal{I}$, all three sides of the right triangle formed by $\mathcal{C}^2$ are integers. Examples are: (a) $\mathcal{R} = 2, \mathcal{I} = 1$: $\mathcal{R}^2 - \mathcal{I}^2 = 3$, $2\,\mathcal{R}\mathcal{I} = 4$, $\mathcal{R}^2 + \mathcal{I}^2 = 5$ with $9 + 16 = 25$; (b) $\mathcal{R} = 3, \mathcal{I} = 2$: $\mathcal{R}^2 - \mathcal{I}^2 = 5$, $2\,\mathcal{R}\mathcal{I} = 12$, $\mathcal{R}^2 + \mathcal{I}^2 = 13$ with $25 + 144 = 169$.

In our case, $\mathcal{R} = \frac{\dot{\alpha}}{\alpha}$ and $\mathcal{I} = \dot{\varphi}$, so the question arises, where do we find a quantization of these quantities in physics? The quantization of $\dot{\varphi}$ can obviously be found considering the HO with constant frequency $\omega = \omega_0$. In this case, and for constant WP width, $\omega_0 = \dot{\varphi}$ is valid and the energy is quantized in terms of $\dot{\varphi}$ according to

$$E_n = \hbar \left(n + \frac{1}{2} \right) \dot{\varphi} \tag{29}$$

with n being an integer, i.e. there is a quantization in terms of $n\,\dot{\varphi}$.

In order to find a quantization of $\mathcal{R} = \frac{\dot{\alpha}}{\alpha}$, one has to look at the currents in the continuity equation. In the TD case for Gaussian WP solutions the velocity field v_-, according to the definition below Eq. (20), can be written as

$$v_- = \dot{\eta} + \frac{\dot{\alpha}}{\alpha} \tilde{x} \tag{30}$$

with the classical contribution $\dot{\eta} = \langle v \rangle$.

In the QHE, the current $j = \varrho\, v \propto \sigma E$ (here only one dimension shall be considered) is proportional to the electric field strength E and the conductivity σ with $\sigma = n\, c \left(\frac{e^2}{c\hbar} \right) = n\, c\, \alpha_{\text{Som}}$ where n = integer, c = velocity of light, e = electic elementary charge and $\alpha_{\text{Som}} = \left(\frac{e^2}{c\hbar} \right) \approx \frac{1}{137}$ is Sommerfeld's fine structure constant (not to be confused with $\alpha(t)$, the WP width as defined in Sec. 2).

For fixed position of the system's centre of mass, $\dot{\eta} = 0$, the conductivity or current displays the following proportionality:

$$j \propto n \left(\frac{e^2}{c\hbar} \right) = n\, \alpha_{\text{Som}} \propto \frac{\dot{\alpha}}{\alpha}, \tag{31}$$

therefore, changes of j by integers n would correspond to integer changes of $\frac{\dot{\alpha}}{\alpha}$.

However, the continuity equation is valid for isolated systems without dissipation, whereas the QHE takes frictional effects into account. Looking at the dissipative scenario described by the Smoluchowski equation (20), the convective velocity field changes into $v_- = \dot{\eta} + \left(\frac{\dot{\alpha}}{\alpha} - \frac{\gamma}{2} \right) \tilde{x}$ and an additional contribution $v_D = -D \frac{\frac{\partial}{\partial x} \varrho}{\varrho} = \frac{\gamma}{2} \tilde{x}$ from the diffusion term shows up, leading to a total velocity field $v_{tot} = v_- + v_D = \frac{\dot{\alpha}}{\alpha} \tilde{x}$ that is again proportional to $\frac{\dot{\alpha}}{\alpha}$, so (for $\dot{\eta} = 0$) also in the dissipative situation that is present in the QHE, $\frac{\dot{\alpha}}{\alpha}$ is the quantity that is quantized. Particularly, $\frac{\dot{\alpha}}{\alpha}$ is proportional to $\frac{e^2}{c\hbar}$ that is dimensionless and is discussed in more detail below.

Apart from the QHE, there is also the FQHE observed where

$$j \propto \frac{n}{k} \left(\frac{e^2}{c\hbar} \right) \tag{32}$$

with n, k both being integers. There are different interpretations of this effect in the literature and not all possible combinations of n and k (being odd, even or both) have been observed. To my knowledge none of the existing theories is able to explain all observed combinations of n and k or even predict yet unobserved ones. Therefore, there might still be room for some unknown in the theory of the FQHE.

One interpretation that had been given by the author and colleagues [18] some time ago assumes that the FQHE is a manifestation of a phenomenon where two effects are involved, each being quantized separately. Since it was already stated in the introduction that *action* is the quantity that is usually quantized in nature, we assumed that in the FQHE one of the effects is quantized in terms of $\hbar$ while the other is in terms of $\frac{e^2}{c}$, so the ratio of both is just Sommerfelds fine structure constant $\alpha_{\text{Som}} = \frac{e^2}{c\hbar}$.

Certainly, as shown above, the action $\hbar$ is well known in quantum mechanics; the action $\frac{e^2}{c}$ was already mentioned by Einstein [19], Schrödinger [20] and Eddington [21], assuming its importance but not knowing its physical relevance. Our interpretation of the FQHE is that this is the first observed physical effect that shows the relevance of a second quantum of action, $\frac{e^2}{c}$.

In [17], we called $\frac{e^2}{c}$ the "*quantum of electrostatic action*". Why this name? For two electrons interacting via Coulombs law, the electrostatic action involved is action = energy × time = $\frac{e^2}{r} \times t$. As the distance between the electrons, r, and their interaction time, t, can be related via the interaction velocity $v = \frac{r}{t}$, one can define the action $= \frac{e^2}{v}$. This action becomes minimum when the velocity attains its maximum value $v = c =$ velocity of light. So, the last Coulombic action is $\frac{e^2}{c}$.

But since nature is usually very economical, as expressed by the various extremal principles, e.g., in mechanics, optics, etc., why does it permit the luxury of two elementary quanta of action? Of course no definite answer shall be given at this point; but the following remarks, based on the afore-mentioned facts shall be allowed. Planck's constant (divided by 2π), $\hbar$, is well known, particularly in the quantum mechanics of oscillating systems where, e.g., the energy is quantized in terms of $\hbar\omega = \hbar\dot{\varphi}$, i.e., the frequency is related to some kind of angular velocity and thus to some kind of *angular* aspect. On the other hand, it has just been shown that $\frac{e^2}{c}$ is somehow related to a physical situation where a *radial* distance plays the important role.

As the radius and circumference of a circle cannot be expressed in terms of the same units ("*quanta*") since they are related via the irrational number π, maybe the "*radial*" *action* and the "*angular*" *action* also cannot be expressed in terms of the same fundamental units.

As seen from our discussion of integer values for $\frac{\dot{\alpha}}{\alpha}$ and $\dot{\varphi}$ related to $\mathcal{C} = \frac{\dot{\lambda}}{\lambda}$, also there the radial aspect is for integer values of $\frac{e^2}{c}$ connected to e = 2.718... whereas the angular aspect, $\dot{\varphi} = \frac{1}{\alpha^2}$, is connected to $\pi = 3.141...$. So, maybe $\frac{e^2}{c}$ and $\hbar$ just offer the possibility to quantize in a (complex?) two-dimensional (phase?) space radial changes as well as angular ones.

5. Conclusions and perspectives

Quantum theory is essentially based on complex quantities that posses an amplitude and a phase, like the wave function solving the linear SE. As has been shown above, this equation can also be considered as a linearized version of a complex Riccati equation. Due to the quadratic nonlinearity occurring in the Riccati equation, real- and imaginary-parts of the complex quantity are no longer independent of each other, but coupled (in our case by some kind of conservation law).

Considering the dissipative case the TDSE has shown that the effect of the the dissipative environment on the linearized complex quantity (i.e., $\lambda(t)$ fulfilling a complex Newtonian equation) is essentially an exponential decay of the amplitude of this quantity. In the TI case, this is by definition not possible, as the linearized quantity is the wave function whose amplitude is always normalized due to the probabilistic interpretation. Nevertheless a description of the same dissipative system was possible, but now requiring a change of the action function S into S' as defined below Eq. (28).

Therefore it seems that radial changes as they occur e.g. in irreversible decay processes can equally well be described by damping of an amplitude, i.e. radial changes, or changing of the phase of a complex quantity, i.e. angular changes. This would suggest that non-unitary transformations that appear to be necessary for the description of open quantum systems can equally well be described by gauge-transformations changing only the phase of the wave function, but leaving the amplitude (and thus the interpretation of the wave function) untouched.

Attempts to link NLSEs with (NL) gauge trandformations have already been studied [22,23], but there is still a wide field to explore.

Also the apparent occurrence of a second quantum of action that seems to be related to some radial changes that are usually not recognized as such in quantum theory needs further investigation.

References

[1] Peitgen H-O and Richter P H 1986 *The Beauty of Fractals* (Berlin: Springer) pp 9, 11 22 48, 84-86
[2] Yang C N 1987 *Schrödinger – Centenary Celebration of a Polymath* ed C W Kilmister (Cambridge: Cambridge University Press) p 53
[3] Wigner E 1960 *Commun. Pure Appl. Math.* **13** 1
[4] Schrödinger E 1926 *Ann. d. Phys.* **79** 361
[5] Schuch D 2008 *SIGMA* **4** 043
[6] Schuch D and Moshinsky M 2006 *Phys. Rev.* A **73** 062111
[7] Schuch D 2011 in: *Latin-American School of Physics XL ELAF; AIP Conf. Proceedings* **1334** 291
[8] Reinisch G 1994 *Physica* A **206** 229
[9] Madelung E 1926 *Z. Physik* **40** 322
[10] Schuch D 2012 *J. Phys. Conf. Ser.* **380** 012009
[11] Schuch D, Guerrero J, López-Ruiz F F and Aldaya V *Phys. Scr.* **90** 045209
[12] Yu L H and Sun C P 1994 *Phys. Rev.* A **49** 592
[13] Sun C P and Yu L H 1995 *Phys. Rev.* A **51** 1845
[14] Madelung E 1950 *Die Mathematischen Hilfsmittel des Physikers* (Berlin: Springer) p 432
[15] Mrowka B 1951 *Z. Phys.* **130** 164
[16] Schuch D, Chung K-M and Hartmann H 1983 *J. Math. Phys.* **24** 1652
[17] Hasse R W 1975 *J. Math. Phys.* **16** 2005
[18] Wadlinger R L P, Hunter G, Kostro L and Schuch D 1990 *Phys. Essays* **3** 194
[19] Einstein A 1909 *Z. Phys.* **6** 185
[20] Schrödinger E 1922 *Z. Phys.* **12** 23
[21] Eddington A S 1935 *New Pathways in Science* (Cambridge: Cambridge University Press) p 234
[22] Doebner H-D and Goldin G A 1996 *Phys. Rev.* A **54** 3764
[23] Schuch D 2000 in *Quantum Theory and Symmetries* (Singapore, World Scientific) p 152

IOP Conf. Series: Journal of Physics: Conf. Series **839** (2017) 012009 doi:10.1088/1742-6596/839/1/012009

The Harmonic Oscillator in the Framework of Scale Relativity

M Bonilla and O Rosas-Ortiz

Physics Department, Cinvestav, A.P. 14-740, 07000, México City, Mexico

E-mail: mbonilla@fis.cinvestav.mx, orosas@fis.cinvestav.mx

Abstract. The dynamical law obeyed by the one-dimensional physical systems in the scale relativity approach is reduced to a Riccati nonlinear differential equation. Applied to the harmonic oscillator potential, we show that such an approach permits the calculation of the solutions of the scale relativity problem in terms of the well known solutions of the Schrödinger equation for the harmonic oscillator.

1. Introduction

Scale relativity is an emergent approach that aims to unify quantum physics and relativity theory [1]. This is based on the notion that any measuring process in physics depends on the scale, so that no absolute measurements can be associated with any coordinate system. The latter would mean that the derivatives introduced by Newton must, at least, be revised. Introducing a new complex derivative which takes into account such scale dependence, the dynamical law of scale relativity is represented by a complex-differential equation with a term that encodes the nonclassicality of the system under study. Thus, by necessity, the velocity is a complex vector in such an approach. Then, besides the potential energy, the scale-dependent Hamiltonian includes a complex 'kinetic energy' plus a term that is proportional to the divergence of the complex velocity. If this last is different from zero, the Hamiltonian of scale relativity differs from the one of the conventional approaches even if the imaginary part of the velocity is zero. Thus, the divergence of the complex velocity is a measure of the nonclassicality of a given system in the scale relativity approach.

In this paper we solve the fundamental equation of scale relativity for the harmonic oscillator. As the latter is the simplest exactly solvable model in any of the physical theories, the results reported here are addressed to obtain a better understanding of the way in which scale relativity works. In particular we show that the solutions of the quantum problem are intimately connected with the ones of the scale relativity approach via the Riccati non-linear differential equation.

2. Basic equations

In the simplest case, preserving the notion of Newtonian time, the space metric is continuous and non-differentiable everywhere [1]. Then, the spatial displacements as well as the velocities are twice-valued. The complex velocity

$$\vec{\mathcal{V}}(\vec{x},t) \equiv \vec{F}(\vec{x},t) - \mathrm{i}\vec{G}(\vec{x},t) \tag{1}$$

IOP Conf. Series: Journal of Physics: Conf. Series **839** (2017) 012009 doi:10.1088/1742-6596/839/1/012009

is introduced to include the two velocities $\vec{v}_\pm$ since its real and imaginary parts are respectively given by $\vec{F}(\vec{x},t) \equiv \frac{\vec{v}_+ + \vec{v}_-}{2}$ and $\vec{G}(\vec{x},t) \equiv \frac{\vec{v}_+ - \vec{v}_-}{2}$. In the Hamiltonian formulation [5], for a stationary system with definite energy W, the dynamical law is represented by the Riccati equation

$$W = \frac{m}{2}\vec{\mathcal{V}}^2 + \Phi - \mathrm{i}\frac{\hbar}{2}(\vec{\nabla}\cdot\vec{\mathcal{V}}). \tag{2}$$

For a particle of mass m subjected to the one-dimensional potential $\Phi = \frac{1}{2}m\omega^2x^2$, the dynamical law (2) takes the form

$$\partial_x \mathcal{V} = -\mathrm{i}\frac{m}{\hbar}\mathcal{V}^2 + \mathrm{i}\frac{2}{\hbar}\Big(W - \Phi(x)\Big). \tag{3}$$

As the complex velocity of a stationary state has no real part, we arrive at the expression

$$-G' = \frac{m}{\hbar}G^2 + \frac{2}{\hbar}\Big(W - \frac{1}{2}m\omega^2x^2\Big). \tag{4}$$

With the appropriate change of variables [2], we rewrite (4) in dimensionless form

$$-g' = g^2 + \epsilon - \mathrm{x}^2. \tag{5}$$

The latter nonlinear differential equation can be linearized by the logarithmic derivative $g(\mathrm{x}) = \frac{\mathrm{d}}{\mathrm{dx}}\ln\xi(\mathrm{x})$. One obtains the Schrödinger-like equation

$$\mathrm{H}\xi = \Big(-\frac{\mathrm{d}^2}{\mathrm{dx}} + \mathrm{x}^2\Big)\xi = \epsilon\,\xi, \tag{6}$$

where $\mathrm{H} = -\frac{\mathrm{d}^2}{\mathrm{dx}^2} + \mathrm{x}^2$ is the related Hamiltonian and ϵ corresponds to the energy eigenvalue. That is, (6) is the eigenvalue equation associated with the (mathematical) harmonic oscillator $\Phi = \mathrm{x}^2$. We are going to take full advantage of the connection between (5) and (6) to solve the dynamical problem of the scale relativity for the harmonic oscillator. That is, we shall obtain the solutions of (5) by solving the eigenvalue problem (6).

3. Solution of the problem

It is well known that the operators

$$\mathsf{a} = \frac{\mathrm{d}}{\mathrm{dx}} + \mathrm{x}, \quad \mathsf{a}^\dagger = -\frac{\mathrm{d}}{\mathrm{dx}} + \mathrm{x}, \quad (\mathsf{a}^\dagger)^\dagger = \mathsf{a}, \tag{7}$$

satisfy the Heisenberg algebra

$$[\mathsf{a}, \mathsf{a}^\dagger] = 2\mathrm{I}, \quad [\mathrm{H}, \mathsf{a}] = -2\mathsf{a}, \quad [\mathrm{H}, \mathsf{a}^\dagger] = 2\mathsf{a}^\dagger, \tag{8}$$

and factorize the Hamiltonian H as follows

$$\mathsf{a}\mathsf{a}^\dagger = -\frac{\mathrm{d}^2}{\mathrm{dx}^2} + \mathrm{x}^2 + 1 = -\frac{\mathrm{d}^2}{\mathrm{dx}^2} + \mathrm{x}^2 - \epsilon \equiv \mathrm{H} - \epsilon. \tag{9}$$

The simplest form of solving the eigenvalue equation (6) considers the factorization (9) and an extremal function $\phi_0(\mathrm{x})$ which is annihilated by the operator a. That is

$$\mathsf{a}\phi_0(\mathrm{x}) = 0 \quad \Rightarrow \quad \mathrm{H}\phi_0(\mathrm{x}) = \phi_0(\mathrm{x}). \tag{10}$$

If $\phi_0(\mathrm{x})$ is of finite norm then it is a normalizable eigenfunction of H with eigenvalue $E_0 = 1$. Using (7) it is immediate to obtain

$$\phi_0(\mathrm{x}) = C_0 e^{-\frac{\mathrm{x}^2}{2}}. \tag{11}$$

Clearly $\phi_0(\mathrm{x}) \in \mathrm{L}^2(\mathbb{R})$, so that the integration constant C_0 is fixed by normalization. On the other hand, from the oscillation theorem we know that there is no square-integrable solution belonging to the eigenvalue $E < E_0$ since $\phi_0(\mathrm{x})$ is free of nodes. Therefore, $\phi_0(\mathrm{x})$ is the wave function of the ground energy of the oscillator.

Iterating n-times the action of $\mathrm{a}^\dagger$ on $\phi_0(\mathrm{x})$ one obtains

$$\phi_n(\mathrm{x}) = \frac{C_n}{\sqrt{\sqrt{\pi}\, 2^n n!}} \left(\mathrm{a}^\dagger\right)^n \phi_0(\mathrm{x}), \tag{12}$$

which is the wave function of the state associated to the energy $E_n = E_0 + 2n$. After the straightforward calculation we have

$$\phi_n(\mathrm{x}) = C_n \exp\left(-\frac{\mathrm{x}^2}{2}\right) \mathrm{H}_n(\mathrm{x}), \tag{13}$$

with C_n the normalization constant and $\mathrm{H}_n(\mathrm{x})$ standing for the Hermite polynomials [3].

A first set of solutions to the Riccati equation (5) are given by the logarithmic derivative of the wave functions

$$g_n(\mathrm{x}) = \frac{\mathrm{d}}{\mathrm{dx}} \ln \phi_n(\mathrm{x}) = -\mathrm{x} + \frac{2n\mathrm{H}_{\mathrm{n}-1}(\mathrm{x})}{H_n(\mathrm{x})} = \mathrm{x} - \frac{\mathrm{H}_{\mathrm{n}+1}(\mathrm{x})}{H_n(\mathrm{x})}. \tag{14}$$

That is,

$$g_0(\mathrm{x}) = -\mathrm{x}, \quad g_1(\mathrm{x}) = -\mathrm{x} + \frac{1}{\mathrm{x}}, \quad g_2(\mathrm{x}) = -\mathrm{x} + \frac{4\mathrm{x}}{2\mathrm{x}^2 - 1}, \quad \ldots \tag{15}$$

Additionally, there is a solution of (5) which cannot be obtained from the functions (13). Namely, for $\epsilon = -1$ the simplest solution is $g_{-1}(\mathrm{x}) = \mathrm{x}$. However, such a function gives rise to a solution of (6) that is not normalizable

$$\phi_{-1}(\mathrm{x}) = \exp\left(\frac{\mathrm{x}^2}{2}\right) \quad \Rightarrow \quad \mathrm{H}\phi_{-1}(\mathrm{x}) = -\phi_{-1}(\mathrm{x}). \tag{16}$$

The same holds for the solutions of the Riccati equation (5) for any other $\epsilon < E_0 = 1$. In this form, among the solutions of the scale relativity equation (5), only the g-functions obeying the transformation (14) admit an interpretation in the quantum approaches [2]. We may write

$$g_n'(\mathrm{x}) + g_n^2(\mathrm{x}) = \mathrm{x}^2 - 2n - 1, \quad n = 0, 1, 2, \ldots \tag{17}$$

as the equation in scale relativity that defines physical solutions in the quantum picture for the harmonic oscillator.

In the panel of Fig. 1 we show some of the solutions g_n of (5) as well as the corresponding solutions ϕ_n of (6). Note that, in all the cases, the g-functions diverge as $|x| \to \infty$. Besides, we can appreciate that the singularities of g_1 and g_2 are associated with the nodes of ϕ_1 and ϕ_2 respectively. In turn, the function g_0 has no singularities since the ground state wave function ϕ_0 is free of nodes. Interestingly, although g_{-1} is free of singularities and diverges at $x = \pm\infty$, it is associated with the non square-integrable function ϕ_{-1}. The latter because the slopes of g_{-1} and g_0 have opposite sign. Another interesting property of the solutions of (17) is that their zeros are in connection with the local maxima and minima of ϕ_n. On the other hand, it is well known that the ϕ_n are nonclassical states for $n \geq 1$ since their P-function is as singular as the derivatives of the delta distribution [4]. In contrast, the ground state ϕ_0 is classical because its P-function is equal to $\delta(x)$. The relationship between the singularities of g_n and the nonclassicality of ϕ_n is in progress and will be reported elsewhere.

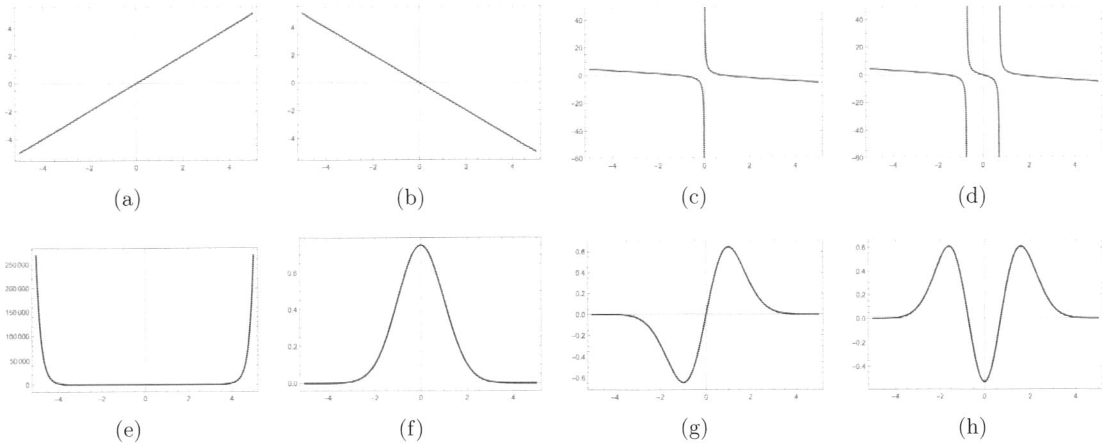

Figure 1. The upper and lower rows include some solutions of the Riccati equation (5) and the Schrödinger equation (6) respectively. From left to right the columns correspond to $\epsilon = -1$, $\epsilon = E_0$, $\epsilon = E_1$, and $\epsilon = E_2$, with $E_n = 2n + 1$, $n = 0, 1, 2, \ldots$, the energy eigenvalues of the quantum harmonic oscillator. Notice that g_{-1} is connected with the function ϕ_{-1} which is not square-integrable, see Eq. (16).

4. Concluding remarks

We have shown that the fundamental equation of the scale relativity for the harmonic oscillator can be reduced to a Riccati nonlinear differential equation. Using the well known relationship between the Riccati and the Schrödinger equations we have solved the scale relativity problem in terms of the oscillator quantum eigenfunctions. In contrast with other works [5], we require no numerical approximations to justify the form of the solutions. We have found some interesting relationships between the singularities of the scale relativity solutions and the nodes of the quantum wave functions. Further progress will be reported elsewhere.

Acknowledgment
MB acknowledges the funding received through a CONACyT scholarship.

References

[1] Notalle L, *Scale Relativity and Fractal Space-Time: A New Approach to Unifying Relativity and Quantum Mechanics*, Imperial College Press, 2011
[2] Bonilla M, *From Scale Relativity to Supersymmetric Quantum Mechanics via the Riccati Equation*, M.Sc. Thesis, Cinvestav, México City, Mexico, 2016.
[3] Abramowitz M and Stegun I A, *Handbook of Mathematical Functions*, Dover Publications, Inc., New York, 1970.
[4] Glauber R J, *Quantum Theory of Optical Coherence*. Selected Papers and Lectures, Wiley-VCH, Weinheim, 2007.
[5] Nottale N,Teh M T and Le Bohec S, Scale Relativistic Formulation of Non-Differentiable Mechanics I: Applications to the harmonic oscillator, arXiv:1601.07778 [physics.gen-ph]

Fractional Driven Damped Oscillator

F Olivar-Romero and O Rosas-Ortiz

Physics Department, Cinvestav, A.P. 14-740, 07000, México City, Mexico

E-mail: folivar@fis.cinvestav.mx

Abstract. The resonances associated with a fractional damped oscillator which is driven by an oscillatory external force are studied. It is shown that such resonances can be manipulated by tuning up either the coefficient of the fractional damping or the order of the corresponding fractional derivatives.

1. Introduction

The simplest oscillating system (a *harmonic oscillator*) can be modeled by a mass at the end of a spring which slides back and forth without friction. The motion is characterized by the natural frequency of oscillation ω_0 and the total stored energy E (which is a constant of motion and defines the amplitude of oscillation) [1]. Actual oscillating systems present some loss of energy due to friction forces so that the amplitude of their oscillations is a decreasing function of time. However, the oscillations can be driven to avoid their damping down by the action of a repetitive force $F(t)$ on the system. Such a system is called *driven damped oscillator* [2]. An special excitation of the system arises when the frequency of the applied force matches the natural frequency of the oscillator since the spectral energy distribution takes its maximum value. The phenomenon, known as *resonance*, is a subject of study in classical mechanics, electromagnetism, optics, acoustics and quantum mechanics, among other physical theories [3].

The present work is addressed to the study of the driven damped oscillator in the context of fractional calculus [4–6]. That is, the second-order differential equation associated with the Newtonian law of motion for a damped oscillator that is driven by an external force will be substituted by a fractional differential equation of order 2α, with $0 < \alpha \leq 1$. Special emphasis will be placed on the resonance phenomenon.

2. Fractional Oscillator with fractional damping

Given an oscillator of natural frequency ω_0, the general expression for the displacement of the mass can be expressed as the integral equation [8, 9]

$$x(t) = x_0 + I_t(\dot{x}_0) + \omega_0^2 I_t\left[I_t(x(t))\right], \tag{1}$$

where x_0 and $\dot{x}_0$ are constants of integration, and

$$I_t(x(t)) := \int x(t)dt \tag{2}$$

represents the Riemman time-integration of $x(t)$. The fractional generalization of (1) is performed in two steps. First we replace I_t with the Riemman-Liouville fractional time-integral operator I^α [4–6], and ω_0 with ω_0^α. The latter for consistency of units. Then we have

$$x(t) = x_0 + I^\alpha(\dot{x_0}) + \omega_0^{2\alpha} I^\alpha\left[I^\alpha(x(t))\right], \quad 0 < \alpha \le 1. \tag{3}$$

Now, a fractional differential form of (3) can be obtained by applying twice the time-fractional derivative operator of Caputo

$$D^\alpha D^\alpha x(t) + \omega^{2\alpha} x(t) = 0 \tag{4}$$

(for details about the operator D^α see, e.g., [5]). Let us introduce a 'fractional damping' which is proportional to the fractional time-derivative of the position $D^\alpha x(t)$. That is

$$D^\alpha D^\alpha x(t) + 2\beta^\alpha D^\alpha x(t) + \omega^{2\alpha} x(t) = 0. \tag{5}$$

One can show that the solution of this last equation is of the form

$$\begin{aligned} x(t) =& \frac{x_0 t^{-\alpha}}{\sqrt{\beta^{2\alpha} - \omega_0^{2\alpha}}}\left[E_{\alpha,1-\alpha}\left(-\Omega_- t^\alpha\right) - E_{\alpha,1-\alpha}\left(-\Omega_+ t^\alpha\right)\right] \\ &+ \frac{2\beta^\alpha x_0 + x_0^{(\alpha)}}{\sqrt{\beta^{2\alpha} - \omega_0^{2\alpha}}}\left[E_{\alpha,1-\alpha}\left(-\Omega_- t^\alpha\right) - E_{\alpha,1-\alpha}\left(-\Omega_+ t^\alpha\right)\right], \end{aligned} \tag{6}$$

where

$$E_{\alpha,\beta}(z) = \sum_{k=0}^{\infty} \frac{z^k}{\Gamma(\alpha k + \beta)}, \quad \mathrm{Re}(\alpha) > 0, \quad \mathrm{Re}(\beta) > 0, \quad z \in \mathbb{C}, \tag{7}$$

is the Mittag-Leffler function [7], and

$$\Omega_\pm = \beta^\alpha \pm \sqrt{\beta^{2\alpha} - \omega_0^{2\alpha}}.$$

3. Driven fractional oscillator with fractional damping

Let us add a driving force at the right hand side of Eq. (5), we have

$$D^\alpha D^\alpha x(t) + \beta^\alpha D^\alpha x(t) + \omega^{2\alpha} x(t) = f_0 \cos(\omega t + \phi), \tag{8}$$

where f_0 is a real constant. Applying the Laplace transform L and solving for $X(s) = L[x(t)]$ we arrive at the expression

$$X(s) = f_0\left[\frac{s\cos\phi - \omega\sin\phi}{\left(s^2+\omega^2\right)\left(s^{2\alpha} + 2\beta^\alpha s^\alpha + \omega_0^{2\alpha}\right)}\right] + \frac{x_0 s^{2\alpha-1} + \left(x_0 + 2\beta^\alpha x_0^{(\alpha)}\right) s^{\alpha-1}}{s^{2\alpha} + 2\beta^\alpha s^\alpha + \omega_0^{2\alpha}}. \tag{9}$$

Making $f_0 = 0$ we see that the second term in (9) corresponds to *transient* oscillations because there is no force present which can ensure their predominance; the corresponding inverse Laplace transform has been evaluated in the previous section. In turn, for the inverse Laplace transform of the firs term one has

$$x(t) = \frac{f_0}{2\pi i}\lim_{T\to\infty}\int_{g-iT}^{g+iT} e^{st}\left[\frac{s\cos\phi - \omega\sin\phi}{\left(s^2+\omega^2\right)\left(s^{2\alpha} + 2\beta^\alpha s^\alpha + \omega_0^{2\alpha}\right)}\right] ds. \tag{10}$$

The integrand contains a branch point at $s = 0$ and simple poles at $s = \pm i\omega$, and $s = \left(\Omega_\pm e^{\pm i\pi}\right)^{1/\alpha}$. Following [10] we find that $x(t)$ is given as the sum of three contributions: $x_1(t)$,

$x_2(t)$ and $x_3(t)$. The first one results from the calculation of (10) along the Hankel-Bromwich path shown in Fig. 1 of Ref. [10], we obtain

$$x_1(t) = \frac{f_0}{\pi}\int_0^\infty \frac{e^{-rt}\left[r^{2\alpha}\sin(2\pi\alpha) + 2\beta^\alpha r^\alpha \sin(\pi\alpha)\right]}{(r^2+\omega^2)\left[r^{4\alpha} + 4\beta^{2\alpha}r^{2\alpha} + \omega^{4\alpha} + 4\beta^\alpha r^{3\alpha}\cos(\alpha\pi) + 4\beta^\alpha r^\alpha \omega_0^{2\alpha} + 2r^{2\alpha}\omega_0^{2\alpha}\cos(2\alpha\pi)\right]}\,dr. \tag{11}$$

The latter expression vanishes as $t \to \infty$. On the other hand, the sum of residues associated with the poles $s = \left(\Omega_\pm e^{\pm i\pi}\right)^{1/\alpha}$ gives

$$\begin{aligned}
x_2(t) = {} & \frac{2e^{t\gamma_+\cos(\pi/\alpha)}}{\gamma_+^{\alpha-1}(\gamma_-^\alpha - \gamma_+^\alpha)}\left[\frac{\omega^2\gamma_+\cos\phi\cos\left(t\gamma_+\sin(\pi/\alpha) - \frac{\pi}{\alpha}(\alpha-2)\right) + \gamma_+^3\cos\phi\cos\left(t\gamma_+\sin(\pi/\alpha)-\pi\right)}{\omega^4+\gamma_+^4+2\omega^2\gamma_+^2\cos(2\pi/\alpha)}\right] \\
& + \frac{2e^{t\gamma_-\cos(\pi/\alpha)}}{\gamma_-^{\alpha-1}(\gamma_+^\alpha - \gamma_-^\alpha)}\left[\frac{\omega^2\gamma_-\cos\phi\cos\left(t\gamma_-\sin(\pi/\alpha) - \frac{\pi}{\alpha}(\alpha-2)\right) + \gamma_-^3\cos\phi\cos\left(t\gamma_-\sin(\pi/\alpha)-\pi\right)}{\omega^4+\gamma_-^4+2\omega^2\gamma_-^2\cos(2\pi/\alpha)}\right] \\
& + \frac{2e^{t\gamma_+\cos(\pi/\alpha)}}{\gamma_+^{\alpha-1}(\gamma_-^\alpha - \gamma_+^\alpha)}\left[\frac{\omega^3\sin\phi\cos\left(t\gamma_+\sin(\pi/\alpha) - \frac{\pi}{\alpha}(\alpha-1)\right) + \omega^2\gamma_+^2\sin\phi\cos\left(t\gamma_+\sin(\pi/\gamma)-\frac{\pi}{\alpha}(\alpha+1)\right)}{\omega^4+\gamma_+^4+2\omega^2\gamma_+^2\cos(2\pi/\alpha)}\right] \\
& + \frac{2e^{t\gamma_-\cos(\pi/\alpha)}}{\gamma_-^{\alpha-1}(\gamma_+^\alpha - \gamma_-^\alpha)}\left[\frac{\omega^3\sin\phi\cos\left(t\gamma_-\sin(\pi/\alpha) - \frac{\pi}{\alpha}(\alpha-1)\right) + \omega^2\gamma_-^2\sin\phi\cos\left(t\gamma_-\sin(\pi/\gamma)-\frac{\pi}{\alpha}(\alpha+1)\right)}{\omega^4+\gamma_-^4+2\omega^2\gamma_-^2\cos(2\pi/\alpha)}\right],
\end{aligned} \tag{12}$$

where $\gamma_\pm = \Omega_\pm^{1/\alpha}$. The term $x_2(t)$ is parameterized by the order of the Caputo operator D^α, the coefficient β of the fractional damping, and the natural frequency ω_0; the appropriate combination of these three parameters produces $x_2 \to 0$ as $t \to \infty$. Further details will be reported elsewhere. On the other hand, the term associated with the poles $s = \pm i\omega$ is of the form

$$x_3(t) = A\cos(\omega t + \delta), \tag{13}$$

where the amplitude and phase are respectively given by

$$A = f_0\frac{\sqrt{\omega^{4\alpha} + \omega_0^{4\alpha} + 2\omega^{2\alpha}\omega_0^{2\alpha}\cos(\pi\alpha - 2\phi) + 4\omega^\alpha\beta^\alpha\left[\omega^\alpha\beta^\alpha + (\omega^{2\alpha}+\omega_0^{2\alpha})\cos\left(\frac{\pi\alpha}{2} - 2\phi\right)\right]}}{\omega^{4\alpha} + \omega_0^{4\alpha} + 2\omega^{2\alpha}\omega_0^{2\alpha}\cos(\pi\alpha) + 4\omega^\alpha\beta^\alpha\left[\omega^\alpha\beta^\alpha + (\omega^{2\alpha}+\omega_0^{2\alpha})\cos\left(\frac{\pi\alpha}{2}\right)\right]}, \tag{14}$$

and

$$\delta = \arctan\left[-\frac{\omega^{2\alpha}\sin(\pi\alpha - \phi) + \omega_0^2\sin(\phi) + 2\beta^\alpha\omega^\alpha\sin\left(\frac{\pi\alpha}{2}\right)}{\omega^{2\alpha}\cos(\pi\alpha-\phi) + \omega_0^2\cos(\phi) + 2\beta^\alpha\omega^\alpha\cos\left(\frac{\pi\alpha}{2}\right)}\right]. \tag{15}$$

As in the conventional case, the amplitude A of the oscillations dictated by $x_3(t)$ is proportional to the amplitude f_0 of the driving force. At zero frequency ω (i.e., for a constant driving force), the quotient $\Lambda = A/f_0$ becomes $\omega_0^{-2\alpha}$ which, in turn, reproduces the (low frequencies) Newtonian result for $\alpha = 1$. At very high frequencies we find $\Lambda \approx \omega^{-2\alpha}$, so that the external driving force is dominant. On the other hand, at $\omega = \omega_0$ with β and ϕ fixed, we find that Λ is as larger as α approaches the value $\alpha = 1$ and becomes smaller for $\alpha \in (0, 1/2)$. That is, given the fractional damping parameter β, the fractional system behaves as an underdamped oscillator for $\alpha \to 1$, and as an overdamped one if α approximates $1/2$ from above.

The behavior of Λ is depicted in Fig. 1 as a function of the frequency ω with ϕ, ω_0 and β fixed, and for different values of α. For $\alpha = 1$ (i.e., for the Newtonian case) we find that Λ reaches its maximum value when the driving force oscillates at the natural frequency ω_0, as expected. Such a large response of the system to the driving force is the fingerprint of a resonance. Notice however that the maximum decreases and shifts to the left as α decreases. That is, for $\alpha \lesssim 1$ the resonance occurs at a frequency ω which is lower than ω_0. The latter means that the resonances can be controlled by fixing the fractional-damping parameter β and tuning up the order of the fractional derivative D^α. The same holds if one fixes the value of α and adjust the fractional-damping parameter β.

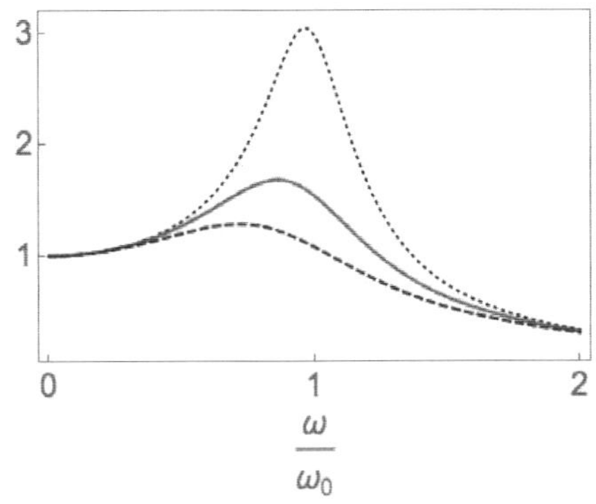

Figure 1. (Color online) The quotient $\Lambda = A/f_0$ of the amplitude of oscillation A defined in (14) and the amplitude f_0 of the driving force introduced in (8) with $\phi = 0$, $\omega_0 = 1$, $\beta = 0.1\omega_0$, for $\alpha = 0.99$ (dotted-black), $\alpha = 0.95$ (solid-red), and $\alpha = 0.90$ (dashed-blue).

4. Conclusions

Using fractional calculus one finds that the classical harmonic oscillator is affected by an 'intrinsic damping' [8,9], such a damping is also present in the quantum-fractional case [11]. The response of the classical fractional oscillator to the presence of a driving force has been already studied in e.g. [10]. In this paper we have presented some preliminary results of our study on a driven fractional oscillator which is affected by a fractional damping of the form $\beta^\alpha D^\alpha x(t)$, with D^α the Caputo time-derivative operator and $0 < \alpha \leq 1$. In particular, we have shown that the resonance phenomenon can be controlled by tuning up either the coefficient β of the fractional-damping or the order α of the Caputo operator. Further results will be reported elsewhere.

5. Acknowledgments

FOR acknowledges the funding received through a CONACyT scholarship.

References

[1] French A P, *Vibrations and waves*, W W Norton, New York, 1971
[2] Taylor J R, *Classical Mechanics* University Science Books 2005
[3] Rosas-Ortiz O, Fernández García N, Cruz y Cruz S, *AIP Conf. Proc.* **1077** (2008) 31
[4] Miller K S and Ross B, *An introduction to the fractional calculus and fractional differential equations*, John Wiley, New York, 1993
[5] Podlubny I, *Fractional Differential Equations*, Academic Press, New York, 1999
[6] Hilfer R, *Applications of Fractional Calculus in Physics*, World Scientific, New York, 2000
[7] Erdélyi A, *Higher Transcendental Functions*, Vol. III, McGraw-Hill, New York, 1955
[8] Narahari A B N, Hanneken J W, Enck T and Clarke T, *Physica A* **297** (2001) 361
[9] Olivar-Romero F, *A first approach to the fractional quantum mechanics*, M.Sc. Thesis (in Spanish), Physics Department, Cinvestav, México City, 2014
[10] Narahari A B N, Hanneken J W and Clarke T, *Physica A* **309** (2002) 275
[11] Olivar-Romero F and Rosas-Ortiz O, *J. Phys. Conf. Ser.* **698** (2016) 012025

IOP Conf. Series: Journal of Physics: Conf. Series **839** (2017) 012011 doi:10.1088/1742-6596/839/1/012011

Branched Hamiltonians for a Class of Velocity Dependent Potentials

B Bagchi[1], S M Kamil[1], T R Tummuru[1], I Semorádová[2] and M Znojil[2]

[1] Department of Physics, Shiv Nadar University, Dadri, UP 201314, India
[2] Nuclear Physics Institute, ASCR, 250 68 Řež, Czech Republic

E-mail: bbagchi123@gmail.com, kamil.syed@snu.edu.in, tt295@snu.edu.in,semoradova@ujf.cas.cz and znojil@ujf.cas.cz

Abstract. Hamiltonians that are multivalued functions of momenta are of topical interest since they correspond to the Lagrangians containing higher-degree time derivatives. Incidentally, such classes of branched Hamiltonians lead to certain not too well understood ambiguities in the procedure of the quantization. Within this framework we pick up a model which samples the latter ambiguities and which, simultaneously, turns out to be amenable to a transparent analytic and perturbative treatment.

1. Introduction

Models of classical systems with branched structures [1], in either coordinate (x) space or in its momentum (p) counterpart, have of late been a subject of active theoretical enquiry [2–9]. The key idea is that classical Lagrangians possessing time derivatives in excess of quadratic powers inevitably lead to p becoming a multi-valued function of velocity (v) thereby yielding a multivalued class of Hamiltonian systems.

Branched Hamiltonians in the classical context, and their quantized forms, have been recently discussed by Shapere and Wilczek [2]. Following it, Curtright and Zachos [3] analyzed certain representative models for a classical Lagrangian described by a pair of convex, smoothly tied functions of v. The underlying v turns out to be a double-valued function of p. Proceeding to the quantum domain shows that the double-valued Hamiltonians thus obtained have the inherent feature of being expressible in a supersymmetric form in the p space. Subsequently, a class of nonlinear systems whose Hamiltonians exhibit branching was explored by Bagchi et al [4] who also considered the possibility of quantization for some specific cases of the underlying coupling parameter.

In this paper we present a class of velocity-dependent Lagrangians which define a canonical momentum that yields exactly a pair of velocity variables in fractional terms. As a consequence, the corresponding Hamiltonians develop a branching character. An interesting aspect of our scheme is that it is well-suited for a perturbative treatment.

IOP Conf. Series: Journal of Physics: Conf. Series **839** (2017) 012011 doi:10.1088/1742-6596/839/1/012011

2. Branched Hamiltonians: A brief review

Let us briefly review the example of a branched system that was put forward in [3]. It was noted that a typical classical model of branched Hamiltonians results from a non-conventional form of the Lagrangian say, for the one given by

$$L = (v-1)^{\frac{2k-1}{2k+1}} - V(x) \tag{1}$$

where the traditional kinetic-energy term features the replacement of a typical quadratic form by a fractional function of "velocity" v while the function $V(x)$ stands for a convenient local interaction potential. The fractional powers of the difference $v-1$ was invoked to make plausible connections to known phenomenology such as the supersymmetric pairing. In detail, the $(2k+1)$−st root was required to be real and positive or negative for $v > 1$ or $v < 1$, respectively. Correspondingly, the quantity v turned out to be a double-valued function of p.

Working out the standard steps leads to the following two branches:

$$H_{\pm} = p \pm \frac{1}{4k-2}\left(\frac{1}{\sqrt{p}}\right)^{2k-1} + V(x)\,. \tag{2}$$

Note that the $k=1$ case speaks of the canonical supersymmetric structure [10] for the difference $H_{\pm} - V(x)$ namely, $p \pm \frac{1}{2\sqrt{p}}$, but in the momentum space if viewed as a quantum mechanical system. The spectral and boundary condition linkages of these Hamiltonians are not difficult to set up.

3. A velocity dependent potential

Against the above background we consider setting up of an extended Lagrangian model having a velocity dependent potential $U(x,v)$ that gives rise to a branched Hamiltonian under Legendre transformation:

$$L(x,v) = C(v-1)^{\frac{2k+1}{2k-1}} - U(x,v) \quad \text{where} \quad C = \frac{2k-1}{2k+1}\left(\frac{1}{4}\right)^{\frac{2}{2k+1}} \tag{3}$$

where we assume $U(x,v)$ to be given in a separable form $U(x,v) = f(v) + V(x)$, $f(v)$ and $V(x)$ are certain functions of v and x respectively.

Using the standard definition of the canonical momentum, we find that it is given by

$$p = \left(\frac{1}{4}\right)^{\frac{2}{2k+1}} (v-1)^{\frac{2}{2k-1}} - f'(v)\,. \tag{4}$$

The above equation is too complicated to put down the multivalued nature of velocity in a tractable closed form.

If we try to determine the associated branches of the Hamiltonian corresponding to this Lagrangian (3), $H_{\pm}$ emerge in a mixed form involving the momentum p, the function $f(v)$ and its derivative.

$$H_{\pm} = p \pm \frac{1}{4}\left[p + f'(v)\right]^{-\frac{2k-1}{2}} \left(\frac{2k+1}{2k-1} - p\left[p+f'(v)\right]^{-1}\right) + U(x,v)\,. \tag{5}$$

Since a Hamiltonian has to be a function of the coordinate and the corresponding canonical momentum, $H_{\pm}$ as derived above is of little use.

IOP Conf. Series: Journal of Physics: Conf. Series **839** (2017) 012011 doi:10.1088/1742-6596/839/1/012011

We note that the case $k = 1$ is particularly interesting to understand the spectral properties of $L(x, v)$. Explicitly, the Lagrangian assumes the simple but a general form

$$L = 3\left(\frac{1}{4}\right)^{\frac{2}{3}} (v-1)^{\frac{1}{3}} - f(v) - V(x)\,. \tag{6}$$

A sample choice for $f(v)$ could be

$$f(v) = \lambda v + 3\delta(v-1)^{\frac{1}{3}} \tag{7}$$

with $\lambda\,(\geq 0)$ and $\delta\left(< 4^{-\frac{2}{3}}\right)$ being suitable real constants. Observe that the presence of δ rescales the kinetic energy coefficient which now enjoys a parametric representation.

The above form of f facilitates determination of the canonical momentum p in a closed form as given by

$$p = \mu(v-1)^{-\frac{2}{3}} - \lambda \tag{8}$$

where $\mu = 4^{-\frac{2}{3}} - \delta > 0$. On inversion, we find a pair of relations for the velocity depending on p :

$$v_{\pm}(p) = 1 \mp \mu^{\frac{3}{2}}(p+\lambda)^{-\frac{3}{2}}\,. \tag{9}$$

As a consequence, we run into two branches of the Hamiltonian which we put down in the form

$$H_{\pm} - V(x) = (p+\lambda) \pm \frac{2\gamma}{\sqrt{p+\lambda}} \tag{10}$$

For the ease of notation, note that we have replaced $\mu^{3/2}$ with γ. In the special case where $\lambda = 0$

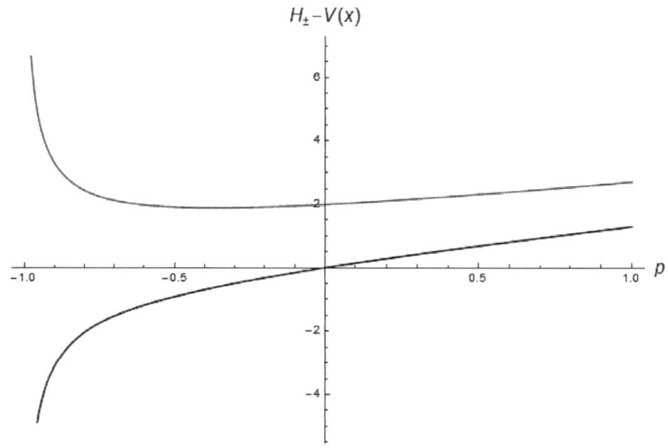

Figure 1. When $\lambda = 1$ and $\gamma = \frac{1}{2}$, $H_{\pm} - V(x)$ branches correspond to the upper and lower curves respectively.

and $\gamma = \frac{1}{4}$, we recover the Hamiltonian derived in [3]. However, the presence of the parameter γ in (10) is nontrivial as our following treatment of perturbative analysis will show. In Figure 1, we have given a graphical illustration (for $\lambda = 1$ and $\gamma = \frac{1}{2}$) of the behavior of the two branches of the Hamiltonian against some typical values of the momentum variable. As in the $\lambda = 0$ case of [3] here also we encounter a cusp asymptotically with regard to p for a fixed γ.

4. Lowest excitations and the Fourier transform

After one decides to consider just small excitations of our quantum system over a local or global minimum (x_0) of a generic analytic potential $V(x)$, one may put the origin of the coordinate axis to this minimum, $x \to y = x - x_0$, and write down the Taylor series

$$V(x) = V(x_0) + (x - x_0)V'(x_0) + \frac{1}{2}(x - x_0)^2 V''(x_0) + \dots \tag{11}$$

Recall that $V'(x_0) = 0$ and the zero of the energy scale can be shifted in such a manner that $V(x_0) = 0$. Finally, the series is truncated after the first non-trivial term yielding, in *ad hoc* units,

$$V(x_0 + y) = y^2. \tag{12}$$

After a Fourier transform to the momentum space, we get a transformed quantum form of the Hamiltonian guided by the second-order differential operator,

$$H = -\frac{d^2}{dp^2} + W(p) \tag{13}$$

containing a one-parametric family of pseudo-potentials

$$W(p) = p + \frac{2\gamma}{\sqrt{p}}. \tag{14}$$

Here, the original subscript $_\pm$ entering Eq. (10) may be perceived as equivalent to an optional switch between positive coupling-type parameter $\gamma > 0$ and its negative alternative $\gamma < 0$. Besides such a freedom of the sign of the dynamical characteristic, the consequent quantum-theory interpretation of the model requires also a few nontrivial mathematical addenda. The form Eq. (14) matches with Eq. (10) for $\lambda = 0$ which will now be our point of inquiry.

First of all, the most natural tentative candidate

$$H\phi_n(p) = E_n\phi_n(p), \qquad p \in (-\infty, \infty) \tag{15}$$

for the quantum Schrödinger equation living on the whole real line of momenta (i.e., with $\phi_n(p) \in L^2(\mathbb{R})$) is characterized by the asymptotically linear *decrease* of the pseudo-potential (14) along the left half-line. Hence, the negative half-axis of momenta p must be excluded, *a priori*, as unphysical. In other words, the acceptable wave functions $\phi_n(p)$ should vanish, identically, whenever $p \in (-\infty, 0)$. The consistent quantization of our model must be based on the modified, half-line version of Eq. (15), viz., on Schrödinger equation

$$H\phi_n(p) = E_n\phi_n(p), \qquad p \in (0, \infty) \tag{16}$$

such that (cf. also [2] and [3])

$$\phi_n(p) \in L^2(\mathbb{R}^+). \tag{17}$$

Still, the discussion is not yet complete. Due care must be also paid to the fact that the inverse-square-root singularity of $W(p)$ in the origin is "weak" (see, e.g., Ref. [11] for a detailed explanation of the rigorous, "extension theory" mathematical contents of this concept). In the language of physics, such a comment means that the information about possible bound states and physics represented by Eq. (16) with constraint (17) is incomplete.

In the rest of this paper (i.e., in sections 5 and 6) we shall, therefore, describe the two alternative versions of the completion of the missing, phenomenology-representing information.

5. Eligible "missing" boundary conditions at small γ and $p = 0$

As we emphasized above, the existence of the usual discrete spectrum of bound states can only be guaranteed via an *additional* physical boundary condition at $p = 0$. Although, from the point of view of pure mathematics, the choice of such a condition is flexible and more or less arbitrary, the necessary suppression of this unwanted freedom can rely upon several forms of the physics-based intuition.

Let us split the problem into two subcategories. In a simpler scenario we shall assume that the central core is repulsive and strong (i.e., that our parameter is positive and large, $\gamma \gg 1$). This possibility will be discussed in the next section 6. For the present, let us admit that the (real) value of γ is arbitrary and that the regular nature of our ordinary differential Schrödinger equation near $p = 0$ implies that the integrability condition (17) itself still does not impose any constraint upon the energy E [11]. A fully explicit and constructive demonstration of such an observation may be based on the routine reduction of (16) to its simplified, leading-order form

$$- \sqrt{p}\,\frac{d^2}{dp^2}\psi(p) + 2\gamma\,\psi(p) = 0\,. \tag{18}$$

Being valid at the very small (though still positive) values of $p \ll 1$ this equation is exactly solvable in terms of Bessel functions [12]. Thus, one may choose either $\gamma > 0$ or $\gamma < 0$.

After some algebra we obtain the respective two-parametric families of the general solutions which depend on two parameters $C_{1,2}$ or $D_{1,2}$ and which remain energy-dependent. At small p they behave, respectively, as follows,

$$\psi\,(p) = C_1\,\sqrt{p}\,I_{2/3}\left(\frac{4\sqrt{2}}{3}\sqrt{\gamma}\,p^{\frac{3}{4}}\right) + C_2\,\sqrt{p}\,K_{2/3}\left(\frac{4\sqrt{2}}{3}\sqrt{\gamma}\,p^{\frac{3}{4}}\right) \tag{19}$$

and

$$\psi\,(p) = D_1\,\sqrt{p}\,J_{2/3}\left(\frac{4\sqrt{2}}{3}\sqrt{-\gamma}\,p^{\frac{3}{4}}\right) + D_2\,\sqrt{p}\,Y_{2/3}\left(\frac{4\sqrt{2}}{3}\sqrt{-\gamma}\,p^{\frac{3}{4}}\right) \tag{20}$$

On this purely analytic background, one of the most natural resolutions of the paradox of the ambiguity of the physical boundary conditions at $p = 0$ may be based on the brute-force choice of the parameters $C_{1,2}$ or $D_{1,2}$ in these formulae.

Finally, let us emphasize that intuitively by far the most plausible requirement of the absence of the jump in the wave functions at $p = 0$, i.e., the Dirichlet boundary condition

$$\lim_{p\to 0} \psi\,(p) = 0 \tag{21}$$

would remove the latter ambiguity of quantization in the most natural manner. The resulting pair of the requirements

$$C_2 = 0, \quad D_2 = 0 \tag{22}$$

may be then recommended as easily derived from the well known approximate formulae for the Bessel functions near the origin [12].

6. Perturbation-theory analysis at large $\gamma \gg 1$

In a purely formal spirit one could complement the above-recommended Dirichlet boundary condition (21) by its Neumann vanishing-derivative analogue

$$\lim_{p\to 0} \psi'\,(p) = 0 \tag{23}$$

IOP Conf. Series: Journal of Physics: Conf. Series **839** (2017) 012011 doi:10.1088/1742-6596/839/1/012011

or, more generally, by a suitable Robin boundary condition. In this context it is worth adding that with a systematic strengthening of the repulsive version of the barrier (i.e., with the growth of the positive coupling constant γ) the specification of the additional boundary conditions at $p = 0$ becomes less and less relevant because the two alternative energy levels will degenerate in the limit $\gamma \to \infty$.

The most immediate explanation of this phenomenon may be provided by perturbation theory. In the dynamical regime, when the parameter is large, a perturbative approach seems to be particularly well suited. With $\gamma \gg 1$, we look at the absolute minimum of the potential $W(p)$ which occurs at p_0, say. This value is, incidentally, unique

$$p_0 = \gamma^{\frac{2}{3}} \gg 1 \tag{24}$$

With the construction of a Taylor series in its vicinity,

$$W(p) = W(p_0) + (p - p_0)W'(x_0) + \frac{1}{2}(p - p_0)^2 W''(p_0) + \dots \tag{25}$$

we observe that the first term, which is given by

$$W(p_o) = 3\gamma^{\frac{2}{3}} \tag{26}$$

in very large in this scenario. In contrast, all of the further Taylor coefficients remain very small and asymptotically negligible,

$$W''(p_o) = \frac{3}{2}\gamma^{-\frac{2}{3}}, \quad W'''(p_o) = -\frac{15}{4}\gamma^{-\frac{4}{3}} \quad \dots \tag{27}$$

Clearly then, with $\gamma \gg 1$, H can be expressed as

$$H = -\frac{d^2}{dp^2} + 3\gamma^{\frac{2}{3}} + \frac{3}{4}\gamma^{-\frac{2}{3}}(p - p_0)^2 + \mathcal{O}(\gamma^{-\frac{4}{3}}(p - p_0)^3)\,. \tag{28}$$

After one re-scales the axis $p = \rho q$, equation (16) acquires the modified form

$$\tilde{H}\tilde{\phi}_n(q) = E_n \rho^2 \tilde{\phi}_n(q) \tag{29}$$

where,

$$\tilde{H} = -\frac{d^2}{dq^2} + 3\rho^2\gamma^{\frac{2}{3}} + \frac{3}{4}\rho^4\gamma^{-\frac{2}{3}}(q - q_0)^2 + \mathcal{O}(\rho^5\gamma^{-\frac{4}{3}}(q - q_0)^3)\,. \tag{30}$$

One may now set

$$\rho = \left(\frac{4}{3}\right)^{\frac{1}{4}} \gamma^{\frac{1}{6}} \tag{31}$$

yielding the very weakly perturbed harmonic-oscillator Hamiltonian

$$\tilde{H} = -\frac{d^2}{dq^2} + (q - q_0)^2 + \gamma\sqrt{12} + \mathcal{O}(\rho^5\gamma^{-\frac{4}{3}}(q - q_0)^3)\,. \tag{32}$$

In full analogy to many models with similar structure (cf., the study [13] containing further references), the exact solvability of the model in the leading-order harmonic-oscillator approximation proves sufficient because in the domain of large $\gamma \gg 1$ the contribution of the anharmonic corrections becomes negligible.

7. Summary

To summarize, we looked at the particular example of a non-conventional form of a velocity-dependent Lagrangian that leads to a double-valued structure of the associated Hamiltonian for some specific choice of the underlying coupling parameter. We showed that our scheme allows for a perturbative analysis by constructing a Taylor series near the vicinity of the absolute minimum of the potential.

Acknowledgments
For one of us (BB), it is a pleasure to thank Prof. Sara Cruz y Cruz and Prof. Oscar Rosas-Ortiz for warm hospitality during Quantum Fest 2016 held at UPIITA-IPN, Mexico City. MZ acknowledges the short-stay hospitality by Shiv Nadar University and the support by the GAČR Grant Nr. 16-22945S. We also all thank Prof. Bhabani Prasad Mandal for fruitful discussions.

References

[1] Henneaux M, Teitelboim C and Zanelli J 1987 *Phys. Rev.* **A36** 4417
[2] Shapere A and Wilczek F 2012 *Phys. Rev. Lett.* **109** 200402
[3] Curtright T L and Zachos C K 2014 *J. Phys. A: Math. Gen.* **47** 145201
[4] Bagchi B, Modak S, Panigrahi P K, Ruzicka F and Znojil M 2015 *Mod. Phys. Lett.* A **30** 1550213
[5] Zhao L, Yu P and Xu W 2013 *Mod. Phys. Lett.* A **28** 1350002
[6] Chi H-H and He H-J 2014 *Nucl. Phys.* B **885** 448
[7] Avraham E and Brustein R 2014 *Phys. Rev. D* **90** 024003
[8] Rosu H C, Mancas S C and Chen P 2015 *Phys. Scp.* **90** 055208
[9] Ruz S, Mandal R, Debnath S and Sanyal A K 2016 *Gen. Rel. Grav.* **48** 1
[10] Witten E 1981 *Nucl. Phys.* B **185** 513
[11] Znojil M 2000 *Phys. Rev.* A **61** 066101
[12] Olver F W J and Maximon L C *Bessel Functions* in the NIST Digital Library of Mathematical Functions. Chapter 10 (Open access at http://dlmf.nist.gov/10)
[13] Znojil M 1994 *J. Phys. A: Math. Gen.* **27** 4945

Extremal Density Matrices for the Expectation Value of a Qudit Hamiltonian

O Castaños, A Figueroa, J López, and R López-Peña

Instituto de Ciencias Nucleares, Universidad Nacional Autónoma de México, Apartado Postal 70-543, 04510 México City, Mexico

E-mail: ocasta@nucleares.unam.mx

Abstract. An algebraic procedure to find extremal density matrices for the expectation value of a finite Hamiltonian matrix is established. The extremal density matrices for pure states provide a complete description of the system, that is, its corresponding energy spectrum and projectors. For density matrices representing mixed states, one gets the most probable eigenstates that yield extremal mean values of the energy. The procedure uses mainly the stationary solutions of the von Neumann equation of motion, the orbits of the Hamiltonian, and the positivity conditions of the density matrix. The method is illustrated for matrix Hamiltonians of dimensions $d = 2$ and $d = 3$.

1. Introduction

There are mathematical problems emerging in quantum information theory and in finite-dimensional quantum systems which can be studied by an algebraic parametrisation of the density matrix. As examples one can consider Hamiltonian systems characterising symmetric qubits [1], which appear in several branches of physics. The two-site Bose-Hubbard model to study Josephson tunnelling between two-mode Bose-Einstein condensates [2, 3]; or Hamiltonians of the Lipkin-Meshkov-Glick type to study many body theories and quantum phase transitions [4, 5]. Additionally for symmetric qubits there are entanglement witnesses, i.e., observables that can be measured in a given experimental set-up and thus provide an experimental verification of entanglement [6, 7]. There are also many properties in quantum information related with optimisation problems, for instance, in the detection and quantification of entanglement [8]. If we do not have information about an ensemble of systems, by maximizing the entropy the probability of finding the system in any of its possible states is the same. On the contrary having knowledge of the mean value of the energy one can determine the density matrix of a system in thermal equilibrium; or by knowing the mean value of the spin operator the description of the density matrix for the spins in terms of the Pauli matrices and the polarization vector is obtained [9]. The definition of the relative entropy can be used to determine test density matrices associated to a canonical distribution of eigenstates of finite dimensional matrix Hamiltonian, which allows the determination of upper bounds to the exact free energy of Helmholtz to describe stretch vibrations of triatomic molecules [10]. In the studies of indivisible systems the nature of correlations is not obvious. However for these systems, the existence of quantum correlations can be detected by different types of inequalities which are violated for entangled states or by entropic and information inequalities written for the density matrices of systems and subsystems [11].

IOP Conf. Series: Journal of Physics: Conf. Series **839** (2017) 012012 doi:10.1088/1742-6596/839/1/012012

In finite quantum systems the density matrices can be formulated into a su(d) description [12]. Actually, there are several parametrizations of finite density matrices: generalizations of the Bloch vector [13], the canonical coset decomposition of unitary matrices [14], the recursive procedures to describe $n \times n$ unitary matrices in terms of those of U$(n-1)$ [15], and by defining generalized Euler angles [16]. Recently we have established a procedure to determine the extremal density matrices of a qudit system associated to the expectation value of any observable [17]. These matrices provide an extremal description of the mean values of the energy; and in the case of restricting them to pure states the energy spectrum is recovered. So, apart from being an alternative tool to find the eigen-system, one has information of mixed states which minimize its mean value.

The aim of this work is to compute extremal density matrices in a qudit space by means of an algebraic approach [18] that uses the stationary solutions of the von Neumann equation of motion, the orbits of the Hamiltonian matrix, and the positivity conditions of the density matrix. This approach complement and simplify the variational procedure as we are going to shown. To establish this new method we review in the second section the variational procedure and apply the method to the $d = 2$ dimensional case. In the third section, the algebraic approach is introduced together with the determination of the orbits of the Hamiltonian through the construction of the Gram matrix. An explicit method to get the eigen-system, for the pure case, associated to the Hamiltonian matrices with or without degeneracy is also presented, together with the determination of the density matrices of mixed states corresponding to the extremal mean values of the energy. In section fourth, the Hamiltonian systems of $d = 3$ are studied. Finally a summary and concluding remarks are presented.

2. Variational approach to extremal density matrices

In a previous work [17] we proposed an approach to obtain information of the mean value of a finite dimensional Hamiltonian, in which a general test density matrix is written in terms of the generators $\hat{\lambda}_k$ of a special unitary algebra in d dimensions, $su(d)$, i.e.,

$$\hat{\rho} = \frac{1}{d}\hat{I} + \frac{1}{2}\sum_{k=1}^{d^2-1} \lambda_k \, \hat{\lambda}_k \, , \tag{1}$$

with $\lambda_k = \mathrm{Tr}(\hat{\rho}\,\hat{\lambda}_k)$. The operators $\hat{\lambda}_k$ are completely characterized by means of the commutation and anti-commutation relations [13]

$$\left[\hat{\lambda}_j, \hat{\lambda}_k\right] = 2\,i \sum_{q=1}^{d^2-1} f_{jkq}\hat{\lambda}_q \, , \quad \{\hat{\lambda}_j, \hat{\lambda}_k\} = \frac{4}{d}\delta_{jk}\hat{I} + 2\sum_{q=1}^{d^2-1} d_{jkq}\hat{\lambda}_q \, , \tag{2}$$

where d_{jkq} and f_{jkq} are the symmetric and antisymmetric structure constants given by

$$d_{jkq} = \frac{1}{4}\mathrm{Tr}(\{\hat{\lambda}_j, \hat{\lambda}_k\}\hat{\lambda}_q) \, , \quad f_{jkq} = \frac{1}{4i}\mathrm{Tr}(\left[\hat{\lambda}_j, \hat{\lambda}_k\right]\hat{\lambda}_q) \, . \tag{3}$$

The density matrix must satisfy the following three properties: (a) It is Hermitian, (b) it has trace one, and (c) all its eigenvalues are positive semidefinite. While for dimension $d = 2$, the condition $\mathrm{Tr}(\hat{\rho}^2) \leq 1$ implies (c), for $d \geq 3$ that is not true. Therefore the positivity conditions of the density matrix are established by the set $\{a_k\}$ of coefficients of its corresponding characteristic polynomial. This set can be obtained by means of the recursive relation known as Newton-Girard formulas [14, 19]

$$a_k = \frac{1}{k}\sum_{j=1}^{k}(-1)^{j-1}\, a_{k-j}\, t_j \, , \tag{4}$$

IOP Conf. Series: Journal of Physics: Conf. Series **839** (2017) 012012 doi:10.1088/1742-6596/839/1/012012

with the definitions $a_0 = a_1 = 1$, $a_d = \det\hat{\rho}$, and $t_j = \text{Tr}(\hat{\rho}^j)$, for $j = 1, \ldots, d$. Therefore the allowed density matrix must be in the region given by [13, 20, 21]

$$0 \leq a_k \leq \frac{1}{d^k}\binom{d}{k}, \tag{5}$$

where $\binom{d}{k}$ denotes a binomial coefficient. The upper bound defines the most mixed state while the lower bound specifies pure states.

The compatible regions bounding the invariants a_k are obtained through the intersection of the positivity conditions of the density matrix (5) with the respective positivity conditions of the symmetric matrix called Bezoutian [22–24].

In order to consider the mean value of the Hamiltonian together with the $d-1$ constraints to guarantee the positivity of the density matrix, we define the function

$$f(\lambda_k, \Lambda_j, h_i, c_l) \equiv \text{Tr}(\hat{H}\,\hat{\rho}) + \sum_{j=2}^{d} \Lambda_j (a_j - c_j)\,, \tag{6}$$

which depends on $d-1$ Lagrange multipliers $\{\Lambda_j\}$, $d-1$ positive real constants $\{c_j\}$ to fix the degree of purity of the density matrix, d^2-1 independent variables $\{\lambda_k\}$, and d^2 real parameters associated to the expansion

$$\hat{H} = \frac{1}{d} h_0 \hat{I} + \frac{1}{2}\sum_{k=1}^{d^2-1} h_k\,\hat{\lambda}_k\,, \tag{7}$$

where $h_0 = \text{Tr}(\hat{H})$ and $h_k = \text{Tr}(\hat{H}\,\hat{\lambda}_k)$.

The extremal values of (6) one has to take the derivatives of $f(\lambda_k, \Lambda_j, h_i, c_l)$ with respect to the variables $\{\lambda_k\}$ and the Lagrange multipliers $\{\Lambda_j\}$. Their respective derivatives give d^2+d-2 algebraic equations,

$$\frac{1}{2}h_q + \sum_{j=2}^{d} \Lambda_j \frac{\partial a_j}{\partial \lambda_q} = 0\,, \qquad q = 1, \ldots d^2-1\,, \tag{8}$$

$$a_p = c_p\,, \qquad p = 2, \ldots d\,. \tag{9}$$

These sets of algebraic equations determine the critical values of the density matrix, i.e., $\lambda_q = \lambda_q^c$ and $\Lambda_q = \Lambda_q^c$ for which the expectation value of the Hamiltonian takes an extremal value. If we restrict the solutions to pure states $\{c_p = 0\}$, we have shown explicitly that the energy spectrum of the Hamiltonian is recovered for $d = 2$ and 3 [17]. Extremal expressions for the mean value of the Hamiltonian can be obtained with density matrices representing mixed quantum states, which determine also the corresponding mixture of eigenstates of the Hamiltonian.

2.1. $d = 2$

The generators $\{\hat{\lambda}_k\}$ are realised in terms of the Pauli matrices, i.e., $\hat{\lambda}_1 = \hat{\sigma}_1$, $\hat{\lambda}_2 = \hat{\sigma}_2$ and $\hat{\lambda}_3 = \hat{\sigma}_3$. Then the density and Hamiltonian matrices can be written as

$$\hat{\rho} = \frac{1}{2}\begin{pmatrix} 1+\lambda_3 & \lambda_1 - i\,\lambda_2 \\ \lambda_1 + i\,\lambda_2 & 1-\lambda_3 \end{pmatrix}, \quad \hat{H} = \frac{1}{2}\begin{pmatrix} h_0+h_3 & h_1 - i\,h_2 \\ h_1 + i\,h_2 & h_0 - h_3 \end{pmatrix}, \tag{10}$$

where $\boldsymbol{\lambda}$ is also called the polarization vector. Substituting the last expressions into the expressions (8) and (9) one obtains the system of equations

$$h_k - 2\,\Lambda\lambda_k = 0\,, \quad \frac{1}{4}\left(1 - \lambda_1^2 - \lambda_2^2 - \lambda_3^2\right) = c_2\,, \tag{11}$$

with $k = 1, 2$ and 3. The solutions of the system of equations are $\lambda_k^c = \pm\delta/h\, h_k$ and $\Lambda^c = \pm\, h/\delta$. with $h = \sqrt{h_1^2 + h_2^2 + h_3^2}$ and $\delta = \sqrt{1 - 4c_2}$. Thus one gets two density matrices:

$$\hat{\rho}_{\pm}^c = \frac{1}{2}\begin{pmatrix} 1 \pm \delta\, h_3/h & \pm\delta\,(h_1 - ih_2)/h \\ \pm\delta\,(h_1 + ih_2)/h & 1 \mp \delta\, h_3/h \end{pmatrix}, \tag{12}$$

which yields the extremal mean values of the energy $\langle H\rangle_{\pm}^c = \frac{1}{2}\,(h_0 \pm \delta\, h)$. These values depend on the parameter c_2 whose value is bounded, $0 \le c_2 \le 1/4$. For $c_2 = 1/4$, one has the maximally mixed state $\hat{\rho}_{\pm}^c = 1/2\hat{I}_2$, which has maximum entropy. We can distinguish two types of solutions: (a) Pure case (when $c_2 = 0$, that is, $\delta = 1$): The eigenvalues $\epsilon_{\pm} = \frac{1}{2}\,(h_0 \pm h)$ of the Hamiltonian, and from (12), the corresponding orthogonal projectors. (b) Mixed case (when $0 < c_2 \le 1/4$): The extremal density matrices for the expectation value of the Hamiltonian are given in terms of the convex sum

$$\hat{\rho}_{\pm}^{\text{mixed}} = \frac{1}{2}(1+\delta)\hat{\rho}_{\pm}^{\text{pure}} + \frac{1}{2}(1-\delta)\hat{\rho}_{\mp}^{\text{pure}}\,, \tag{13}$$

where $p_{\pm} = \frac{1}{2}(1+\delta)$ indicates the probability of finding the system with eigenvalue ϵ_+ while $p_{\mp} = \frac{1}{2}(1-\delta)$ the corresponding probability of finding an energy ϵ_-.

3. Algebraic approach

The stationary solutions of the von Neumann equation of motion, which satisfy $[\,\hat{\rho}, \hat{H}\,] = 0$, are equivalent to the critical values of the generalised Bloch vector $\boldsymbol{\lambda}$ that make the expectation value of the Hamiltonian an extremal [18]. By replacing Eqs. (1) and (7) into the last expression, using the commutation relations of the generators $\hat{\lambda}_k$, multiplying by $\hat{\lambda}_q$, and tracing over the result; one gets the homogeneous system of equations

$$\sum_j \boldsymbol{M}_{qj}\,\lambda_j = 0\,, \quad q = 1, 2, \cdots, d^2 - 1\,, \tag{14}$$

where we have defined the $d^2 - 1$-dimensional skew symmetric matrix

$$\boldsymbol{M}_{i,j} = \sum_{k=1}^{d^2-1} f_{i\,j\,k}\, h_k\,. \tag{15}$$

The solution through the Gauss-Jordan elimination method of the underdetermined system of equations (14) leads to the critical values for the components of the Bloch vector, $\boldsymbol{\lambda}^c$. Notice that it has n free components, whose number is associated to the dimension of the null space of $\boldsymbol{M}$. This implies that maximal mixed states (all the components of the Bloch vector equal to zero) are always critical for any observable because the null space always contains the zero vector.

On the other hand, notice that $\hat{w}_q \equiv i\,[\hat{H}\,,\,\hat{\lambda}_q]$ are hermitian vectors spanning the tangent space of the orbits associated with $\hat{H}$, with $q = 1, 2, \ldots, d^2 - 1$. Then by substituting the Hamiltonian expression (7) one gets

$$\hat{w}_q = \sum_{k,l} f_{qkl}\, h_k\, \hat{\lambda}_l\,. \tag{16}$$

Note that these vectors give the rows of $\boldsymbol{M}$, and the number of independent vectors r is determined by the rank of the Gram matrix

$$\boldsymbol{G}_{q,p} = \mathrm{Tr}(\hat{w}_q\, \hat{w}_p) = \sum_{k_1,k_2,j} f_{q\,j\,k_1}\, f_{p\,j\,k_2}\, h_{k_1}\, h_{k_2}\,. \tag{17}$$

Because the rank $\boldsymbol{G}$ is an invariant under unitary transformations [25], one has the freedom to calculate the Gram matrix in the diagonal representation of the Hamiltonian, i.e., by means of the diagonal components of the generalised Bloch vector of the Hamiltonian $\boldsymbol{h} = (h_1, h_2, \cdots, h_{d^2-1})$. Thus, if all the eigenvalues of $\hat{H}$ are equal, any density matrix commutes with $\hat{H}$, implying that $r = 0$. The other extreme is when all the eigenvalues are different [26, 27]; then the dimension of the orbit is associated to the quotient space $U(d)/[U(1) \times \cdots \times U(1)]$ and thus one has that $r = d(d-1)$. Thus one has in this case $n = d-1$ free components of the Bloch vector of the critical density matrix. These components can be obtained by establishing the positivity conditions of the density matrices given in expression (9).

3.1. *Non degenerate case of* $\hat{H}$

In this case, the rank of the matrix $\boldsymbol{M}$ is given by $r = d(d-1)$ and the n free variables reach its minimum number, i.e., $n = d-1$. Therefore, $\boldsymbol{\lambda}^c$ is given by

$$\boldsymbol{\lambda}^c = (\lambda_1^c, \ldots, \lambda_{d(d-1)}^c, \lambda_{d(d-1)+1}, \cdots \lambda_{d^2-1})\,, \tag{18}$$

where the $d(d-1)$ components $\{\lambda_q^c\}$ are functions of the parameters of the Hamiltonian, the antisymmetric structure constants and a set of $d-1$ independent free variables.

The substitution of $\boldsymbol{\lambda}^c$ in (1) gives its associated critical density matrix denoted as $\hat{\rho}^c$. Therefore, the determination of the $d-1$ free variables is done by solving the system of $d-1$ polynomial equations (9), which has at most $d!$ different solutions [28, 29]. They satisfy the same polynomial system (9) but give different mean values of $\hat{H}$. Therefore, the critical density matrices can be written in the form

$$\hat{\rho}_m^c \;=\; \frac{1}{d}\hat{I} + \frac{1}{2}\sum_{k=1}^{d^2-1} \lambda_{m,k}^c\, \hat{\lambda}_k\,, \tag{19}$$

with m labelling the independent critical solutions. The variables $\{\lambda_{m,k}^c\}$ are function only of the known quantities, i.e., the structure constants, the parameters of the Hamiltonian (h_0, h_k), and $d-1$ constants $\{c_k\}$. The number m of solutions decreases up to d when (19) represent pure states (all $\{c_k = 0\}$) and the extremal density matrices are one-dimensional orthogonal projectors.

In general the expectation value of the Hamiltonian is given by

$$\langle \hat{H} \rangle_m^c \equiv \mathrm{Tr}(\hat{H}\, \hat{\rho}_m^c)\,, \tag{20}$$

for each critical (or extremal) density matrix $\hat{\rho}_m^c$, with $m = 1, 2, \ldots d!$. For the pure case the expectation values yield the energy spectrum of the system and the extremal density matrices are orthogonal projectors.

3.2. *Degenerate case of* $\hat{H}$

In this case the rank of the matrix $\boldsymbol{M}$ satisfies that $r < d(d-1)$, whose value is associated to the orbits of the Hamiltonian. As the critical Bloch vector $\boldsymbol{\lambda}^c$ has n free variables with $n = d^2 - 1 - r$, one then has that $n > d-1$. Thus, if N denotes the number of variables appearing in the expression for the mean value of the Hamiltonian in the state $\hat{\rho}^c$, there are two cases to consider:

i) When $1 \leq N \leq d-1$, one has to select $n-N$ components of the density matrix Bloch vector to have $d-1$ free variables and then solve the polynomial system of equations (9).

ii) When $d-1 < N \leq n$, one has only to pick up $d-1$ components of the density matrix Bloch vector from the set of N elements, and again to solve the mentioned polynomial equation.

In both cases $d-1$ free variables are determined by the polynomial system (9). The remaining $n-(d-1)$ components can be taken equal to zero because they do not affect the commutator of $\hat{H}$ with $\hat{\rho}^c$. Specifically, if we are interested in the eigensystem, i.e., all the set of $\{c_k = 0\}$, one can apply the following method recursively:

1) Make zero the $n-(d-1)$ components, to solve the polynomial system (9), whose solution give at least two extremal density matrices.
2) Ask the orthogonality of the first set of solutions $\hat{\rho}^c_k$ with the general density matrix $\hat{\rho}^c$, i.e., $\mathrm{Tr}(\hat{\rho}^c\,\hat{\rho}^c_k) = 0$, where k labels the number of solutions.
3) Substitute the solutions of the linear system, to express the new critical density matrix in terms of the free variables, where $d-1$ are fixed by means of the positivity conditions (all $\{c_k = 0\}$) and the rest of the components can be taken equal to zero.
4) Return to step 1 and repeat the procedure again, until one gets d orthogonal projectors.

Of course, one has in this case several solutions related with the degeneracy of the Hamiltonian in similar form as in the standard diagonalization procedure of a finite Hamiltonian matrix.

4. Case d=3.

For the qutrit case, the generators $\hat{\lambda}_k$, with $k = 1, 2, \ldots 8$ can be realized in terms of the Gell-Mann matrices. Thus, an arbitrary density matrix is given by

$$\hat{\rho} = \frac{1}{2}\begin{pmatrix} \lambda_7 + \frac{\lambda_8}{\sqrt{3}} + \frac{2}{3} & \lambda_1 - i\lambda_4 & \lambda_2 - i\lambda_5 \\ \lambda_1 + i\lambda_4 & \frac{\lambda_8}{\sqrt{3}} - \lambda_7 + \frac{2}{3} & \lambda_3 - i\lambda_6 \\ \lambda_2 + i\lambda_5 & \lambda_3 + i\lambda_6 & \frac{2}{3} - \frac{2}{\sqrt{3}}\lambda_8 \end{pmatrix}. \tag{21}$$

We will illustrate the procedure taking special cases. For the non-degenerate case we take

$$\hat{H} = \begin{pmatrix} b & \frac{c}{\sqrt{2}} & 0 \\ \frac{c}{\sqrt{2}} & 0 & \frac{c}{\sqrt{2}} \\ 0 & \frac{c}{\sqrt{2}} & b \end{pmatrix}, \tag{22}$$

where b and c are real parameters. The Bloch vector for the Hamiltonian is given by

$$\boldsymbol{h} = (\sqrt{2}\,c,\ 0,\ \sqrt{2}\,c,\ 0,\ 0,\ 0,\ b,\ -\sqrt{3}\,b)\,.$$

Substituting the components of $\boldsymbol{h}$ into (15), one finds that the rank of $\boldsymbol{M}$ is 6, implying that the Hamiltonian is non degenerate. Solving the system of equations (14) one obtains the extremal Bloch vector for the density matrix

$$\boldsymbol{\lambda} = \left(\frac{c\,\lambda_2}{\sqrt{2}\,b} - \frac{\sqrt{6}\,c\,\lambda_8}{b}, \lambda_2, \frac{c\,\lambda_2}{\sqrt{2}\,b} - \frac{\sqrt{6}\,c\,\lambda_8}{b}, 0, 0, 0, -\sqrt{3}\,\lambda_8, \lambda_8\right).$$

Thus the associated critical density matrix is found by replacing the components of $\boldsymbol{\lambda}^c$ into (21), which is denoted by $\hat{\rho}^c$. For this case, to guarantee the positivity of the density matrix, one must consider

$$c_2 = \frac{1}{2}\left(1 - \mathrm{Tr}(\hat{\rho}^c)^2\right. \quad , \quad c_3 = \det\hat{\rho}^c\,. \tag{23}$$

IOP Conf. Series: Journal of Physics: Conf. Series **839** (2017) 012012 doi:10.1088/1742-6596/839/1/012012

For the case $d = 3$, the positivity conditions (5) of the density matrix are

$$0 \leq c_2 \leq \frac{1}{3}, \quad 0 \leq c_3 \leq \frac{1}{27}, \tag{24}$$

while the Bezoutian matrix is

$$\boldsymbol{B}_3 = \begin{pmatrix} 3 & 1 & t_2 \\ 1 & t_2 & t_3 \\ t_2 & t_3 & t_4 \end{pmatrix}.$$

By applying the Cayley-Hamilton theorem and the formula (4) one obtains

$$t_2 = 1 - 2\,c_2\,, \quad t_3 = 1 - 3c_2 + 3c_3\,, \quad t_4 = 1 - 4c_2 + 2c_2^2 + 4c_3\,. \tag{25}$$

Similarly to the case of $d = 2$, $\det \boldsymbol{B}_3 \geq 0$ is the only relevant positivity condition of $\boldsymbol{B}_3$. Expressed in terms of $\{c_2,\, c_3\}$, it gives

$$c_2^2 - 4c_2^3 + 18c_2c_3 - c_3(4 + 27c_3) \geq 0\,. \tag{26}$$

Thus, the inequalities system formed by (24) and (26) produces the compatible region between c_2 and c_3. This is shown in Fig. 1(a). The bottom line is associated with one eigenvalue zero (yielding the condition on c_2 for the case $d = 2$), while the other curves imply two equal eigenvalues for the density matrix. The $(c_2, c_3) = (0, 0)$ case is associated to density matrices of pure states and the highest is the maximal mixed state (all the eigenvalues are equal).

Again one has two types of solutions: (a) Pure case (taking $c_2 = c_3 = 0$): Solving the system of equations $c_2 = 0$ and $c_3 = 0$, one arrives to 3 different solutions for λ_2 and λ_8, denoted by

$$(\lambda_2, \lambda_8)_0 = \left(-1, -\frac{1}{2\sqrt{3}}\right), \quad (\lambda_2, \lambda_8)_\pm = \frac{1}{2}\left(1 \pm \frac{b}{\Delta}, \frac{\sqrt{3}}{6} \mp \frac{\sqrt{3}\,b}{2\,\Delta}\right), \tag{27}$$

where we defined $\Delta = \sqrt{b^2 + 4\,c^2}$, which yield three independent Bloch vectors of the density matrix, namely

$$\boldsymbol{\lambda}_0 = \left(0, -1, 0, 0, 0, 0, \frac{1}{2}, -\frac{1}{2\sqrt{3}}\right), \tag{28}$$

$$\boldsymbol{\lambda}_\pm = \left(\pm\frac{\sqrt{2}\,c}{\Delta}, \frac{1}{2} \pm \frac{b}{2\,\Delta}, \pm\frac{\sqrt{2}\,c}{\Delta}, 0, 0, 0, -\frac{1}{4} \pm \frac{3\,b}{4\,\Delta}, \frac{\sqrt{3}}{12} \mp \frac{\sqrt{3}\,b}{4\,\Delta}\right), \tag{29}$$

whose norm is equal to $4/3$ and the scalar products between them are equal to $-2/3$. Thus it is straightforward to check that the corresponding extremal density matrices are orthogonal projectors associated to the energy eigenvalues of the Hamiltonian $\epsilon_0 = b$, $\epsilon_\pm = 1/2(b \pm \Delta)$. (b) Mixed case: For any other values for c_2 and c_3 in the region shown in Fig. 1(a), one can solve the polynomial system given by (23). As an example we take $c_2 = 29/100$ and $c_3 = 1/50$. There are 6 different solutions for λ_2 and λ_8 which give rise to 6 Bloch vectors:

$$\begin{aligned}
\boldsymbol{\lambda}^{(1)}{}_\pm &= \left(\pm\frac{\sqrt{2}\,c}{10\Delta}, \frac{7}{20} \pm \frac{b}{20\,\Delta}, \pm\frac{\sqrt{2}\,c}{10\,\Delta}, 0, 0, 0, -\frac{7}{40} \pm \frac{3\,b}{40\,\Delta}, \frac{\sqrt{3}}{120} \mp \frac{\sqrt{3}\,b}{40\,\Delta}\right), \\
\boldsymbol{\lambda}^{(2)}{}_\pm &= \left(\pm\frac{2\sqrt{2}\,c}{5\,\Delta}, -\frac{1}{10} \pm \frac{b}{5\,\Delta}, \pm\frac{2\sqrt{2}\,c}{5\,\Delta}, 0, 0, 0, +\frac{1}{20} \pm \frac{3\,b}{10\,\Delta}, -\frac{\sqrt{3}}{60} \mp \frac{\sqrt{3}\,b}{10\,\Delta}\right), \\
\boldsymbol{\lambda}^{(3)}{}_\pm &= \left(\pm\frac{3\sqrt{2}\,c}{10\,\Delta}, -\frac{1}{4} \pm \frac{3\,b}{20\,\Delta}, \pm\frac{3\sqrt{2}\,c}{10\,\Delta}, 0, 0, 0, +\frac{1}{8} \pm \frac{9\,b}{40\,\Delta}, -\frac{5\sqrt{3}}{120} \mp \frac{9\sqrt{3}\,b}{120\,\Delta}\right).
\end{aligned} \tag{30}$$

The corresponding extremal expectation values of the Hamiltonian are given by

$$\langle\hat{H}\rangle_\pm^{(1)} = \frac{11\,b}{20} \pm \frac{\Delta}{20}, \quad \langle\hat{H}\rangle_\pm^{(2)} = \frac{7\,b}{10} \pm \frac{\Delta}{5}, \quad \langle\hat{H}\rangle_\pm^{(3)} = \frac{3\,b}{4} \pm \frac{3\,\Delta}{20}. \tag{31}$$

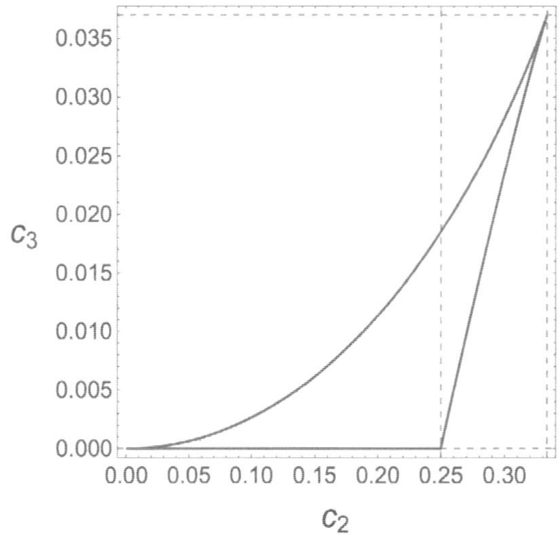

Figure 1. Regions of c_2, and c_3, where the positivity conditions of density matrix are satisfied. The horizontal and vertical dashed lines taking the values $c_3 = \{0, 1/27\}$ and $c_2 = \{1/4, 1/3\}$, respectively.

We find the expansion of the extremal density matrices for the mixed case in terms of the pure case described before,

$$\begin{aligned} \hat{\rho}_+^{(1)} &= \frac{1}{10}\,\hat{\rho}_0 + \frac{1}{2}\,\hat{\rho}_+ + \frac{2}{5}\,\hat{\rho}_- , & \hat{\rho}_-^{(1)} &= \frac{1}{10}\,\hat{\rho}_0 + \frac{2}{5}\,\hat{\rho}_+ + \frac{1}{2}\,\hat{\rho}_- , \\ \hat{\rho}_+^{(2)} &= \frac{2}{5}\,\hat{\rho}_0 + \frac{1}{2}\,\hat{\rho}_+ + \frac{1}{10}\,\hat{\rho}_- , & \hat{\rho}_-^{(2)} &= \frac{2}{5}\,\hat{\rho}_0 + \frac{1}{10}\,\hat{\rho}_+ + \frac{1}{2}\,\hat{\rho}_- , \\ \hat{\rho}_+^{(3)} &= \frac{1}{2}\,\hat{\rho}_0 + \frac{2}{5}\,\hat{\rho}_+ + \frac{1}{10}\,\hat{\rho}_- , & \hat{\rho}_-^{(3)} &= \frac{1}{2}\,\hat{\rho}_0 + \frac{1}{10}\,\hat{\rho}_+ + \frac{2}{5}\,\hat{\rho}_- . \end{aligned} \tag{32}$$

Note that the expressions (31) can be checked by calculating the expectation value of the Hamiltonian with the expansions given in the last expression.

4.0.1. Degenerate case. Now we consider the Hamiltonian matrix given by

$$\hat{H} = \begin{pmatrix} 2 & -1+i & -1-\frac{i}{3} \\ -1-i & \frac{13}{3} & 1+i\,2 \\ -1+\frac{i}{3} & 1-i\,2 & 3 \end{pmatrix} . \tag{33}$$

In this case the Bloch vector characterising the Hamiltonian is

$$\boldsymbol{h} = \left(-2,\ -2,\ 2,\ -2,\ \frac{2}{3},\ -4,\ -\frac{7}{3},\ \frac{\sqrt{3}}{9}\right) \tag{34}$$

with $h_0 = \frac{28}{3}$. Replacing this values into the matrix (15), the rank of $\boldsymbol{M}$ is $r = 4$, which according to [26] exhibits a double degeneracy. Thus, if $\alpha > \beta$ the diagonal representation is $\text{diag}(\alpha, \beta, \beta)$, or in opposite way, if $\beta > \alpha$ then $\text{diag}(\beta, \beta, \alpha)$.

Applying the Gauss-Jordan method to (14), one gets the solutions

$$\begin{aligned} \lambda_1^c &= \frac{1}{14}\left(6\,\lambda_5 + 8\,\lambda_6 + \lambda_7 + 7\sqrt{3}\,\lambda_8\right), & \lambda_2^c &= \frac{1}{7}\left(3\,\lambda_5 - 3\,\lambda_6 + 11\,\lambda_7 - 7\sqrt{3}\,\lambda_8\right), \\ \lambda_3^c &= \frac{1}{42}\left(24\,\lambda_5 - 24\,\lambda_6 + 11\,\lambda_7 - 7\sqrt{3}\,\lambda_8\right), & \lambda_4^c &= \frac{1}{42}\left(-30\,\lambda_5 + 30\,\lambda_6 - 19\,\lambda_7 + 35\sqrt{3}\,\lambda_8\right). \end{aligned}$$

IOP Conf. Series: Journal of Physics: Conf. Series **839** (2017) 012012 doi:10.1088/1742-6596/839/1/012012

Hence, the critical Bloch vector is given by $\boldsymbol{\lambda}^c = (\lambda_1^c, \lambda_2^c, \lambda_3^c, \lambda_4^c, \lambda_5, \lambda_6, \lambda_7, \lambda_8)$, with 4 free parameters and its associated critical density matrix is denoted as $\hat{\rho}^c$.

In order to obtain the eigen-system of $\hat{H}$, we use the procedure established before for the degenerated case:

• We select $\lambda_5 = \lambda_6 = 0$, solve the polynomial condition (23) with $c_2 = c_3 = 0$, and get the following Bloch vectors of the density matrix

$$\boldsymbol{\lambda}^{(1)} = \frac{1}{11}\left(6,\,0,\,0,\,6,\,0,\,0,\,7,\,11/\sqrt{3}\right), \quad \boldsymbol{\lambda}^{(2)} = \frac{1}{46}\left(12,\,36,\,6,\,0,\,0,\,0,\,35,\,19/\sqrt{3}\right). \tag{35}$$

These yield two density matrices $\hat{\rho}^{(1)}$, and $\hat{\rho}^{(2)}$ which are not independent. Both give an energy eigenvalue $\epsilon = 4/3$.

• We establish the algebraic system of equations,

$$\left\{\mathrm{Tr}\left(\hat{\rho}^{(1)}\,\hat{\rho}^c\right), \mathrm{Tr}\left(\hat{\rho}^{(2)}\,\hat{\rho}^c\right)\right\} = 0\,, \tag{36}$$

whose solution, together with the positivity condition, gives another independent Bloch vector

$$\boldsymbol{\lambda}^{(3)} = \frac{1}{16}\left(-6,\,-6,\,6,\,-6,\,2,\,-12,\,-7,\,1/\sqrt{3}\right). \tag{37}$$

Thus we have obtained another extremal density matrix orthogonal to $\hat{\rho}^{(1)}$, and $\hat{\rho}^{(2)}$ and the expectation value of the Hamiltonian yields the eigenvalue $\epsilon_2 = 20/3$. Until now we have obtained 2 independent and orthogonal projectors, namely $\hat{\rho}^{(1)}$ and $\hat{\rho}^{(3)}$.

• We repeat the procedure by establishing the algebraic system of equations

$$\left\{\mathrm{Tr}\left(\hat{\rho}^{(1)}\,\hat{\rho}^c\right), \mathrm{Tr}\left(\hat{\rho}^{(3)}\,\hat{\rho}^c\right)\right\} = 0\,, \tag{38}$$

whose solution gives the Bloch vector

$$\boldsymbol{\lambda}^{(4)} = \left(-\frac{15}{88},\,\frac{3}{8},\,-\frac{3}{8},\,-\frac{15}{88},\,-\frac{1}{8},\,\frac{3}{4},\,-\frac{35}{176},\,-\frac{17}{16\sqrt{3}}\right). \tag{39}$$

One gets then another orthogonal projector $\hat{\rho}^{(4)}$ and the corresponding expectation value of the Hamiltonian is $\epsilon_3 = 4/3$.

We have obtained the complete eigen-system of the degenerated Hamiltonian. For the eigenvalue $\epsilon = 4/3$, we indeed have a family of projectors yielding the same eigenvalue.

5. Summary and Conclusions

The main contribution of our work is to give an algebraic procedure to find extremal density matrices for a given Hamiltonian. Our approach applies to both the degenerate and non degenerate cases of the Hamiltonian. The examples of the procedure are given for dimensions $d = 2,\,3$ and show that the Hamiltonian spectrum for the pure case is recovered. For the mixed case, we have verified that the extremal values of the expectation value of the Hamiltonian is a convex sum of the corresponding results for the pure case. We want to enhance that the method can be applied by replacing the Hamiltonian for any observable acting on a qudit space.

We established that an extremal density matrix commutes with the Hamiltonian operator and optimizes its mean value. We demonstrated that at most $d-1$ variables are necessary to find extremal density matrices with appropriate positivity conditions, for the non-degenerated case of the finite matrix Hamiltonian. In the degenerate pure case, one has more free components of the extremal density matrix which can be selected by asking orthogonality between the projectors, which allow to obtain the energy spectrum.

Finally, for the examples with $d = 2$, and $d = 3$ following the method given in [24], we find also the compatible regions between the coefficients of the characteristic polynomial of the density matrix in terms of the positivity conditions of the Bezoutian matrix in order to provide a self-contained approach.

Acknowledgement

This work was partially supported by CONACyT-México (under Project No. 238494). The authors thank G. Marmo, M. A. Man'ko and V. I. Man'ko for valuable comments.

References

[1] Amico L, Fazio R, Osterloh A, and Vedral V 2008 *Rev. Mod. Phys.* **80** 517.
[2] Links J, Foerster A, Tonel A, and Santos G 2006 *Ann. Henri Poincare* **7** 1591-1600.
[3] Pérez-Campos C, González-Alonso J R, Castaños O, López-Peña R, 2010 Ann. Phys. (N.Y.) **325** 325.
[4] Lipkin H J, Meshkov N and Glick A J 1965 *Nucl. Phys.* **62** 188.
[5] Romera E, Calixto M and Castaños O 2014 *Phys. Scr.* **89** 095103.
[6] Günhe O and Tóth G 2009 *Phys. Reports* **474** 1-75.
[7] Bertlmann R A and Krammer P 2008 *J. Phys. A* **41** 235303.
[8] Spengler C, Huber M, and Hiesmayr B C 2010 *J. Phys.* A **43** 385306.
[9] Louisell W H 1973 *Quantum Statistical Properties of Radiation* John Wiley & Sons.
[10] Benjamin I, Alhassid Y, and Levine R D 1985 *Chem. Phys. Letters* **115** 113-118.
[11] Chernega V N, and Man'ko V I 2007 *J. Russ. Laser Res.***28** 103.
[12] Mahler G, Waberruss V A 1995 *Quantum Networks: Dynamics of Open Nanostructures.* Springer.
[13] Kimura G, 2003 Phys. Lett. A **314** 339.
[14] Brüning E, Mäkelä H, Messina A, Petruccione F, 2012 J. Mod. Opt. **59** 1.
[15] Jarlskog C, 2006 J. Math. Phys. **47** 013507.
[16] Tilma T, Sudarshan E C G, 2002 J. Phys A: Math. Gen. **35** 10467.
[17] Figueroa A, López J, Castaños O, López-Peña R, Man'ko M A, Man'ko V I, 2015 J. Phys. A: Math. Theor. **48** 065301.
[18] Figueroa A, López J, Castaños O, López-Peña R, ArXiv:1609.09835.
[19] Seroul R, 2000 *Programming for Mathematicians.* Springer.
[20] Byrd M S, Khaneja N, 2003 Phys. Rev. A **68** 062322.
[21] Deen S M, Kabir P K, 1971. Phys. Rev. D **4** 1662.
[22] Procesi C. 2007 *Lie Groups: An Approach through Invariants and Representations* (New York: Springer).
[23] Procesi C, Schwarz G, 1985 Invent. Math. **81** 539.
[24] Gerdt V P, Khvedlidze A M, Palii Yu G, 2014 J. Math. Sci. **200** 682.
[25] Kus M, Zyczkowski, 2001Phys. Rev. A **63** 032307.
[26] Ercolessi, E, Marmo, G, Morandi, G, 2001 Int. Jour. Mod. Phys. A **16** 5007.
[27] Schirmer S G, Zhang T, Leahy J V, 2004 J. Phys. A: Math. Gen. **37** 1389.
[28] Van der Waerden B L, 1991 *Modern Algebra II* (New York: Springer).
[29] Gelfand I M, Kapranov M M, Zelevinsky A V, 2008 *Discriminants Resultants and Multidimensional Determinants* (Germany: Birkhäuser).

Non Linear Response of a Harmonic Diatomic Molecule: an Exact Algebraic Approach[1]

J Récamier

Instituto de Ciencias Físicas, Universidad Nacional Autónoma de México
C.P. 62210 Cuernavaca, Morelos, México

E-mail: pepe@fis.unam.mx

Abstract. Even harmonic molecules display a nonlinear behavior when driven by an inhomogeneous field. Here we present the response of a single harmonic molecule to a monochromatic time and space dependent electric field $E(r,t)$ of frequency ω employing exact algebraic methods. We evaluate the responses at the fundamental frequency ω and at succesive harmonics 2ω, 3ω, etc., as a function of the intensity of the field and compare the results with those of the first and second order perturbation theory.

1. Introduction

Even though the harmonic oscillator is the prototype of a linear system, its response to inhomogeneous time dependent fields contains nonlinear contributions. The study of the optical linear and nonlinear responses of harmonic systems to periodic electromagnetic fields has been done both with classical [2–4] and quantum mechanical treatments [5]. It has been shown that when perturbation theory is applied both methods yield the same response at the fundamental frequency and at its second harmonic 2ω. Here we present an algebraic treatment to study nonperturbatively the time evolution of a quantum mechanical harmonic system forced by a time and space dependent electric field.

Consider a dynamic charge q with mass m displaced a distance x from its equilibrium position r_0, to which it is bound by a harmonic force with resonant frequency ω_0 driven by a classical electric field $E(r,t)$. For simplicity, we assume that the system is one-dimensional. The interaction potential is then, up to second order in the displacement x:

$$V = -qxE(r_0,t) - \frac{q}{2}x^2\partial_r E(r_0,t) \tag{1}$$

the first term is the usual dipolar interaction while the second term is the quadrupolar interaction. The classical equation of motion is:

$$m\frac{d^2x}{dt^2} = qE(r_0,t) - m\omega_0^2 x + qx\partial_r E(r_0,t) \tag{2}$$

where the field

$$E(r,t) \simeq E(r_0,t) + x\partial_r E(r_0,t)$$

[1] Most of the information presented in this contribution has already been published in Ref. [1].

is evaluated at the position of the charge $r = r_0 + x$. Notice that the terms linear in the displacement are time dependent, thus the spatial dependence of the field transforms the harmonic oscillator into a forced parametric oscillator. Notice also that the sign of the contribution arising from the inhomogeneity of the field has a sign opposite to that of the harmonic oscillator contribution, then there will be a competition between the confining term and the quadrupolar term. Now assume a monochromatic field

$$E(r,t) = Re[E_\omega(r)e^{-i\omega t}]$$

Using Ref. [2] we see that the induced dipole moment is given by $p^{(1)}(t) = Re(p_\omega e^{i\omega t})$ with

$$p_\omega = qx_\omega = \alpha(\omega)E_\omega(r_0)$$

to first order in the driving field, and the linear polarizability is:

$$\alpha(\omega) = \frac{q^2/m}{\omega_0^2 - \omega^2}. \tag{3}$$

The equation of motion at 2ω is:

$$-m(2\omega)^2 x_{2\omega} = -m\omega_0^2 x_{2\omega} + qx_\omega \partial_r E_\omega(r_0)$$

whose solution is:

$$x_{2\omega} = \frac{1}{2q^2}\alpha(\omega)\alpha(2\omega)[\partial_r E_\omega^2(r_0)]. \tag{4}$$

and the second order dipole moment at 2ω is, therefore,

$$p_{2\omega} = \frac{1}{2q}\alpha(\omega)\alpha(2\omega)[\partial_r E_\omega^2(r_0)]. \tag{5}$$

As mentioned before, the nonlinear response derived above classically using a perturbative solution, coincides with the quantum mechanical expressions obtained with perturbation theory.

2. Algebraic approach: harmonic oscillator

Here we present a Lie algebraic approach that yields the exact solution to the problem [6–8]. First we write the system's Hamiltonian as:

$$H = H_0 + V, \tag{6}$$

where H_0 is the unperturbed harmonic oscillator Hamiltonian and V is the interaction given in Eq. 1. We go into the interaction picture where the time evolution operator satisfies the equation:

$$i\hbar \partial_t U_I(t,t_0) = H_I(t)U_I(t,t_0), \quad U_I(t_0,t_0) = 1 \tag{7}$$

with the interaction picture Hamiltonian given as:

$$H_I(t) = U_0^\dagger(t,t_0)VU_0(t,t_0),$$

and where $U_0(t,t_0)$ is the free time evolution operator. We have chosen $t = t_0$ as our reference time and assumed that the interaction is negligible for times $t < t_0$.
The interaction picture Hamiltonian is then:

$$H_I(t) = -qE(r_0,t)\sqrt{\frac{\hbar}{2m\omega_0}}(\hat{a}e^{-i\omega_0(t-t_0)} + \hat{a}^\dagger e^{i\omega_0(t-t_0)})$$
$$+\frac{q}{2}\partial_r E(r_0,t)\left(\frac{\hbar}{2m\omega_0}\right)(\hat{a}^2 e^{-2i\omega_0(t-t_0)} + 2\hat{a}^\dagger\hat{a} + 1 + \hat{a}^{\dagger 2}e^{2i\omega_0(t-t_0)}) \tag{8}$$

which can be written as:

$$H_I(t) = \sum_{n=1}^{6} F_n(t)\hat{X}_n \tag{9}$$

with known, time dependent functions $F_n(t)$ and where the operators $\hat{X}_n$ are given in the Schrödinger picture. Since the set of operators in $H_I(t)$ is closed under commutation, we can use the Wei-Norman Theorem [6] and write the *exact* time evolution operator as:

$$U_I(t) = \prod_{n=1}^{6} e^{\alpha_n(t)\hat{X}_n} \tag{10}$$

For simplicity, we choose normal ordering in the product of exponentials, that is: $\hat{X}_1 = \hat{n}$, $\hat{X}_2 = \hat{a}^{\dagger 2}$, $\hat{X}_3 = \hat{a}^\dagger$, $\hat{X}_4 = \hat{a}$, $\hat{X}_5 = \hat{a}^2$ and $\hat{X}_6 = 1$. After substitution of Eq. 10 into Eq. 7 we get the following set of first order, coupled, nonlinear differential equations for the complex functions $\alpha_n(t)$ which is generally solved by numerical means:

$$\partial_t\alpha_1 = \frac{i}{\hbar}(F_1 - 4F_5\alpha_2 e^{-2\alpha_1}), \tag{11}$$

$$\partial_t\alpha_2 = \frac{i}{\hbar}(F_2 e^{2\alpha_1} - 4F_5\alpha_2^2 e^{-2\alpha_1}), \tag{12}$$

$$\partial_t\alpha_3 = \frac{i}{\hbar}(F_3 e^{\alpha_1} - 2F_4\alpha_2 e^{-\alpha_1}), \tag{13}$$

$$\partial_t\alpha_4 = \frac{i}{\hbar}(F_4 e^{-\alpha_1} - 2F_5\alpha_3 e^{-2\alpha_1}), \tag{14}$$

$$\partial_t\alpha_5 = \frac{i}{\hbar}F_5 e^{-2\alpha_1}, \tag{15}$$

$$\partial_t\alpha_6 = \frac{i}{\hbar}(F_6 - F_4\alpha_3 e^{-\alpha_1} + (\alpha_3^2 - 2\alpha_2)F_5 e^{-2\alpha_1}), \tag{16}$$

with initial conditions $\alpha_i(t_0) = 0$, $i = 1, \ldots 6$.
The temporal evolution of any observable can be obtained from the creation, annihilation operators in the Heisenberg picture, these are given by:

$$\hat{a}(t) = U_I(t)^\dagger U_0(t)^\dagger \hat{a} U_0(t) U_I(t) = t_{11}(t)\hat{a} + t_{12}(t)\hat{a}^\dagger + t_{13}(t) \tag{17}$$

$$\hat{a}^\dagger(t) = U_I(t)^\dagger U_0(t)^\dagger \hat{a}^\dagger U_0(t) U_I(t) = t_{21}(t)\hat{a} + t_{22}(t)\hat{a}^\dagger + t_{23}(t) \tag{18}$$

where the functions $t_{ij}(t)$ are known functions of the $\alpha_n(t)$. The transformation amounts to a rotation due to the presence of the quadratic operators and a displacement due to the linear operators in $H_I(t)$.

The average value of the displacement coordinate taken between number states is:

$$\langle n|\hat{x}(t)|n\rangle = \sqrt{\frac{\hbar}{2m\omega_0}}(t_{13}(t) + t_{23}(t)). \tag{19}$$

Using an algebraic approach we have reduced the problem of calculating the quantum mechanical response of a nonlinearly driven harmonic diatomic molecule forced by a time and space dependent field, to the solution of a small set of coupled nonlinear ordinary differential equations.

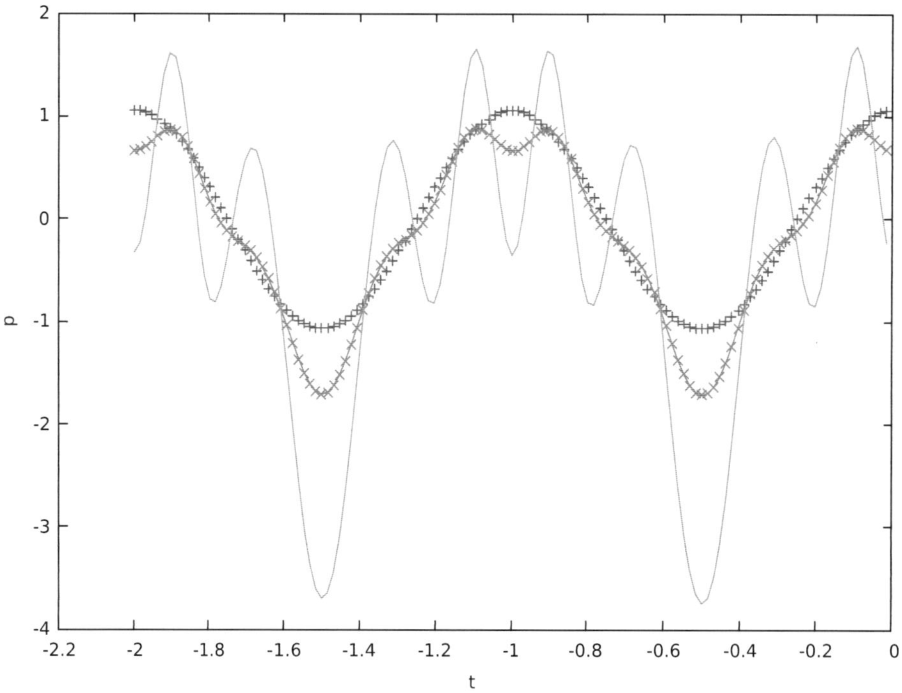

Figure 1. Normalized dipole moment induced in a harmonic molecule by a inhomogeneous field ($\beta = 0.05/l$) as a function of time for $\omega = 0.24\omega_0$ and E_ω=0.1E (dots), E_ω=5E (line-points), and E_ω=10E (full line). The dipole moment is normalized to $\alpha_0 E_\omega$ where α_0 is the static polarizability of the oscillator, and the time is given in units of the fundamental period $\tau = 2\pi/\omega$

3. Numerical results

In order to be specific we now consider an adiabatically switched monochromatic field of the form

$$E(r_0, t) = E_\omega(r_0)\cos(\omega t)e^{\eta t} \tag{20}$$

where η is a small switching rate and the initial time is such that $e^{\eta t_0} \ll 1$. In Fig. 1 we show the time-dependent dipole moment induced by an inhomogeneous field. The inhomogeneity is characterized by the parameter $\beta \equiv \partial_r E_\omega(r_0)/E_\omega(r_0)$ which we set as $\beta = 0.05/l$ where $l = \sqrt{\hbar/m\omega}$ is a length scale, of the order of the amplitude of a molecular vibration in the ground state. We also introduce a field scale $\mathcal{E} = \hbar\omega_0/ql$ which is the field that would produce an appreciable energy change, of the order of a vibrational quantum when a charge q is displaced a distance of the order of l. To produce Fig. 1 we have chosen $\eta = 2\times 10^{-4}$ and an initial time $|t_0| \gg 1/\eta$ such that $e^{\eta t_0} \leq 10^{-5}$. In this figure we have also set $\omega = 0.24\omega_0$ so that the fourth harmonic is nearly resonant with the natural frequency ($4\omega \simeq \omega_0$) and we consider several values for the field intensity. For a small value of the field intensity $E_\omega = 0.1\mathcal{E}$ the response of the system is dominated by the linear contribution (line with points), response that coincides with that of first order perturbation theory. When we increase the field intensity to $E_\omega = 5\mathcal{E}$ the nonlinear contributions start to play a role as can be seen from the figure (points). Finally, when the intensity of the field is large $E_\omega = 10\mathcal{E}$ the response is dominated by the fourth harmonic (full line). This conduct cannot be obtained from a perturbative treatment.

In Fig. 2 we have chosen $\omega = 0.325\omega_0$ so that the third harmonic is nearly resonant with the natural frequency ($3\omega \simeq \omega_0$) and we consider several values for the field intensity. For a small

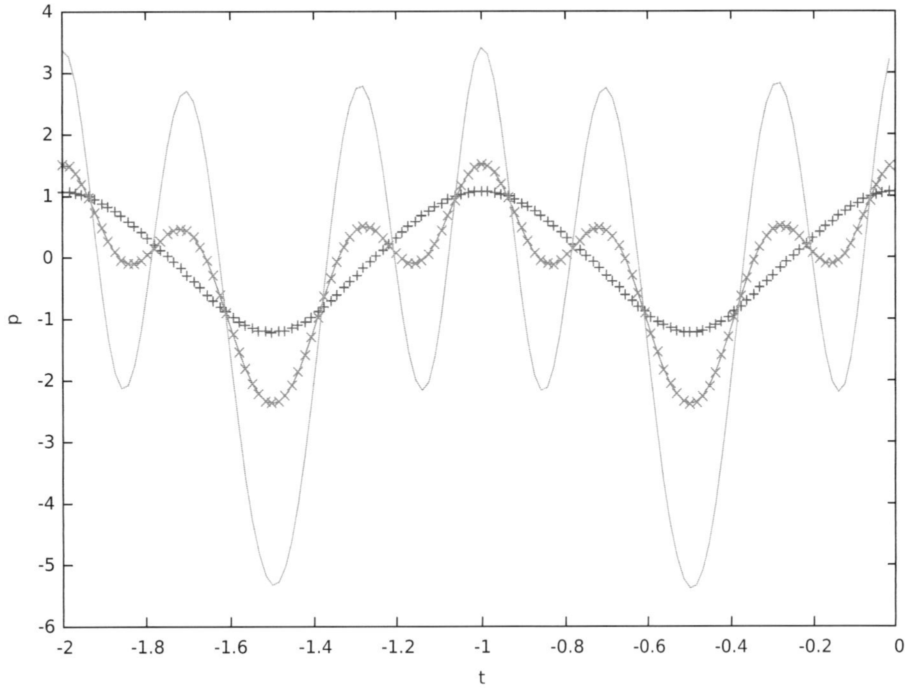

Figure 2. Normalized dipole moment induced in a harmonic molecule by a inhomogeneous field ($\beta = 0.05/l$) as a function of time for $\omega = 0.325\omega_0$ and E_ω=0.1E (dots), E_ω=5E (line-points), and E_ω=8E (full line). The dipole moment is normalized to $\alpha_0 E_\omega$ where α_0 is the static polarizability of the oscillator, and the time is given in units of the fundamental period $\tau = 2\pi/\omega$

value of the field intensity $E_\omega = 0.1\mathcal{E}$ the response of the system is again dominated by the linear contribution (line with cruces). When we increase the field intensity to $E_\omega = 5\mathcal{E}$ the nonlinear effects are evident and we can see the appearance of structure at $\omega/3$ (line with points) and at even larger values of the field intensity $E_\omega = 8\mathcal{E}$ the structure is much more evident with a much larger amplitude.

4. Discussion

Here we have presented an algebraic approach which allows the calculation of the dipole moment nonlinearity induced by a time and space dependent electric field acting upon a harmonic diatomic molecule. The nonlinearity is due to the quadrupolar coupling between the system and the field transforming the problem of a forced harmonic diatomic into that of a forced parametric one. The method used here does not depend on the specific form of the temporal dependence of the field, however, we specified its functional form as adiabatically switched monochromatic in order to analyze the Fourier components of the polarization. We found that, for small field amplitudes $E_\omega \ll \mathcal{E}$ the $n'th$ order polarization $p_{n\omega}$ α E_ω^n as expected from perturbation theory. At high enough fields this power law is no longer valid. When the field's frequency is close to a subharmonic of the resonance frequency ($\omega \simeq \omega_0/n$) the response is dominated by its $n'th$ harmonic component. Since the quadrupolar contribution has an oposite sign to the harmonic one, for large enough fields the nonlinear coupling may overwhelm the restoring force causing the braking of the molecule.
We introduced a length unit l of the order of a fraction of an atomic unit and the gradient of

the field was characterized by the parameter β; for an H_2 molecule $l \simeq 2$Å. For a free wave interacting with an isolated molecule we would expect the nonlinearity contribution of the order of βl, thus, a very small quantity. However, for a molecule adsorbed on a surface or interacting with nearby neighbors, the local field that polarizes the molecule might have a longitudinal character and may be a few orders of magnitude larger than that for a free wave. Thus, the values used in this work for $\beta l = 0.05$ might be attainable. There are other nonperturbative approaches to analyze the response of a molecular system to an applied field; the most common being Floquet theory [9], which allows calculations for time-periodic fields.

References

[1] Récamier J, Mochán W L and Maytorena J 2005 *Phys. Rev.* A **72** 023805
[2] Mendoza B S and Mochán W L 1996 *Phys. Rev.* B **53**, 4999
[3] Maytorena J A, Mendoza B S and Mochán W L 1998 *Phys. Rev.* B **55**, 2569
[4] Mendoza B S and Mochán W L 1996 *Phys. Rev.* B **53**, R10473.
[5] Schaich W L and Mendoza B S 1992 *Phys. Rev.* B **45**, 14279.
[6] Wei J and Norman E 1963 *Jour. Math. Phys.* **4**(4) 575
[7] Gilmore R and Yuan J M 1987 *J. Chcm. Phys.* **86** 130
[8] Gilmore R 1974 *Lie Groups, Lie Algebras and Some of Their Applications* (Wiley, New York).
[9] Shirley J H 1965 *Phys. Rev.* **138**, B979.

IOP Conf. Series: Journal of Physics: Conf. Series **839** (2017) 012014 doi:10.1088/1742-6596/839/1/012014

Universal Quantum Gates for Quantum Computation on Magnetic Systems Ruled by Heisenberg-Ising Interactions

F Delgado

Departamento de Física y Matemáticas, Escuela de Ingeniería y Ciencias, Tecnológico de Monterrey, Campus Estado de México, Atizapán, Estado de México, CP. 52926, México.

E-mail: `fdelgado@itesm.mx`

Abstract. The gate version of quantum computation exploits several quantum key resources as superposition and entanglement to reach an outstanding performance. In the way, this theory was constructed adopting certain supposed processes imitating classical computer gates. As for optical as well as magnetic systems, those gates are obtained as quantum evolutions. Despite, in certain cases they are attained as an asymptotic series of evolution effects. The current work exploits the direct sum of the evolution operator on a non-local basis for the driven bipartite Heisenberg-Ising model to construct a set of equivalent universal gates as straight evolutions for this interaction. The prescriptions to get these gates are reported as well as a general procedure to evaluate their performance.

1. Introduction

Quantum computation and quantum information are modern developments taking advantage from the Quantum Mechanics features to propose technological applications. In their quantum gate version, the gate construction on specific physical systems is a central aspect to develop those applications. The requirement of entanglement as resource implies the introduction of physical interactions between the parts of the system. This requirement introduces in the dynamics other natural basis than computational basis, which is only natural for single and isolated qubits. For this reason, in general, it is not easy fit the evolution into the theoretical gate constructions inspired on the classical computational elements [1, 2]. The same is true for more complex gates involving multiqubit systems. Thus, several attempts to define general ways to construct gates are in order. These approaches uses unitary factorization, asymptotic approximations, etc. [3–6]. Still the most common approach is the construction of universal sets of gates whose composition allows to generate any other gate.

In this work, we propose and develop a set of universal gates easily constructed for bipartite magnetic systems ruled by the anysotropic Heisenberg-Ising interaction with strengths J_k along each direction and including driven magnetic fields B_{i_h} on each qubit $i = 1, 2$ in only one of the x, y, z directions ($h = 1, 2, 3$):

$$\mathcal{H}_h = \sum_{k=1}^{3} J_k \sigma_{1_k} \sigma_{2_k} - B_{1_h} \sigma_{1_h} - B_{2_h} \sigma_{2_h} \tag{1}$$

this model comprises several models reported in the literature [7–13]. Their $SU(4)$ dynamics exhibits a block form when it is expressed on the non-local basis of Bell states [14]. As it has been proved, the general blocks in that decomposition has the form $U(2) = U(1) \times SU(2)$:

$$s_{h_j} = e^{i\Delta_{h\alpha}^{+}} \begin{pmatrix} e_{h\alpha}^{\beta *} & -qi^h d_{h\alpha} \\ qi^{*h} d_{h\alpha} & e_{h\alpha}^{\beta} \end{pmatrix} = e^{i\Delta_{h\alpha}^{+}} \left(\cos\Delta_{h\alpha}^{-} \mathbf{I}_{h_j} - i \sin\Delta_{h\alpha}^{-} \mathbf{n} \cdot \mathbf{S}_{h_j}\right) \quad (2)$$

$$\text{with}: \quad e_{h\alpha}^{\beta} = \cos\Delta_{h\alpha}^{-} + i\beta j_{h_{-\alpha}} \sin\Delta_{h\alpha}^{-}, \quad d_{h_\alpha} = b_{h_{-\alpha}} \sin\Delta_{h\alpha}^{-} \quad (3)$$

Note $e_{h\alpha}^{\beta}$ and d_{h_α} are related with the reduced transversal strength and magnetic fields $j_{h_{-\alpha}}, b_{h_{-\alpha}}$ and with the Rabi frequencies involved $\Delta_{h\alpha}^{\pm}$, as they are presented in [14]. At the same time, these parameters are finally expressed in terms of the physical parameters $t, J_i, B_k, i = 1,2,3; k = 1,2$. While, $\mathbf{n} = (qb_{h_{-\alpha}} \sin\frac{h\pi}{2}, qb_{h_{-\alpha}} \cos\frac{h\pi}{2}, \beta j_{h_{-\alpha}}), \alpha = (-1)^{h+j+1}, \beta = (-1)^{j(h+l_j-k_j+1)}, q = \beta(-1)^{h+1}$. h is the direction of magnetic field and $j = 1,2$ a position label for each block in the whole evolution matrix. k_j, l_j are the labels for its rows.

There, $\mathbf{I}_{h_j}$ and $\mathbf{S}_{h_j}$ are a set of extended Pauli matrices stating a basis for each $SU(2)$ block j applied on definite pairs of selectable Bell states as function of h. Those matrices and their algebra have been widely discussed in [14,15].This basis could be understood too as noise or error basis for deviations from the evolution prescriptions (such as the traditional X, Y and Z are noise effects in the computational basis, dephasing and flipping noise). Thus $U(t) = \bigoplus_{j=1}^{2} s_{h_j}$, stating a semi-direct product $SU(4) = U(1) \times SU(2)^2$ for $U(t)$ with the Bell states as general basis for all cases $h = 1,2,3$. This dynamics mixes the selectable pairs of Bell states in a programmed way. Thus, in this work we explode last property to show that an alternative set of universal gates can be easily constructed. These gates operate on these subspaces defined by the pairs of Bell states as a privileged grammar instead or alternatively to the computational basis (through the block forms depicted there). The second section presents the Boykin et al universal gates and the proposed alternative gates for the current systems. The third section shows the prescriptions to generate the set of universal gates, discussing some possible issues. The fourth section presents a general strategy to evaluate the performance for all gates proposed. Last section states the conclusions and extensions.

2. Boykin set of universal gates and alternative gates for the Heisenberg-Ising interactions

It has been shown that two level quantum channel processing is universal in the quantum gate version of quantum computation [16]. In addition, a universal set of two level gates for the computational basis was given by Boykin et al [17]: $\mathcal{B} \equiv \{S_{\pi/8}, S_{\pi/4}, H, C^a NOT_b\}$ (Table 1). Boykin et al gates are really universal for $U(4)$ operations on quantum information, independently from the basis being used. Then, their form can be used for another arbitrary basis on two levels where quantum computation is being settled. For the Bell states basis, an analog set of gates could be alternative to the last universal set. These gates operates on the entire quantum information space of the bipartite system: $\mathcal{D} = \{(\mathbf{1}_1 \otimes S_{\pi/8_2})_{\mathrm{B}}, (\mathbf{1}_1 \otimes S_{\pi/4_2})_{\mathrm{B}}, (S_{\pi/8_1} \otimes \mathbf{1}_2)_{\mathrm{B}}, (S_{\pi/4_1} \otimes \mathbf{1}_2)_{\mathrm{B}}, (\mathbf{1}_1 \otimes H_2)_{\mathrm{B}}, (H_1 \otimes \mathbf{1}_2)_{\mathrm{B}}, (C^1 NOT_2)_{\mathrm{B}}, (C^2 NOT_1)_{\mathrm{B}}\}$. Gate subscript remarks that those forms are written for the Bell basis. Thus, they need be considered covering the two quantum channels simultaneously. For this reason, the graph of any gate in $\mathcal{D}$ comprises always the two channels (Table 2). In particular, the classical symbol for $CNOT$ gate appears horizontally (as a symbolic issue) in the corresponding graphs, due to they are controlled with respect to the quantum information states, not with respect to the single qubit states.

Table 1. Universal set of 1 and 2 level gates for the computational basis.

Levels	Gate	Symbol	Graph	Matrix form
1-Level	$\frac{\pi}{8}$ gate	$S_{\pi/8}$	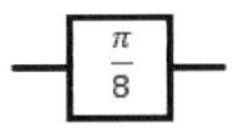	$\begin{pmatrix} 1 & 0 \\ 0 & e^{i\pi/4} \end{pmatrix} \simeq \begin{pmatrix} e^{-i\pi/8} & 0 \\ 0 & e^{i\pi/8} \end{pmatrix}$
	$\frac{\pi}{4}$ gate	$S_{\pi/4}$	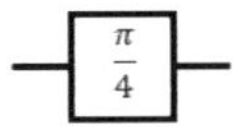	$\begin{pmatrix} 1 & 0 \\ 0 & e^{i\pi/2} \end{pmatrix} \simeq \begin{pmatrix} e^{-i\pi/4} & 0 \\ 0 & e^{i\pi/4} \end{pmatrix}$
	Hadamard	H	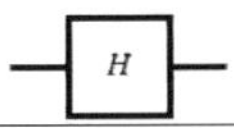	$\frac{1}{\sqrt{2}}\begin{pmatrix} 1 & 1 \\ 1 & -1 \end{pmatrix}$
2-Level	Controlled	C^1NOT_2	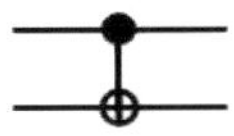	$\begin{pmatrix} 1 & 0 & 0 & 0 \\ 0 & 1 & 0 & 0 \\ 0 & 0 & 0 & 1 \\ 0 & 0 & 1 & 0 \end{pmatrix}$
		C^2NOT_1	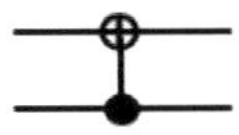	$\begin{pmatrix} 1 & 0 & 0 & 0 \\ 0 & 0 & 0 & 1 \\ 0 & 0 & 1 & 0 \\ 0 & 1 & 0 & 0 \end{pmatrix}$

This structure particularly shows how these gates operate on the quantum information being settled in the system and not properly on the physical system. S_ϕ gate includes $S_{\pi/8}$ and $S_{\pi/4}$ gates for the corresponding values of ϕ. In particular note that $(H_1 \otimes \mathbf{1}_2)_{\mathrm{B}}$ works as a translator between the computational and the Bell basis, closing the equivalence with the $\mathcal{B}$ set. The relevant aspect is the simple construction of these gates for the current Hamiltonian in this work. At the first glance, one can see the correspondence between their block structure of $\mathcal{D}$ elements and the general block in (2). In the next section we state the main concrete prescriptions for each gate.

Table 2. Alternative universal set of 2-level gates for the Bell basis.

Gate	Subspaces	Graph	Matrix form on Bell basis
$(\mathbf{1}_1 \otimes S_{\phi 2})_{\mathrm{B}}$	1	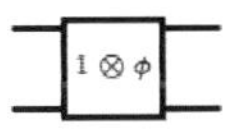	$\begin{pmatrix} e^{-i\phi} & 0 & 0 & 0 \\ 0 & e^{i\phi} & 0 & 0 \\ 0 & 0 & e^{-i\phi} & 0 \\ 0 & 0 & 0 & e^{-i\phi} \end{pmatrix}$
$(S_{\phi 1} \otimes \mathbf{1}_2)_{\mathrm{B}}$	1	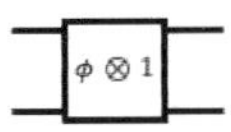	$\begin{pmatrix} e^{-i\phi} & 0 & 0 & 0 \\ 0 & e^{-i\phi} & 0 & 0 \\ 0 & 0 & e^{i\phi} & 0 \\ 0 & 0 & 0 & e^{i\phi} \end{pmatrix}$
$(\mathbf{1}_1 \otimes H_2)_{\mathrm{B}}$	1	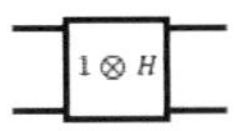	$\frac{1}{\sqrt{2}}\begin{pmatrix} 1 & 1 & 0 & 0 \\ 1 & -1 & 0 & 0 \\ 0 & 0 & 1 & 1 \\ 0 & 0 & 1 & -1 \end{pmatrix}$
$(H_1 \otimes \mathbf{1}_2)_{\mathrm{B}}$	1	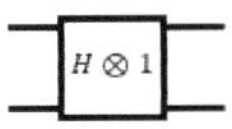	$\frac{1}{\sqrt{2}}\begin{pmatrix} 1 & 0 & 1 & 0 \\ 0 & 1 & 0 & 1 \\ 1 & 0 & -1 & 0 \\ 0 & 1 & 0 & -1 \end{pmatrix}$
$(C^1NOT_2)_{\mathrm{B}}$	2	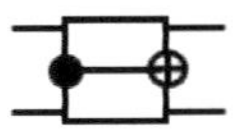	$\begin{pmatrix} 1 & 0 & 0 & 0 \\ 0 & 1 & 0 & 0 \\ 0 & 0 & 0 & 1 \\ 0 & 0 & 1 & 0 \end{pmatrix}$
$(C^2NOT_1)_{\mathrm{B}}$	2	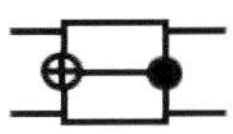	$\begin{pmatrix} 1 & 0 & 0 & 0 \\ 0 & 0 & 0 & 1 \\ 0 & 0 & 1 & 0 \\ 0 & 1 & 0 & 0 \end{pmatrix}$

The demonstration that $\mathcal{D}$ is universal is trivial, which is based on the universality of $\mathcal{B}$ [17]. Note that due any 2-level gate G can be expressed as a product of other unitary operations in

IOP Conf. Series: Journal of Physics: Conf. Series **839** (2017) 012014 doi:10.1088/1742-6596/839/1/012014

$\mathcal{B}$: $G = \Pi_{i=1}^{N} g_i, g_i \in \mathcal{B}$. Then, as $\mathcal{T} \equiv (H_1 \otimes \mathbf{1}_2)_{\mathrm{B}}$ is an operation fulfilling: a) $\mathcal{T}^\dagger = \mathcal{T}$, b) $\mathcal{T}|\beta_{ij}\rangle = |i, i \oplus j\rangle, \mathcal{T}|\beta_{i,i\oplus j}\rangle = |i,j\rangle$. Thus, clearly $G = \mathcal{T}(\Pi_{i=1}^{N} \mathcal{T} g_i \mathcal{T})\mathcal{T}$ and each $\mathcal{T} g_i \mathcal{T}$ is some gate expressed in Bell basis, a gate operating on two levels and then able to be expressed (at least asymptotically [17]) as a product of elements of $\mathcal{D}$: $\mathcal{T} g_i \mathcal{T} = \Pi_{j=1}^{M} u_j^i, u_j^i \in \mathcal{D}$. With this, $G = \mathcal{T}(\Pi_{i=1}^{N} \Pi_{j=1}^{M} u_j^i)\mathcal{T}$. The process of translation is shown in the Figure 1.

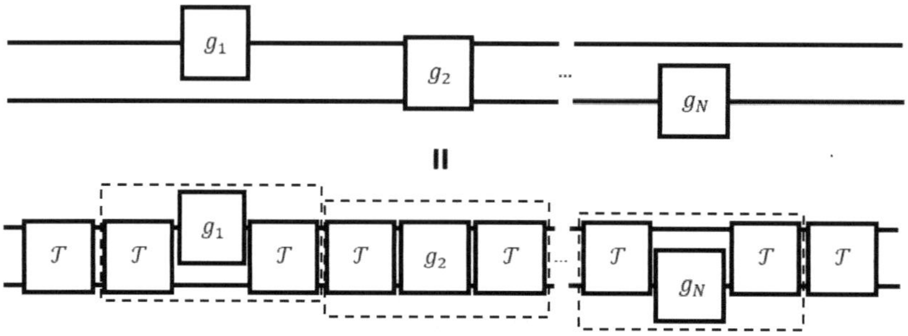

Figure 1. Quantum gate array showing the equivalence between a traditional circuit using the computational basis and another using the translator gate $\mathcal{T}$. Each block in the dotted boxes can be expressed in terms of elements in $\mathcal{D}$.

3. Heisenberg-Ising gate's realization and possible issues

All last gates are achievable for the Hamiltonian being considered and its evolution matrix reducible to the blocks (2), through the prescriptions given in the Table 3. This table contains only the immediate prescriptions.

Table 3. Basic design parameters for the universal gates in $\mathcal{U}$. There $m, m' \in \mathbf{Z}$.

Gate	$(\mathbf{1}_1 \otimes S_{\phi_2})_{\mathrm{B}}$	$(S_{\phi_1} \otimes \mathbf{1}_2)_{\mathrm{B}}$	$(\mathbf{1}_1 \otimes H_2)_{\mathrm{B}}$	$(H_1 \otimes \mathbf{1}_2)_{\mathrm{B}}$	$(C^1NOT_2)_{\mathrm{B}}$	$(C^2NOT_1)_{\mathrm{B}}$
h	1	1	1	3	1	3
$\Delta^+_{h\,\alpha}$	2π	ϕ	$\pi/2$	$\pi/2$	$\pi/4$	$\pi/4$
$\Delta^-_{h\,\alpha}$	ϕ	2π	$\pi/2$	$\pi/2$	$2m\pi$	$2m\pi$
$\Delta^-_{h\,-\alpha}$	ϕ	2π	$\pi/2$	$\pi/2$	$\pi/2 + 2m'\pi$	$\pi/2 + 2m'\pi$
$b_{h-\alpha}$	0	-	$q\beta j_{h-\alpha}$	$-q\beta j_{h-\alpha}$	-	-
$b_{h\alpha}$	0	-	$q\beta j_{h\alpha}$	$-q\beta j_{h\alpha}$	$\|b_{h\alpha}\| \to 1$	$\|b_{h\alpha}\| \to 1$
$j_{h-\alpha}$	β	-	-	-	0	0
$j_{h\alpha}$	β	-	-	-	0	0
Other	-	-	-	-	$m, m' \to \infty$	$m, m' \to \infty$

Specific prescriptions should be still written in a more complicated way for physical parameters $t, B_{h\pm\alpha}, B_{h\pm\alpha}$, nevertheless all them are possible and compatible. Some variations could be possible or necessary, but last prescriptions resume the general conditions to reproduce the universal gates set $\mathcal{D}$ on the Bell states. Due to the extension of these prescriptions they are omitted here. α corresponds always to the first block $j = 1$. Note particularly than prescriptions for $(\mathbf{1}_1 \otimes S_{\phi_2})_{\mathrm{B}}$ could be used to generate $(S_{\phi_1} \otimes \mathbf{1}_2)_{\mathrm{B}}$ too when $h = 1$ is changed by $h = 3$ and

IOP Conf. Series: Journal of Physics: Conf. Series **839** (2017) 012014 doi:10.1088/1742-6596/839/1/012014

$\beta \to -\beta$ (as for other gates). Nevertheless, we report in the Table 1 an easier implementation for $h = 1$. Condition $m \to \infty$ is optional, but it has been used in the current analysis, indicating the necessity to increase sufficiently the magnetic field on the qubits to approach the evolution matrix asymptotically to $(C^a NOT_b)_{\mathrm{B}}$.

4. Error estimation strategy based on quantum fidelity

In order to round the gates proposal, an strategy to evaluate their performance should be delivered. Thus, to set a stability degree for each element of the universal set $\mathcal{D}$, we consider the initial multipartite state (bipartite is sufficient for our purposes):

$$|\Psi_0\rangle = \bigoplus_k \sum_j \alpha_{k,j} |\psi_{k,j}\rangle \tag{4}$$

which is explicitly exhibiting the direct sum structure of the Hilbert space in the current case. Note that $|\psi_{k,j}\rangle$ is the basis being used in the representation (some arrangement of Bell states, dependent on h in our case), with each element labeled by the block k and a label j for each element of the pair. Under the effect of some gate on $SU(4)$: $U = \bigoplus_k s_{h_k} \in \mathcal{D}$, where s_{h_k} is the $U(2) = U(1) \times SU(2)$ block (2). Here, h is an additional label identifying different decompositions and arrangements of $|\psi_{k,j}\rangle$ (the direction in which the control field is applied, in our case) and k is the block number. Then, the transformed state $|\Psi_f\rangle$ under the gate U is:

$$|\Psi_f\rangle = U |\Psi_0\rangle = \bigoplus_k \sum_j \alpha_{k,j} s_{h_k} |\psi_{k,j}\rangle . \tag{5}$$

Now, we are interested in a tiny failure in the gate prescriptions, generating an alternative effect but very near from the proposed gate, $U' = \bigoplus_k s'_{h_k} = \bigoplus_k s_{h_k} + \delta s_{h_k}$, where δs_{h_k} is a tiny variation of s_{h_k} generated by the variation of some of their parameter prescriptions $\mathbf{p} = (p_1, p_2, ..., p_N)$ ($\mathbf{p} = (t, B_{1_h}, B_{2_h}, J_1, J_2, J_3)$ in our case). This gate generates on $|\Psi_0\rangle$:

$$\left|\Psi'_f\right\rangle = U' |\Psi_0\rangle = |\Psi_f\rangle + \bigoplus_k \sum_j \alpha_{k,j} \delta s_{h_k} |\psi_{k,j}\rangle . \tag{6}$$

Gate stability can be quantified by its fidelity [15]. Developing δs_{h_k} to second order:

$$\delta s_{h_k} \approx \mathcal{D} s_{h_k} + \frac{1}{2} \mathcal{D}^2 s_{h_k}, \quad \mathcal{D} = d\mathbf{p} \cdot \nabla_{\mathbf{p}}$$

and if $\rho_f = |\Psi_f\rangle \langle \Psi_f|$, $\rho'_f = \left|\Psi'_f\right\rangle \left\langle \Psi'_f\right|$, then:

$$\mathcal{F}^2 = \mathrm{Tr}(\rho_f \rho'_f) = |1 + \sum_{k,j,j'} \alpha^*_{k,j'} \alpha_{k,j} \left\langle \psi_{k,j'} | s^\dagger_{h_k} \delta s_{h_k} | \psi_{k,j} \right\rangle|^2 \tag{7}$$

Developing the last expression using (7) and considering $s_{h_k} s^\dagger_{h_k} = \mathbf{1}_k$, it gives:

$$\mathcal{F}^2 = 1 - \sum_{k,j,j'} \alpha^*_{k,j'} \alpha_{k,j} a_{k,j,j'} + |\sum_{k,j,j'} \alpha^*_{k,j'} \alpha_{k,j} b_{k,j,j'}|^2 \tag{8}$$
$$\text{with: } a_{k,j,j'} = \left\langle \psi_{k,j'} | \mathcal{D} s^\dagger_{h_k} \mathcal{D} s_{h_k} | \psi_{k,j} \right\rangle, b_{k,j,j'} = \left\langle \psi_{k,j'} | s^\dagger_{h_k} \mathcal{D} s_{h_k} | \psi_{k,j} \right\rangle$$

last ones are simply the entries of $\mathcal{D} s^\dagger_{h_k} \mathcal{D} s_{h_k}$ and $s^\dagger_{h_k} \mathcal{D} s_{h_k}$ in each block. Outstandingly, $\mathcal{F}^2$ depends quadratically from $d\mathbf{p}$, a property due to s_{h_j} is unitary. This expression set a procedure to evaluate the performance of any element in $\mathcal{D}$ for any state $|\Psi_0\rangle$ being processed around of their specific prescriptions given in the Table 3. This task implies evaluate exhaustively $\mathcal{F}^2$ for each gate in $\mathcal{D}$ and any state $|\Psi_0\rangle$ able to be transferred in the Hilbert space $\mathcal{H}^2$.

5. Conclusions

Exact control for time independent magnetic fields has been used here to reduce the evolution blocks to pertinent blocks to reproduce the gates $\mathcal{D}$ alternative to $\mathcal{B}$, but other kinds of control (optimal, non-resonant, etc.) could be introduced [10,18], whose control schemes are well known for the $SU(2)$ dynamics. More achievable forms for magnetic fields pulses are in order to reproduce the proposed gates, particularly avoiding resonance effects.

Nevertheless gate fidelity has been theoretically quantified, additional concrete research is necessary to analyze this fidelity for the current constructions when the prescriptions are slightly modified due to the main uncontrollable factors. In particular, as time and strength of magnetic fields are currently well controlled [19], possibly the most sensible factors in the model are the strengths J_i for the non-local interactions in the Heisenberg-Ising model [20]. A complete analysis based on the formula (8) for each gate is currently in progress.

This proposal is based on Bell states instead of single qubits states, which are technologically a challenge in terms of experimental stability and coherence. Despite, this states appears as better candidates to set a computational grammar on magnetic systems easing quantum control and quantum gate engineering. In addition, the approach is clearly settled on the quantum information manipulation more than in the physical systems where it lays.

Acknowledgments

The support from Tecnológico de Monterrey to develop this research is acknowledged.

References

[1] Deutsch D 1985 *Proc. R. Soc. London*, Ser. A **400** 97
[2] Steane A M 1996 *Phys. Rev. Lett.* **77** 793
[3] Schende V, Bullock S and Markov I 2005 *Proc. Asia and South Pacific Design Automation Conference* 272
[4] Li C K, Jones R and Yin X 2013 *Int. J. Quant. Inf.* **11** 1350015
[5] Delgado F 2014 *Factorization of unitary matrices induced by 3D anisotropic Ising interaction* (quant-ph/1410.5891)
[6] Paige C and Wei M 1994 History and Generality of the CS Decomposition *Linear Algebra and Appl.* **208** 303
[7] Berman G P, Doolen G D, López G V and Tsifrinovich V I 1998 *Generalized Quantum Control-Not Gate in Two-Spin Ising System* (quant-ph/9802013v1)
[8] Wang X 2001 *Phys. Lett.* A **281** 101
[9] D'Alessandro D and Dahleh M 2001 *IEEE Transactions on Automatic Control* **46** 866
[10] Dalessandro D 2007 *Introduction to Quantum Control and Dynamics* (Boca Raton: Chapman Hall Applied Mathematics Nonlinear Science)
[11] Kamta G L and Starace A F 2002 *Phys. Rev. Lett.* **88** 107901
[12] Sun Y, Chen Y and Chen H 2003 *Phys. Rev.* A **68** 044301
[13] Zhou L, Song H S, Guo CY Q and Li C 2003 *Phys. Rev.* A **68** 024301
[14] Delgado F 2015 *Int. J. Quant. Info.* **13** 1550055
[15] Delgado F 2016 *J. Phys.: Conf. Ser.* **648** 012024
[16] Reck M, Zeilinger A, Bernstein H J and Bertani P 1994 *Phys. Rev. Lett.* **73** 58
[17] Boykin P O, Mor T, Pulver M, Roychowdhury V and Vatan F 1999 *On universal and fault tolerant quantum computing* (quant-ph/9906054)
[18] Boscain U and Mason P 2006 *J. Math. Phys.* **47** 062101
[19] Sekatski P, Skotiniotis M, Kolodynski J and Dur W 2016 *Quantum metrology with full and fast quantum control* (quant-ph/1410.5515)
[20] Delgado F 2010 *Rev. Mex. Fis.* **56**(1) 30

IOP Conf. Series: Journal of Physics: Conf. Series **839** (2017) 012015 doi:10.1088/1742-6596/839/1/012015

Disentangling the Time-Evolution Operator of a Single Qubit

M Enríquez[1,2] **and S Cruz y Cruz**[1]
[1]Instituto Politécnico Nacional, UPIITA, Av. Instituto Politécnico Nacional 2580 CP 07340 Ciudad de México, Mexico
[2]Departamento de Física, Cinvestav, AP 14-740, 07000 Ciudad de México, Mexico

E-mail: `menriquezf@fis.cinvestav.mx`, `sgcruzc@ipn.mx`

Abstract. We show that the time-evolution equation for a two-level system can be turned into a system of non-linear differential equations when it is written in the disentangled form. The solutions are determined by solving a parametric-oscillator-like equation with a certain time-dependent frequency. The formalism is used to deal with the problem of a two-level atom interacting with a circularly polarized field. The possibility of generating exactly solvable Hamiltonians is briefly discussed.

1. Introduction

The search of exact solutions to the time-dependent Schrödinger equation represents a long-standing problem in quantum mechanics. Yet, the state of the system at any time $|\psi(t)\rangle$ fulfills the dynamical law

$$i\frac{d}{dt}|\psi(t)\rangle = H(t)|\psi(t)\rangle, \tag{1}$$

meaning that the evolution process is encoded in the family of unitary operators $U(t,t_0)$ defining the dynamics of the system from an initial moment t_0 to an arbitrary time t in the form $|\psi(t)\rangle = U(t,t_0)|\psi(t_0)\rangle$. Then, the evolution operator satisfies the equation

$$i\frac{dU(t,t_0)}{dt} = H(t)\cdot U(t,t_0), \quad U(t_0,t_0) = \mathbb{I}, \tag{2}$$

where $\mathbb{I}$ stands for the identity operator. The simplest case arises when the Hamiltonian H does not depend explicitly on time. In such a case the solution to equation (2) is readily written: $U(t,t_0) = \exp[-iH(t-t_0)]$. In general, if $H(t)$ does not commute with itself at different times one has to sum up the *continuous* Baker-Campell-Hausdorff exponent [1], which is not a trivial problem. Therefore, it is essential the construction of different technics to solve the evolution equation (2).

The main goal of the present work is to develop a framework to approach the problem of the time-evolution of a two-level system. For this purpose it results convenient to write the Hamiltonian in terms of the Hubbard operators. Such set of operators is suitable to deal with the algebraic properties of the involved observables and represents an appropriate way to generalize our results to the case of multipartite systems [2]. The corresponding time-evolution operator is then written in the so-called disentangled form and we show that the operator equation (2) is turned into a set of three non-linear differential equations. This way of approaching the problem is related to the well-known Wei-Norman theorem [3], which, in particular, has been applied to the case of the Lie groups $SU(2)$ [4] and $SU(1,1)$ [5]. In this sense, we discuss the equivalence of our results with those previously reported.

This contribution is organized as follows. In Section 2 some useful preliminaries are presented and the main results are given. This approach is applied to the case of one qubit interacting with a circularly polarized field in Section 3. Finally, in Section 4 we give some conclusions and discuss the possibility of obtaining exactly solvable time-dependent Hamiltonians for one qubit.

2. Two-level system driven by a time-dependent field

The Hubbard operators $X^{i,j}$ (also called X-operators) satisfy the following properties:

(i) $X^{i,j}X^{k,m} = \delta_{jk}X^{i,m}$ (multiplication rule)

(ii) $\sum_k X^{k,k} = \mathbb{I}$ (completeness)

(iii) $(X^{i,j})^\dagger = X^{j,i}$ (non-hermiticity)

In the case of a qubit with eigenstates $|p\rangle$ and $|q\rangle$ and energies ϵ_p and ϵ_q, the Hubbard operator of order 2 is defined as

$$X_2^{k,\ell} := |k\rangle\langle\ell|, \quad k,\ell = p,q. \tag{3}$$

It is a simple matter to verify that the so-defined operators satisfy the properties (i)-(iii). Moreover, from the former definition it is straightforward to find the action of the X-operators on the basis vectors reads:

$$X_2^{k,\ell}|n\rangle = \delta_{\ell,n}|k\rangle, \quad k,\ell,n = p,q, \tag{4}$$

where $\delta_{i,j}$ stands for the Kronecker delta.

Now consider a qubit (*e.g.*, a two-level atom) in a time-dependent driving field. The corresponding Hamiltonian in the Hubbard representation has the form

$$H_2(t) = \epsilon_p X_2^{p,p} + \epsilon_q X_2^{q,q} + V(t)X_2^{p,q} + \overline{V}(t)X_2^{q,p}, \quad \epsilon_p, \epsilon_q \in \mathbb{R}, \tag{5}$$

where V is a complex-valued function and it will be referred to as the driving field. The over bar stands for complex conjugation. Since the energies ϵ_p, ϵ_q are real it is easy to verify that the Hamiltonian (5) is indeed Hermitian. Additionally, if the zero level energy is fixed at the middle of both levels, that is to say $\frac{1}{2}(\epsilon_p + \epsilon_q) = 0$, the Hamiltonian (5) can be written as

$$H_2(t) = \frac{\Delta}{2}\left(X_2^{p,p} - X_2^{q,q}\right) + V(t)X_2^{p,q} + \overline{V}(t)X_2^{q,p},$$

where $\Delta = \epsilon_p - \epsilon_q$ corresponds to the atomic transition frequency and the terms

$$H_0 = \frac{\Delta}{2}\left(X_2^{p,p} - X_2^{q,q}\right), \quad H_i = V(t)X_2^{p,q} + \overline{V}(t)X_2^{q,p}$$

are, respectively, the (free) atomic Hamiltonian and the interaction term. Remark that the operators

$$J_0 = \frac{1}{2}\left(X_2^{p,p} - X_2^{q,q}\right), \quad J_+ = X_2^{p,q} \quad \text{and} \quad J_- = X_2^{q,p} \tag{6}$$

constitute a representation of the $su(2)$ algebra generators [2].

IOP Conf. Series: Journal of Physics: Conf. Series **839** (2017) 012015 doi:10.1088/1742-6596/839/1/012015

The main problem of this work is to find the time-evolution operator $U(t) = U(t,0)$ (we have fixed $t_0 = 0$ for the sake of simplicity) fulfilling the initial value problem (2). We assume that this operator can be written in the *disentangled* normal form

$$U(t) = \exp[\alpha(t)X_2^{p,q}]\exp[\Delta f(t)J_0]\exp[\beta(t)X_2^{q,p}], \tag{7}$$

where α, β and f are complex functions to be determined. Inserting (7) and (5) into (2) we get

$$\begin{aligned}\frac{dU}{dt} &= \alpha' X_2^{p,q} e^{\alpha X_2^{p,q}} e^{\Delta f J_0} e^{\beta X_2^{p,q}} \\ &+ \Delta f' e^{\alpha X_2^{p,q}} J_0 e^{\Delta f J_0} e^{\beta X_2^{p,q}} + \beta' e^{\alpha X_2^{p,q}} e^{\Delta f J_0} X_2^{q,p} e^{\beta X_2^{p,q}} = -iH_2 U.\end{aligned} \tag{8}$$

Here the explicit time dependence of the functions α, f and β has been omitted for shortness and the prime stands for time derivation. The right multiplication of this expression by $U^\dagger$ produces

$$-iH_2 = \alpha' X_2^{p,q} + \Delta f' e^{\alpha X_2^{p,q}} J_0 e^{-\alpha X_2^{p,q}} + \beta' e^{\alpha X_2^{p,q}} e^{\Delta f J_0} X_2^{q,p} e^{-\Delta f J_0} e^{-\alpha X_2^{p,q}}. \tag{9}$$

The similarity transformations in the right-hand side of (9) can be accomplished by using the well-known Baker-Campbell-Haussdorff formula [1]

$$e^{\lambda A} B e^{-\lambda A} = \sum_{n=0}^{\infty} \frac{\lambda^n}{n!} \{A^n, B\}, \tag{10}$$

where the bracket operation $\{\ ,\ \}$ is defined by

$$\{\mathbb{I}, B\} := B, \quad \{A^{n+1}, B\} = [A, \{A^n, B\}], \quad \{A^n, B\} := \underbrace{[A,[A,\cdots[A}_{\text{n-times}}, B\,]]],$$

leading to

$$\begin{aligned}-i\Delta J_0 - iVX_2^{p,q} - i\overline{V}X_2^{q,p} &= \alpha' X_2^{p,q} + \Delta f'\left(J_0 - \alpha X_2^{p,q}\right) \\ &+ \beta' e^{-\Delta f}\left(X_2^{q,p} + 2\alpha J_0 - \alpha^2 X_2^{p,q}\right).\end{aligned}$$

Thus, the linear independence of the X-operators allows to reduce the problem to the following set of non-linear equations

$$\begin{aligned}\alpha' - \alpha\Delta f' - \alpha^2\beta' e^{-\Delta f} &= -iV, \\ \Delta f' + 2\alpha\beta' e^{-\Delta f} &= -i\Delta, \\ \beta' e^{-\Delta f} &= -i\overline{V},\end{aligned} \tag{11}$$

with the initial conditions $\alpha(0) = \beta(0) = f(0) = 0$ in order to fulfill the requirement that $U(0) = \mathbb{I}$. Substituting the second and third equations into the first one produces the following complex Ricatti equation for α

$$\alpha' + i\Delta\alpha - i\overline{V}\alpha^2 + iV = 0. \tag{12}$$

The introduction of the new variable η such that $\alpha = i(\ln\eta)'/\overline{V}$ turns (12) into the second-order differential equation

$$\eta'' - \left(-i\Delta + \frac{d}{dt}\ln\overline{V}\right)\eta' + |V|^2\eta = 0. \tag{13}$$

In view of the forthcoming calculations, it is convenient to define the function $R(t) = e^{-i\Delta t}\overline{V}(t)$. Note that this function differs from the complex conjugate of V only in a phase factor, therefore, we will refer to both V and R, indistinctly, as the driving field. In these terms equation (13) becomes

$$\eta'' - \left(\frac{d}{dt}\ln R\right)\eta' + |R|^2\eta = 0. \tag{14}$$

Finally, the change $\eta(t) = Q(t)\varphi(t)$, with $Q^2 = R$, turns this equation into the parametric oscillator equation

$$\varphi''(t) + \Omega^2(t)\varphi(t) = 0, \tag{15}$$

where the time-dependent frequency, given by

$$\Omega^2(t) = -\frac{1}{4}\left[\frac{d}{dt}\ln R(t)\right]^2 + \frac{1}{2}\frac{d^2}{dt^2}\ln R(t) + |R(t)|^2, \tag{16}$$

is, in general, a complex-valued function.

Now the solution to the equation (12) can be readily written by undoing the previous variable changes. Indeed, in terms of φ the function α has the form

$$\alpha(t) = i\frac{e^{-i\Delta t}}{R(t)}\left[\frac{\varphi'(t)}{\varphi(t)} + \frac{1}{2}\frac{R'(t)}{R(t)}\right]. \tag{17}$$

The functions f and β, in turn, are easily obtained from (11) by quadratures

$$\Delta f(t) = -2\ln\left(\frac{\varphi(t)}{\varphi_0}\right) - \ln\left(\frac{R(t)}{R_0}\right) - i\Delta t, \quad \varphi_0 = \varphi(0), \quad R_0 = R(0), \tag{18}$$

and

$$\beta(t) = -iR_0\varphi_0^2\int_0^t \frac{ds}{\varphi^2(s)}. \tag{19}$$

Remark that $f(t)$ and $\beta(t)$ were explicitly constructed to fulfill the initial conditions $f(0) = \beta(0) = 0$. Moreover, the condition $\alpha(0) = 0$ fixes the initial conditions of φ and its first derivative through

$$\lim_{t\to 0}\frac{1}{R(t)}\left[\frac{\varphi'(t)}{\varphi(t)} + \frac{1}{2}\frac{R'(t)}{R(t)}\right] = 0. \tag{20}$$

On the other hand, from the Hubbard operators properties one can see that $(X_2^{p,q})^2 = (X_2^{q,p})^2 = 0$, $(X_2^{p,p})^2 = X_2^{p,p}$ and $(X_2^{q,q})^2 = X_2^{q,q}$. Then, using the completeness property (iii), we may write

$$\begin{aligned}
&\exp[\alpha(t)X_2^{p,q}] = 1 + \alpha(t)X_2^{p,q}, \quad \exp[\beta(t)X_2^{q,p}] = 1 + \beta(t)X_2^{q,p} \\
&\exp\left[\tfrac{\Delta f(t)}{2}X_2^{p,p}\right] = X_2^{q,q} + e^{\Delta f(t)/2}X_2^{p,p}, \quad \exp\left[\tfrac{-\Delta f(t)}{2}X_2^{q,q}\right] = X_2^{p,p} + e^{-\Delta f(t)/2}X_2^{q,q} \\
&\exp\left[\Delta f(t)J_0\right] = e^{\Delta f(t)/2}X_2^{p,p} + e^{-\Delta f(t)/2}X_2^{q,q}.
\end{aligned} \tag{21}$$

The time-evolution operator of the system follows immediatly:

$$\begin{aligned}
U(t) = &\left[e^{\Delta f(t)/2} + e^{-\Delta f(t)/2}\alpha(t)\beta(t)\right]X_2^{p,p} + e^{-\Delta f(t)/2}X_2^{q,q} \\
&+ e^{-\Delta f(t)/2}\left[\alpha(t)X_2^{p,q} + \beta(t)X_2^{q,p}\right].
\end{aligned} \tag{22}$$

IOP Conf. Series: Journal of Physics: Conf. Series **839** (2017) 012015 doi:10.1088/1742-6596/839/1/012015

Summarizing, we have reduced the problem of constructing the evolution operator for a driven two-level system, to the problem of finding out the solutions of the parametric oscillator (15) with the appropriate initial conditions. Equivalent results can be found, *e.g.*, in the works by Dattoli *et al.* [4,5] for a different ordering in the factorization of the time-evolution operator. It is important to point out that the introduction of Hubbard operators [2] in our approach allows the possibility of extending the method to the case of multipartite systems [6].

3. A simple example

In this section we deal with a physical example to illustrate the convenience of our approach. To begin with, observe that a simple inspection to equation (16) allows us to conclude that the driving field of the form $R(t) = \overline{g}e^{i\delta t}$, with g and δ constants and $\delta \in \mathbb{R}$, produce $\Omega =$ constant. In this case $V(t) = ge^{-i\omega t}$, with $\omega = \delta + \Delta$. The corresponding interaction Hamiltonian

$$H_i = ge^{-i\omega t}X_2^{p,q} + \overline{g}e^{i\omega t}X_2^{q,p}, \tag{23}$$

can be identified as the interaction term of a two-level atom driven by a classical circularly polarized electromagnetic field, of frequency ω, in the dipole approximation: $H_i = -\mathbf{p}\cdot\mathbf{E}(t)$ with $\mathbf{p} = e\mathbf{r}$ the electric dipole operator and $\mathbf{E}(t) = (\mathcal{E}_0\cos\omega t, \mathcal{E}_0\sin\omega t, 0)$ the electric field amplitude, provided that the coupling constant g takes the value (see, *e.g.*, Klimov and Chumakov [7], pp. 26)

$$g = -\frac{e\mathcal{E}_0}{2}\langle p|x+iy|q\rangle.$$

We now follow the formalism of the previous section to construct the time-evolution operator in the form (7). The functions α, f and β are expressed in terms of the solution of the parametric oscillator equation (15) with the frequency given by

$$\Omega^2(t) = -\frac{1}{4}\left[\frac{d}{dt}\ln R(t)\right]^2 + \frac{1}{2}\frac{d^2}{dt^2}\ln R(t) + |R(t)|^2 = \frac{\delta^2}{4} + |g|^2 \equiv \Omega_0^2, \tag{24}$$

where we have used the fact that

$$R(0) = R_0 = \overline{g}, \qquad \frac{d}{dt}\ln R(t) = i\delta = i(\omega - \Delta). \tag{25}$$

Here $\delta = \omega - \Delta$ is the *detuning* given by the difference between the atomic and the driving field frequencies. In the same sense, the quantity Ω_0 is nothing but the *Rabi frequency* of the atomic driven system. Accordingly, the parametric oscillator equation becomes

$$\varphi''(t) + \Omega_0^2\,\varphi(t) = 0, \tag{26}$$

whose general solution is given by $\varphi(t) = c_1\cos(\Omega_0 t) + c_2\sin(\Omega_0 t)$, with c_1 and c_2 two complex constants related to each other through (20) as $c_2 = -i\delta c_1/2\Omega_0$. By setting, without loss of generality $\varphi(0) = c_1 = 1$ we get

$$\varphi(t) = \cos(\Omega_0 t) - \frac{i\delta}{2\Omega_0}\sin(\Omega_0 t), \tag{27}$$

and the solutions of the non-linear system are then computed from (17-19)

$$\begin{aligned}
\alpha(t) &= \frac{-2ige^{-i\omega t}\sin(\Omega_0 t)}{2\Omega_0\cos(\Omega_0 t) - i\delta\sin(\Omega_0 t)},\\
\Delta f(t) &= -2\ln\left[\cos(\Omega t) - \frac{i\delta}{2\Omega_0}\sin(\Omega_0 t)\right] - i\omega t,\\
\beta(t) &= \frac{-2i\overline{g}\sin(\Omega_0 t)}{2\Omega_0\cos(\Omega_0 t) - i\delta\sin(\Omega_0 t)}.
\end{aligned} \tag{28}$$

A similar approach has been done by Klimov and Chumakov [7] for a system consistent on A qubits interacting with a circularly polarized field. Our results are equivalent to theirs for the lowest representation, or, equivalently, for $A = 1$, as the dynamical group in both cases is $SU(2)$. Since in that work the authors focus in the case of exact resonance, their results are retrieved by setting $\delta = 0$ in our expressions (28).

Let us assume that the system is, initially, in the upper-level energy state, *i.e.*, $|\psi(0)\rangle = |p\rangle$. The state of the system at any time reads

$$|\psi(t)\rangle = U(t)|p\rangle = [e^{\Delta f(t)/2} + e^{-\Delta f(t)/2}\alpha(t)\beta(t)]|p\rangle + e^{-\Delta f(t)/2}\beta(t)|q\rangle. \tag{29}$$

After some algebra we retrieve the known expressions for the time-evolution

$$|\psi(t)\rangle = e^{-i\omega t/2}\left[\cos(\Omega_0 t) + \frac{i\delta}{2\Omega_0}\sin(\Omega_0 t)\right]|p\rangle - \frac{i\,\overline{g}}{\Omega_0}e^{i\omega t/2}\sin(\Omega_0 t)|q\rangle. \tag{30}$$

The atomic population inversion $\mathcal{P}$ is also computed to give

$$\mathcal{P}(t) = |c_p|^2 - |c_q|^2 = \frac{|g|^2}{\Omega_0^2}\cos(2\Omega_0 t) + \frac{\delta^2}{4\Omega_0^2}. \tag{31}$$

We stress that this problem is exactly solvable since the general solution to the parametric oscillator equation (26) is well-known. However, in general, the problem is not so simple to deal with and there are a limited number of cases for which exact solutions to (15) can be explicitly constructed (see for instance [4, 5]).

4. Conclusions and perspectives

The evolution operator of one qubit interacting with a time-dependent driving field has been constructed, in the disentangled form, by using the Hubbard representation. First, the evolution equation is reduced to a set of three coupled nonlinear equations whose solution can be expressed in terms of the solution of a single complex Riccati-type one. Next, with some appropriate changes of variable, this equation is transformed into a classical parametric oscillator equation with a time-dependent frequency defined by the form of the driving field. As a particular case, we have determined the time-evolution operator for a two-level atom interacting with a circularly polarized field. The well-known expressions for the time evolution of the state of the system and the population inversion were also obtained. Yet, it is widely recognized that there is only a limited number cases for which the solution of the general problem can be explicitly constructed. In this sense, an alternative to generate exactly solvable time-dependent Hamiltonians is constituted by the inverse technic: assuming that the solutions of the equation (15) for a given $\Omega(t)$ are known, the problem then is to find the driving field $R(t)$ which yields such time-dependent frequency. The results on this matter will be reported elsewhere [8].

Acknowledgments

The financial support of CONACyT through ME's postdoctoral scholarship (233036) and from the Instituto Politécnico Nacional through project SIP20170233 is acknowledged.

References

[1] Mielnik B and Plebanski J 1970 *Ann. Inst. H. Poincare* A **12** 215
[2] Enríquez M and Rosas-Ortiz O 2013 *Ann. Phys.* **339** 218
[3] Wei J and Norman E 1963 *J. Math. Phys.* **4** 575
[4] Dattoli G, Gallardo J and Torre A 1986 *J. Math. Phys.* **27** 772
[5] Dattoli G and Torre A 1987 *J. Math. Phys.* **28** 618

[6] Enríquez M 2016 *J. Phys.: Conf. Ser.* **698** 012021

[7] Klimov A B and Chumakov S M 2009 *A Group-Theoretical Approach to Quantum Optics* (Weinheim: Wiley-VCH) pp 33-35

[8] Enríquez M and Cruz y Cruz S 2017 Exactly solvable one-qubit Hamiltonians generated via nonlinear equations. *In preparation*

IOP Conf. Series: Journal of Physics: Conf. Series **839** (2017) 012016 doi:10.1088/1742-6596/839/1/012016

Stability of Quantum Loops and Exchange Operations in the Construction of Quantum Computation Gates

D Bermúdez and F Delgado

Departamento de Física y Matemáticas, Escuela Nacional de Ingeniería y Ciencias, Tecnológico de Monterrey, Campus Estado de México, Atizapán, Estado de México, CP. 52926, México.

E-mail: fdelgado@itesm.mx

Abstract. Quantum information and quantum computation is a rapidly emergent field where quantum systems and their applications play a central role. In the gate version of quantum computation, the construction of universal quantum gates to manipulate quantum information is currently an intensive arena for quantum engineering. Specific properties of systems should be able to reproduce such idealized gates imitating the classically inspired computational gates. Recently, for magnetic systems driven by the bipartite Heisenberg-Ising model a universal set of gates has been realized, an alternative easy design for the Boykin set but using the Bell states as grammar. Exact control can be then used to construct specific prescriptions to achieve those gates. Physical parameters impose a challenge in the gate control. This work analyzes, based on the worst case quantum fidelity, the associated instability for the proposed set of gates. An strong performance is found in those gates for the most of quantum states involved.

1. Introduction

Quantum engineering for quantum information and quantum computation applications are currently reaching their technology realizations. In the quantum gate implementation of these theories, the construction of efficient quantum algorithms is being developed to implement and to solve high complexity problems. There, quantum engineering is requiring reproduce, on specific physical systems, the theoretical quantum gates idealized. Control and a performance quantification of those realizations are fundamental for a stable functioning of those elements.

This work presents a stability analysis for a universal set of gates based on the non-local grammar of Bell states for magnetic systems presented in [1]. The system assumes:

$$H_h = \sum_{k=1}^{3} J_k \sigma_{1_k} \sigma_{2_k} - B_{1_h} \sigma_{1_h} - B_{2_h} \sigma_{2_h} \quad (1)$$

a near approach to magnetic systems considered as potential implementation. It exhibits a $SU(2)$ block decomposition for its $SU(4)$ dynamics when it is expressed on the Bell states basis in $\mathcal{H}^2$ [2]. It gives a convenient structure to set alternative quantum gates in those theories. In spite time and external magnetic fields involved are highly controllable, the analysis of stability should be centered on the J_i strength values, the more sensible error source here [3, 4]. Based on the universal set of gates proposed in [1], this work presents a stability analysis on the worst fidelity transmission on the entire $\mathcal{H}^2$ states involved. The second section presents a brief review of the evolution blocks involved in the gates analyzed. The third section depicts a convenient projective quantum space for $\mathcal{H}^2$ to shown the performance outcomes. The fourth section presents the corresponding analysis for each gate based on the fidelity presented in [1] to evaluate the performance. The last section states the conclusions.

2. Heisenberg-Ising universal set of gates

Current analysis considers the set of universal gates $\mathcal{D} = \{(\mathbf{1}_1 \otimes S_{\pi/8_2})_{\mathrm{B}}, (\mathbf{1}_1 \otimes S_{\pi/4_2})_{\mathrm{B}}, (S_{\pi/8_1} \otimes \mathbf{1}_2)_{\mathrm{B}}, (S_{\pi/4_1} \otimes \mathbf{1}_2)_{\mathrm{B}}, (\mathbf{1}_1 \otimes H_2)_{\mathrm{B}}, (H_1 \otimes \mathbf{1}_2)_{\mathrm{B}}, (C^1NOT_2)_{\mathrm{B}}, (C^2NOT_1)_{\mathrm{B}}\}$ as it is presented in [1]. Those gates are equivalent to the Boykin set [5], but expressed for the Bell basis instead of the traditional computational basis. All them are reducible from the block form:

$$s_{h_j} = e^{i\Delta_{h\alpha}^+} \begin{pmatrix} e_{h\alpha}^{\beta*} & -qi^h d_{h\alpha} \\ qi^{*h} d_{h\alpha} & e_{h\alpha}^{\beta} \end{pmatrix} \tag{2}$$

as part of the evolution operator $U = \bigoplus_{j=1}^2 s_{h_j}$, where the parameters $d_{h\alpha}, b_{h\alpha}$ are extensively presented in [2] in terms of the physical parameters J_i, B_{k_h} and t. Note that the most of prescriptions are based on the settlement of Evolution Loops or Exchange operations on (2) [6]. In the current analysis, a deviation dJ_i from the original prescriptions for the Heisenberg-Ising strengths is considered as error source.

3. Hilbert space $\mathcal{H}^2$ on S^7 and projection on S^3 as representation space

To plot the gate stability analysis, we will construct an adequate projective space as follows. By using the generalized spherical coordinates on S^3 (S^n is the $n-$dimensional sphere embed in $(n+1)-$dimensional euclidean space), a general physical state in the current problem on $\mathcal{H}^2$ can be written as:

$$|\Psi\rangle = \sin\alpha \sin\beta \cos\gamma\,|00\rangle + e^{i\iota} \sin\alpha \sin\beta \sin\gamma\,|01\rangle + e^{i\kappa} \sin\alpha \cos\beta\,|11\rangle + e^{i\lambda} \cos\alpha\,|10\rangle \tag{3}$$

with: $\alpha, \beta \in [0, \pi], \iota, \kappa, \lambda \in [0, \pi), \gamma \in [0, 2\pi)$. These states lay on a S^7 sphere, but clearly they do not fill it because we restrict the $|00\rangle$ component to be real. Thus, the space been represented by (3) is isomorphic to S^6. This space is a two-folded representation of all physical states (with the proper rearrangement of α, β and γ) because a global sign can be still removed. The space could be projected on S^3 in several ways to be partially plotted. One of them, $P : |\Psi\rangle \to |\Psi\rangle\,|_{\iota,\kappa,\lambda=0}$, lets understand $S^6 = S^3 \times (H^1)^{\times 3}$ (H^n is a hemisphere of S^n), where H^1 is the fiber bundle corresponding to each phase in (3). This selection sets a S^3 subsphere on S^6 (similarly as the Bloch sphere S^2 for a pure qubit can be projected on its equatorial ring S^1). On this space we can additionally restrict $\gamma \in [0, \pi)$ to avoid the two-folding, settling the component of $|01\rangle$ as positive (signs in other components become still independent). This space will be called the representation space, $\mathcal{R}$ and its motivation is represent the result of the analysis on it. There, Bell states can be identified in $\mathcal{R}$ as precise points (Figure 1a). An issue of $\mathcal{R}$ should be remarked. The concurrence $\mathcal{C}$ of (3) is:

$$\mathcal{C}^2 = 4\sin^2\alpha \sin^2\beta |\cos\alpha \cos\gamma - e^{i(\iota+\kappa-\lambda)} \sin\alpha \sin\gamma \cos\beta|^2. \tag{4}$$

If $\mathcal{C}_{\mathcal{R}}^2 \equiv \mathcal{C}^2|_{\iota,\kappa,\lambda=0}$, then it is possible show $1 \geq \mathcal{C}^2 - \mathcal{C}_{\mathcal{R}}^2 = 8\sin^3\alpha \cos\alpha \sin^2\beta \cos\beta \sin\gamma \cos\gamma(1 - \cos(\iota + \kappa - \lambda)) \geq -1$, then partial entanglement is not injective under the projection. Only some separable states clearly are, those projected on the lateral edges of $\mathcal{R}$, with $\alpha, \beta = 0, \pi$.

The Figure 1a exhibits level surfaces for the concurrence, but they are applicable only if $\iota + \kappa - \lambda = 2n\pi, n \in \mathbf{Z}$. Inner surfaces are approaching to the maximal entanglement line, $\mathcal{C} = 1$ where Bell states lie in a twisted helix, repeating it vertically. Another inner surface, $\cos\alpha\cos\gamma - \sin\alpha\sin\gamma\cos\beta = 0$ (separating the two tube-like structures), contains the separable states when $\iota + \kappa - \lambda = 2n\pi, n \in \mathbf{Z}$, but other projected states laying there for $\iota + \kappa - \lambda \neq 2n\pi, n \in \mathbf{Z}$ are not separable. Then, the stability analysis could be represented in $\mathcal{R}$, but there is not an identification with $\mathcal{C}$ there, so this property should be analyzed separately (an alternative projection for S^7 can be achieved in terms of the complex projective map in [7], identifying submanifolds of constant entanglement). Finally, the state (3) should be still expressed in the Bell basis for the analysis.

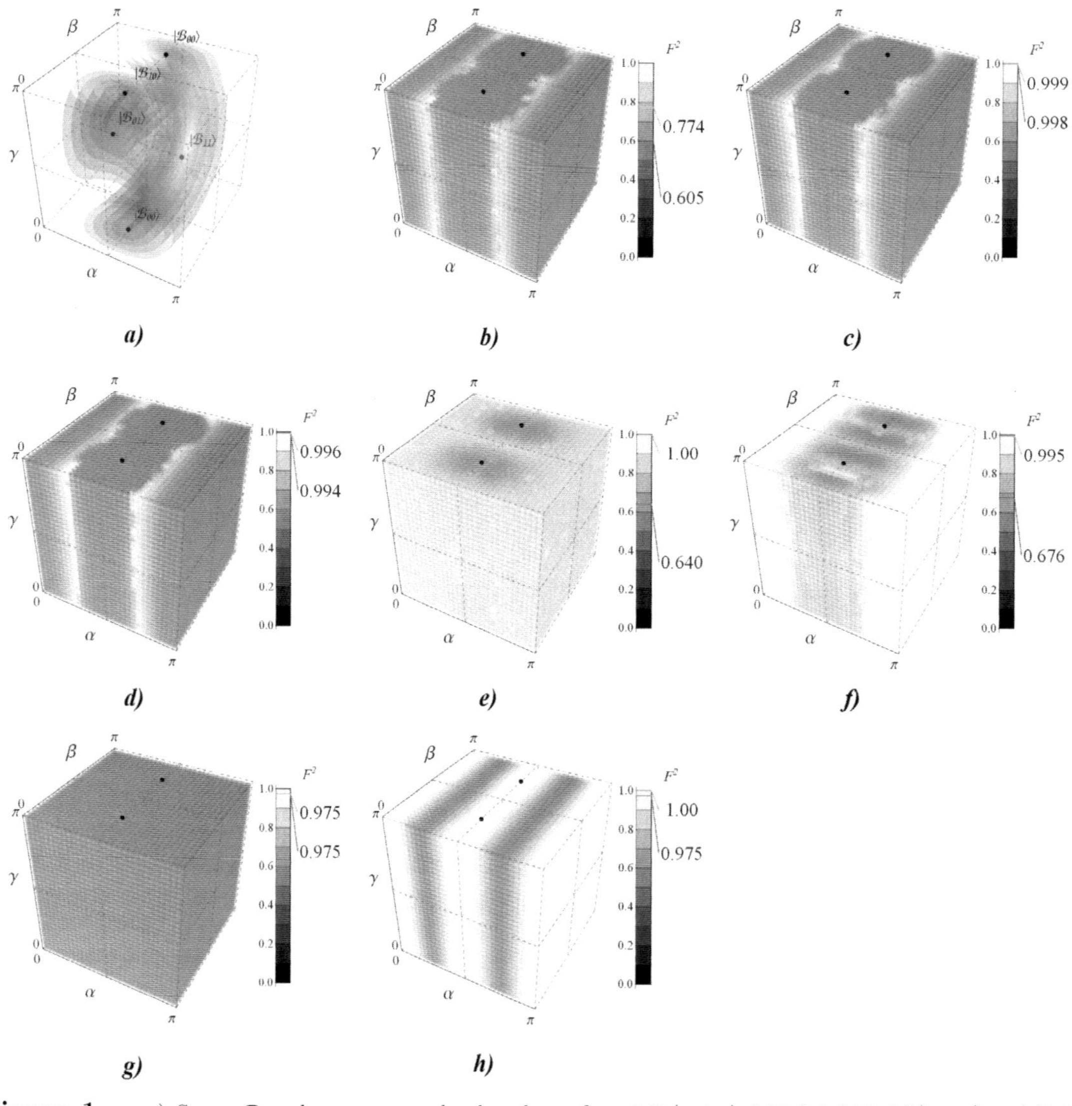

Figure 1. a) Space $\mathcal{R}$ and concurrence level-surfaces $\mathcal{C} = 0.01(\text{outer}), 0.25, 0.5, 0.75, 0.9(\text{inner})$ and Bell states: 3D-density plots for $\mathcal{F}^2$ and their extremal values for b) $(\mathbf{1}_1 \otimes S_{\phi_2})_{\rm B}$, c) $(S_{\pi/8_1} \otimes \mathbf{1}_2)_{\rm B}$, d) $(S_{\pi/4_1} \otimes \mathbf{1}_2)_{\rm B}$, e) $(\mathbf{1}_1 \otimes H_2)_{\rm B}$, f) $(H_1 \otimes \mathbf{1}_2)_{\rm B}$, g) $(C^1NOT_2)_{\rm B}$ and h) $(C^2NOT_1)_{\rm B}$.

4. Error estimation based on quantum fidelity for the elements in $\mathcal{D}$

The stability of each gate can be quantified through the quantum fidelity for each state being processed. Fidelity can be analyzed for each block s_{h_j} in $U(t)$ under the prescriptions of each gate following the development in [1]. If $\left|\Psi_{f_j}\right\rangle$ and $\left|\Psi'_{f_j}\right\rangle$ are the desired and the real states obtained with s_{h_j} and some little variation $s_{h_j}+\delta s_{h_j}$ respectively, departing from any original state $\left|\Psi_{oj}\right\rangle$ (each one in the subspace j of $\mathcal{H}^2$), then (where $s'_{hk} \approx s_{hk} + \mathcal{D}s_{hk} + \frac{1}{2}\mathcal{D}^2 s_{hk}, \quad \mathcal{D} \equiv d\mathbf{J} \cdot \nabla_{\mathbf{J}}$):

$$\mathcal{F}^2 = \left|\left\langle \Psi'_f | \Psi_f \right\rangle\right|^2 = 1 - \sum_{k,j,j'} \alpha^*_{k,j'} \alpha_{k,j} a_{k,j,j'} + \left| \sum_{k,j,j'} \alpha^*_{k,j'} \alpha_{k,j} b_{k,j,j'} \right|^2 \quad (5)$$

$$\text{with}: \; a_{k,j,j'} = \left\langle \beta_{f^h_{k,j'}, f^h_{j',k}} | \mathcal{D}s^\dagger_{h_k} \mathcal{D}s_{h_k} | \beta_{f^h_{k,j}, f^h_{j,k}} \right\rangle, b_{k,j,j'} = \left\langle \beta_{f^h_{k,j'}, f^h_{j',k}} | s^\dagger_{h_k} \mathcal{D}s_{h_k} | \beta_{f^h_{k,j}, f^h_{j,k}} \right\rangle$$

the $\mathcal{D}s^\dagger_{h_k}\mathcal{D}s_{h_k}$ and $s^\dagger_{h_k}\mathcal{D}s_{h_k}$ entries, with: $f^h_{a,b} = \frac{1}{2}((3-h)(a-1)+(h-1)(b-1))$ setting the rules to fix the Bell states in each block. Thus, for a uniform set of 10^6 points in $\mathcal{R}$, a numerical algorithm was used to find the minimum of $\mathcal{F}^2$ with respect to $|\delta J_i| = |dJ_i/J_i| \leq 0.05, i = 1,2,3; \iota, \kappa, \lambda \in [0, 2\pi)$ around the gate prescriptions for J_i, B_i, t [1]. Minimum is reported in the Figures 1b-h as density plots on $\mathcal{R}$ for each gate in $\mathcal{D}$. In this analysis, the minimum was calculated on each entire fiber bundle for each projection.

For $(\mathbf{1}_1 \otimes S_{\phi_2})_\mathrm{B}$, the fidelity is independent of ϕ, including the interest cases $\phi = \pi/4, \pi/8$. There, $0.778 \leq \mathcal{F} \leq 0.880$ for all states (3). Density plot (Figure 1b) exhibits lower $\mathcal{F}$ values for the most of separable states in the lateral boundaries of $\mathcal{R}$. The same analysis is true for $(S_{\phi_1} \otimes \mathbf{1}_2)_\mathrm{B}$ obtained for $h=3$ [1]. For the $(S_{\phi_1} \otimes \mathbf{1}_2)_\mathrm{B}$ gates obtained with $h=1$, $\mathcal{F}$ is ϕ dependent. Despite that for $(S_{\pi/8_1} \otimes \mathbf{1}_2)_\mathrm{B}$ and $(S_{\pi/4_1} \otimes \mathbf{1}_2)_\mathrm{B}$ the density plots are similar to those for $(\mathbf{1}_1 \otimes S_{\phi_2})_\mathrm{B}$, they show improved values of fidelity (Figures 1c-d): $0.999 \leq \mathcal{F} \leq 1.000$ and $0.997 \leq \mathcal{F} \leq 0.998$. Meanwhile, $(\mathbf{1}_1 \otimes H_2)_\mathrm{B}$ and $(H_1 \otimes \mathbf{1}_2)_\mathrm{B}$ exhibit a wider dispersion in the $\mathcal{F}$ values $0.800 \leq \mathcal{F} \leq 1.000$ and $0.822 \leq \mathcal{F} \leq 0.998$ respectively (Figures 1e-f). An unclear behavior in terms of separability appears, which is analyzed below. $(C^1NOT_2)_\mathrm{B}$ gives an almost uniform value $\mathcal{F} \approx 0.987$ (Figure 1g). Finally, $(C^2NOT_1)_\mathrm{B}$ has $0.987 \leq \mathcal{F} \leq 1.000$ (Figure 1h) with some limited slices of low fidelity involving both separable and entangled states.

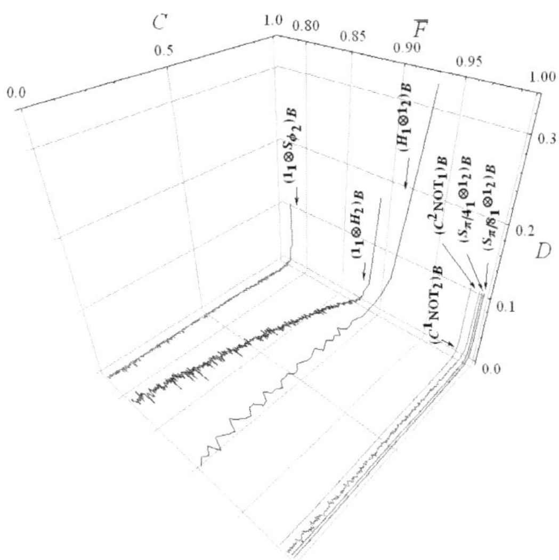

Figure 2. Gates performance summarized showing the mean value of fidelity $\overline{F}$ for each concurrence C value. Vertical axis shows the density of states D in such situation.

Due to density plots shadow some inner details of $\mathcal{F}$ in $\mathcal{R}$ and the unclear relation with $\mathcal{C}$, the Figure 2 summarizes some complementary interesting aspects of the analysis. This figure compares the concurrence $\mathcal{C}$ with the average fidelity $\overline{F}$ for the states in such case, together with their density D (the relative proportion with respect to the overall states considered in the analysis). Note the improved fidelity values for entangled states and the growing number of states in that situation. Hadamard-like gates show the lowest fidelity values together with $(\mathbf{1}_1 \otimes S_{\phi_2})_{\text{B}}$ and a notable dispersion in any case for $|\delta J_i| \leq 0.05$. Despite, $\mathcal{F}$ shows values above of 0.77.

5. Conclusions

Quantum engineering in quantum information is a growing development field relating Quantum Mechanics, Control Theory and Computer Science on a common basis of system's physical properties and achievable technologies. Gate design should combine these fields to construct procedures and physical realizations, including quantification of possible failures as was developed here for the set of universal quantum gates proposed for systems interacting via the Heisenberg-Ising model. Last quantification shows considerably stability when sensible and uncontrolled design parameters could be detuned. There, additional error correction procedures can be still applied on the syndromes defined by $\mathbf{X}_j$ and $\mathbf{Z}_j$ [1] in each block.

Acknowledgments
The support from Tecnológico de Monterrey to develop this research is acknowledged.

References

[1] Delgado F 2016 *Universal quantum gates for Quantum Computation on magnetic systems ruled by Heisenberg-Ising interactions* (quant-ph/1611.05890)
[2] Delgado F 2015 *Int. J. Quant. Info.* **13** 1550055
[3] Sekatski P, Skotiniotis M, Kolodynski J and Dur W 2016 *Quantum metrology with full and fast quantum control* (quant-ph/1410.5515)
[4] Delgado F 2010 *Rev. Mex. Fis.* **56** (1) 30
[5] Boykin P O, Mor T, Pulver M, Roychowdhury V and Vatan F 1999 *On universal and fault tolerant quantum computing* (quant-ph/9906054)
[6] Delgado F 2016 *J. Phys.: Conf. Ser.* **648** 012024
[7] Bengtsson I, Brännlund J and Życzkowski K 2002 *Int. J. of Mod. Phys.* A **17** 4675

Coherent States for Supersymmetric Partners of the Infinite Well

V Hussin[1,2] and V S Morales-Salgado[3]

[1]Département de mathématiques et de statistique, Université de Montréal, C. P. 6128, Succ. Centre-ville, Montréal (Québec) H3C 3J7, Canada.
[2] Centre de recherches mathématiques, Université de Montréal, C. P. 6128, Succ. Centre-ville, Montréal (Québec) H3C 3J7, Canada.
[3]Departamento de Física, Cinvestav, A.P. 14-740, 07000 México City, Mexico

E-mail: veronique.hussin@umontreal.ca, vmorales@fis.cinvestav.mx

Abstract. We define linear and quadratic coherent states for the supersymmetric partners of the quantum infinite well through formal series expansions of the energy eigenfunctions of the systems and we study the appropriateness of this definitions as coherent states by means of their properties. In particular, we examine the localization in position and time evolution, minimum uncertainty relations and the behavior of the Wigner function.

1. Introduction

The concept of coherent state was introduced since the appearance of quantum mechanics on rather heuristic grounds, namely that a coherent state is a quantum state whose behavior is classical nonetheless [1]. A unique definition of such states remains elusive and a set of inequivalent definitions for a general system have been proposed in several works [2–6].

These definitions have lead to the construction of states that have shown to be most fruitful when studying quantum systems. A reason for this is the good localization in position that they tend to posses. The use of coherent states has included systems obtained through the supersymmetric technique [7–9]. This technique allows one to obtain a multi-parametric family of solvable Hamiltonians, departing from an initial one for whose stationary Schrödinger solution is known [10–19]. The thus obtained Hamiltonians are called supersymmetric (SUSY) partners of the initial one. Although the energy spectra of the SUSY partners are known through the technique, several other properties need to be understood once the Hamiltonian is obtained. This is where coherent states provide a better understanding of the dynamics of the system.

The SUSY partners are particularly interesting systems in that their spectrum is not connected in general by the action of the natural ladder operators given immediately by the technique [20–23]. Thus it is challenging to use those definitions of coherent states which rely on ladder operators. Several treatments have been developed to deal with this difficulty mainly focusing on the action of the various alternatives for constructing ladder operators of the system and in studying the formal series expansions of such states. As we will shortly see the SUSY partners of the infinite well [8, 9] (also known as the problem of a particle in a box) are paradigmatic yet simple examples of this behavior.

In this work we will define coherent states for the SUSY partners of the one-dimensional infinite well based on formal series expansions and the idea of good localization in position of these states.

This paper is organized as follows. In Sec. 2 we describe the supersymmetric partners of the one-dimensional infinite well. In Sec. 3 we give a brief enumeration of the alternatives for defining the coherent states of a quantum system. We thus concentrate our attention on two possible choices, i.e. the linear and quadratic coherent states for the SUSY infinite well and we examine their properties. In Sec. 4 we present our concluding remarks.

2. A family of SUSY partners of the 1D infinite well

It is well-known that the quantum infinite well may be described by the Hamiltonian $H = -\frac{\hbar^2}{2M}\frac{d^2}{dx^2} + V(x)$ where the potential $V(x)$ is taken to be

$$V(x) = \begin{cases} 0, & 0 < x < \pi \\ \infty, & \text{otherwise.} \end{cases} \tag{1}$$

The stationary eigenstates and the discrete energies of this system are

$$\psi_n(x) = \sqrt{\frac{2}{\pi}} \sin nx, \qquad E_n = \frac{\hbar^2}{2M}n^2, \quad n = 1, 2, \ldots \tag{2}$$

In the following, we will use dimensionless units, setting $\hbar = 1$, $M = 1/2$.

SUSY partners of the infinite well have been constructed [7] starting from the usual intertwining relations

$$\tilde{H}Q = QH, \quad Q^\dagger \tilde{H} = HQ^\dagger, \tag{3}$$

which involve the supercharges

$$Q = \frac{d^2}{dx^2} + \eta(x)\frac{d}{dx} + \epsilon + \frac{1}{2}(\eta^2(x) - \eta'(x)) \tag{4}$$

and its adjoint $Q^\dagger$ where ϵ is an arbitrary constant. In the so-called confluent case [8], the technique produces a family of new potentials

$$\tilde{V}(x; k, \omega) = 2\eta'(x) = \begin{cases} \frac{32k^2 \sin(kx)[\sin(kx) + k(\pi\omega - x)\cos(kx)]}{[\sin(2kx) + 2k(\pi\omega - x)]^2}, & 0 < x < \pi \\ \infty, & \text{otherwise.} \end{cases} \tag{5}$$

with two real parameters k, ω. The first must satisfy the condition $\epsilon = k^2$ with $k = 1, 2, \ldots$ while ω is an arbitrary constant.

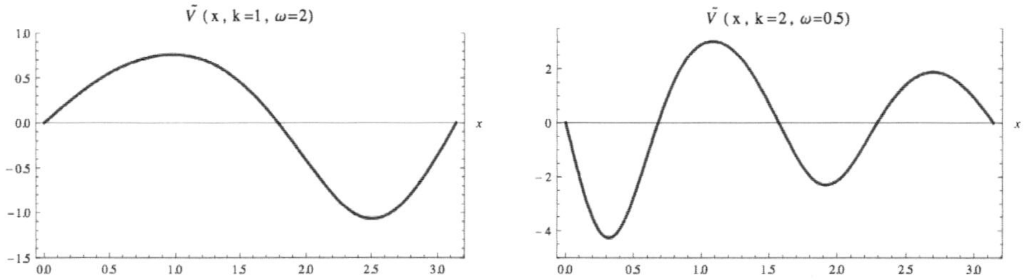

Figure 1. SUSY potential as a function of x.

Notice that the parameter k takes only positive integer values. For these potentials to be non singular, the admissible values of ω are $\omega \in]-\infty, 0[\cup]1, \infty[$. The parameter k controls the amount of oscillations and the potentials are invariant under the combined change $\omega \to 1-\omega$ and $x \to \pi - x$. Two instances are illustrated on Fig. 1.

The normalized SUSY eigenstates $\tilde{\psi}_n(x)$ are obtained from the intertwining relations (3) as:

$$\tilde{\psi}_n(x;k,\omega) = (k^2-n^2)^{-1} Q\psi_n(x), \quad (n \neq k). \tag{6}$$

For $n \neq k$, they are physical states, i.e. they are normalizable and such that $\tilde{\psi}_n(0;k,\omega) = \tilde{\psi}_n(\pi;k,\omega) = 0$, since $\eta(0;k,\omega) = \eta(\pi;k,\omega) = 0$. The corresponding energies are $E_n = n^2$ as in the non SUSY case.

For $n = k$, we get $Q\psi_k(x) = 0$. The missing state is found [8] to be

$$\tilde{\psi}_k(x;k,\omega) = \sqrt{\frac{2}{\pi}} \sin(kx) \frac{2\pi k\sqrt{\omega(\omega-1)}}{\sin(2kx) + 2k(\pi\omega - x)}. \tag{7}$$

It is normalizable and such that $\tilde{\psi}_k(0,k,\omega) = \tilde{\psi}_k(\pi,k,\omega) = 0$. With this additional state the spectrum of $\tilde{H}$ is complete.

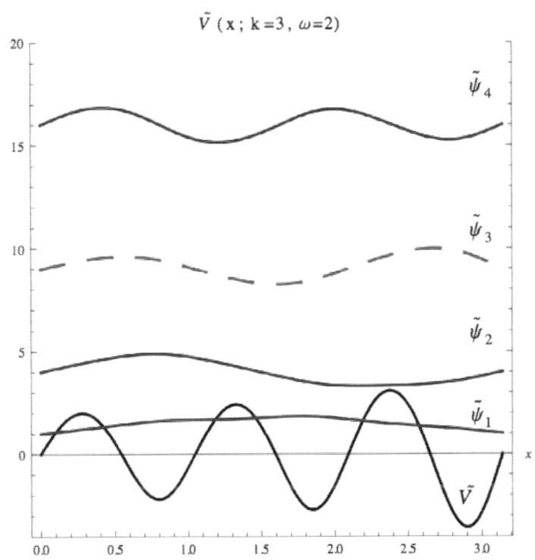

Figure 2. Potential of a SUSY infinite well with its first four energy eigenstates.

Fig. 2 shows the spectrum of a SUSY partner of the infinite well where the second excited state $k = 3$ was used to produce the transformation. The base state and the first three excited states are shown too. For these states we have displaced the axis to their corresponding eigen-energies.

It is worth noticing that, since the infinite well possesses ladder operators A^+ and A^- [24], then there are natural ladder operators for $\tilde{H}$ given by

$$L^{\pm} = QA^{\pm}Q^{\dagger}. \tag{8}$$

3. Different sets of coherent states for the SUSY partners

For an arbitrary system, it is well-known there are distinct definitions of coherent states which in general yield to inequivalent states. Among these definitions we find the following ones.

- A Barut-Girardello coherent state $\Phi_{\rm Ge}$ is that which satisfies the following eigenvalue equation for an annihilation operator a of the system:

$$a\Phi_{\rm Ge}(z;x) = z\Phi_{\rm Ge}(z;x), \; z \in \mathbb{C}. \tag{9}$$

• A Gilmore-Perelomov coherent state is obtained through the action of the displacement operator $D(z) = \exp(za^\dagger - z^*a)$ on an extremal state ψ_0

$$D(z)\psi_0(x) = \exp(za^\dagger - z^*a)\psi_0(x), \; z \in \mathbb{C}. \tag{10}$$

• A coherent state $\Phi_{\rm MUR}$ is such that it minimizes the uncertainty relation

$$\sigma_x\sigma_p = \frac{1}{2}, \quad \sigma_O^2 = \langle\Phi_{\rm MUR}|O^2|\Phi_{\rm MUR}\rangle - \langle\Phi_{\rm MUR}|O|\Phi_{\rm MUR}\rangle^2. \tag{11}$$

where x is the position and p is the momentum of the particle described by the system and $\langle\Phi|O|\Phi\rangle$ is the mean value of the operator O when the system is in the state $|\Phi\rangle$.

Notice that the first two definitions require that the system possesses ladder operators.
For a solvable quantum system with Hamiltonian H_0 having discrete energy spectrum (finite or infinite), ladder operators may be defined such that they act on the energy eigenstates $\psi_n(x)$ as

$$a^-\psi_n(x) = \sqrt{f(n-1)}\,\psi_{n-1}(x), \quad a^+\psi_n(x) = \sqrt{f(n)}\,\psi_{n+1}(x), \quad n = 1, 2, ..., \tag{12}$$

which leads to a generalized Heisenberg algebra ($N\psi_n(x) \equiv n\psi_n(x)$):

$$[a^-, N] = a^-, \quad [a^+, N] = -a^+, \quad [a, a^+] = f(N) - f(N-1), \tag{13}$$

for a positive function $f(n)$ such that $f(0) = 0$. Operatorial definitions of such ladder operators are well-known. For the case of the infinite well, see for example [25].

Departing from the action of the ladder operators on the energy eigenstates we can obtain the proposed coherent states for the system of interest as the formal series

$$\Phi_{\rm Ge}(z; x) = \sum_{n=1}^{d} c_n(z)\psi_n(x), \tag{14}$$

where d is the dimension of the Hilbert space that may be finite or infinite.

The coefficients $c_n(z)$ are different in the case of the Barut-Girardello definition from those in the Gilmore-Perelomov one and in both cases these coefficients depend on the function $f(n)$.

When dealing with SUSY transformations it is common to find that the usual construction for the ladder operators of the resulting system is disconnected through the action of these operators. This yields the problem of having a definition of coherent states that takes into account only a proper subset of the basis of the Hilbert space. An alternative that has been proven [7] to work appropriately in this cases is a construction of the ladder operators with a linear function $f(n) = n$. In such cases the coherent states are defined in each subset which is disconnected by the action of the ladder operators.

For the SUSY partners of the infinite well there are three subsets which are not connected by the action of the ladder operators $L^\pm = QA^\pm\tilde{Q}^\dagger$. The first is composed by all $\tilde{\psi}_n$ with $n < k$, the second has one element, the eigenfunction $\tilde{\psi}_k$, and the third is given by all $\tilde{\psi}_n$ with $n > k$. Nonetheless, let us assume that we can define formally a new set of ladder operators $\mathcal{L}^\pm$ for these systems such that they satisfy (12) and (13), then a generalized coherent state is formally given by

$$\Phi_{\rm Ge}(z; x) = \frac{1}{\sqrt{N(z)}}\sum_{n=1}^{\infty}\frac{z^{n-1}}{\sqrt{\rho(n-1)}}\psi_n(x), \tag{15}$$

where $1/\sqrt{N(z)}$ is a normalization constant and $\rho(n)$ is the generalized factorial of the function $f(n)$ given by

$$\rho(n) = \begin{cases} 1, & \text{if } n = 0 \\ \prod_{i=1}^{n} f(n), & \text{if } n > 0. \end{cases} \tag{16}$$

Thus fixing the function $f(n)$ will fix the coherent states given by (15). In what follows we will study the set of generalized coherent states given by two relevant choices for $f(n)$, say the linear case $f(n) = n$ and the quadratic case $f(n) = n^2$. Indeed, the first choice leads to the usual action (12) of the oscillator-like ladder operators. The second choice leads to the factorization of the SUSY Hamiltonian $\tilde{H} = \mathcal{L}^-\mathcal{L}^+$.

3.1. *Linear coherent states*

When $f(n) = n$, the linear coherent states of the SUSY partners of the infinite well are

$$\tilde{\Phi}_{\mathrm{L}}(x;k,\omega;z) = \frac{1}{\sqrt{N_{\mathrm{L}}}} \sum_{n=1}^{\infty} \frac{z^{n-1}}{\sqrt{(n-1)!}} \tilde{\psi}_n(x;k,\omega), \tag{17}$$

where $|N_{\mathrm{L}}| = \exp\left(|z|^2\right)$.

This series can be associated to the standard series for the coherent states of the harmonic oscillator, although in this case the energy eigenfunctions $\tilde{\psi}_n(x;k,\omega)$ are those of the SUSY partners of the infinite well.

3.2. *Quadratic coherent states*

When $f(n) = n^2$, the quadratic coherent states of the SUSY partners of the infinite well are

$$\tilde{\Phi}_{\mathrm{Q}}(x;k,\omega;z) = \frac{1}{\sqrt{N_{\mathrm{Q}}}} \sum_{n=1}^{\infty} \frac{z^{n-1}}{(n-1)!} \tilde{\psi}_n(x;k,\omega), \tag{18}$$

where $|N_{\mathrm{Q}}| = I(0, 2|z|)$ and $I(a,b)$ is the modified Bessel function of the first kind. Due to the commutation relations (13), these coherent states are related to SU(1,1)-coherent states.

3.3. *Properties*

Let us recall that the time evolution of an arbitrary state is given by the action of the evolution operator $U_t = \exp(-i\tilde{H}t)$. We thus get for our coherent states

$$U_t\left(\tilde{\Phi}_{\mathrm{Ge}}\right) = \frac{1}{\sqrt{N(z)}} \sum_{n=1}^{\infty} \frac{z^{n-1}\,\mathrm{e}^{-in^2 t}}{\sqrt{\rho(n-1)}} \tilde{\psi}_n(x;k,\omega), \tag{19}$$

possibly up to a phase factor.

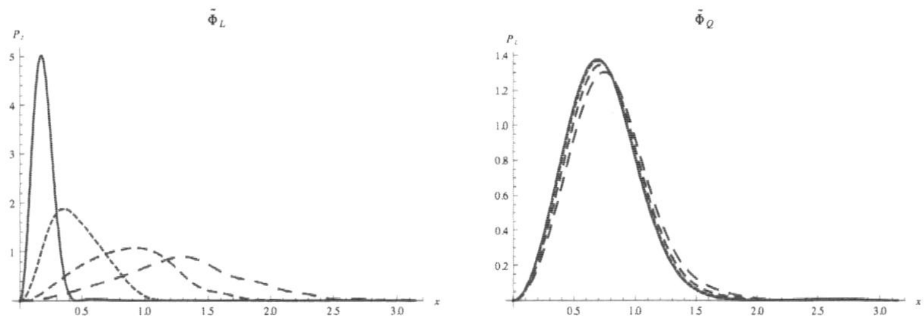

Figure 3. Probability density of the linear and quadratic coherent states for the case $\tilde{V}(x; k = 3, \omega = 2)$ at times $t = 0, 0.03, 0.06$ and $t = 0.09$ from left to right, respectively.

Fig. 3 gives the probability density $P_z = |\tilde{\Phi}|^2$ of both types of coherent states at different times. At $t = 0$, the linear coherent states show a better localization in the position than the quadratic ones. As the states evolve in time the initial localization is lost, however, this effect is stronger in the case of the linear coherent states whose probability density rapidly spreads over the domain of the well.

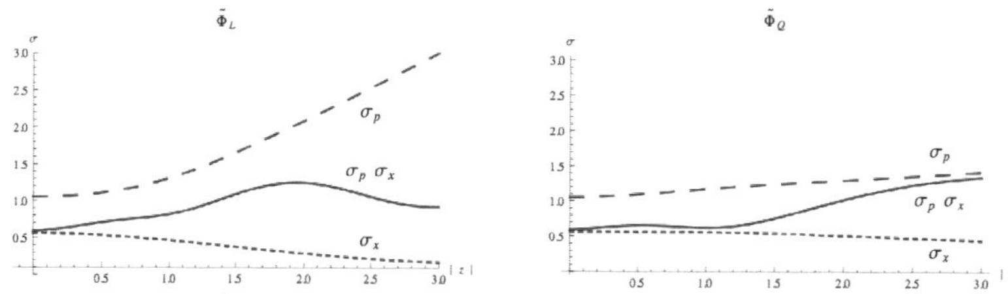

Figure 4. Dispersion in position σ_x (- - -) and momentum σ_p (— — —) and their product $\sigma_x\sigma_p$ (——) for a linear and a quadratic coherent state for $\tilde{V}(x; k = 3, \omega = 2)$.

With respect to the third definition given at the beginning of the section, we compute the dispersion in position σ_x and momentum σ_p, as well as their product, for the linear and quadratic coherent states. In order to do the calculations, we make an approximation of the states keeping 100 terms in the series. Fig. 4 shows that the product $\sigma_x\sigma_p$ obeys the uncertainty relation, however it is not minimal for an any value of $|z|$ other than $z = 0$. As expected for $z = 0$ the dispersions for the linear and the quadratic coherent states are those of the ground state.

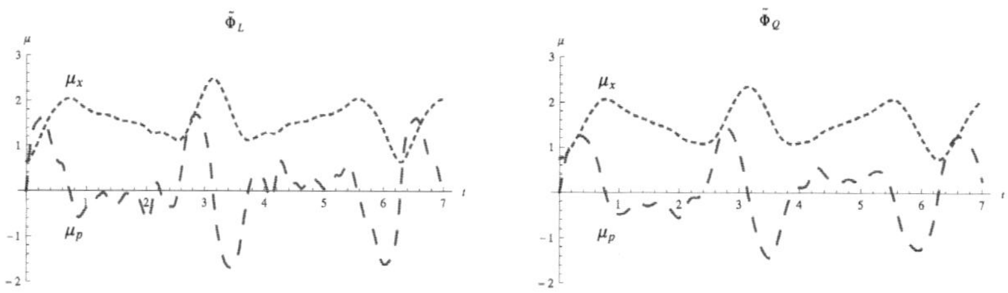

Figure 5. Mean value of x (- - -) and p (— — —) for $\tilde{\Phi}_{\rm L}$ and $\tilde{\Phi}_{\rm Q}$ for $\tilde{V}(x; k = 3, \omega = 2)$ and $z = 1$

On the other hand, the mean values of the observables corresponding to position and momentum evolve in time as shown in Fig. 5. We see that they evolve periodically with period equal to 2π, similar to a bouncing between the walls of the well. Although the temporal stability is not fully attained because the eigen-energies are quadratic functions of n, periodicity is still expected since the time evolution is a phase shift on these states. The oscillations of these mean values behave similarly in both cases.

Finally, Fig. 6 shows the Wigner function of these coherent states, namely

$$W(x,p) = \frac{1}{\pi^2} \int_{-\pi}^{\pi} \tilde{\Phi}^* \left(x + \frac{y}{2}, t; k, \omega; z\right) \tilde{\Phi} \left(x - \frac{y}{2}, t; k, \omega; z\right) \mathrm{e}^{-ipy} dy, \tag{20}$$

where we use fixed values of t, k, ω and z. Notice that in both cases the Wigner function has regions in its domain where it is negative. On the other hand, notice that these functions are not displaced versions of the Wigner function of the ground state $\tilde{\psi}_1$ which can be seen in Fig 7. These features are in contrast with the usual coherent states of the harmonic oscillator whose Wigner function is non-negative and is found to be that of its ground state translated in the phase space.

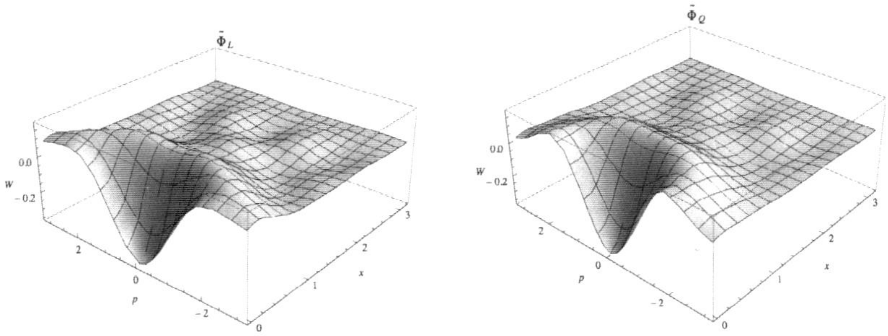

Figure 6. Wigner function of $\tilde{\Phi}_{\rm L}$ and $\tilde{\Phi}_{\rm Q}$ for the case $\tilde{V}(x; k = 3, \omega = 2)$, $z = 1$ and $t = 0$

.

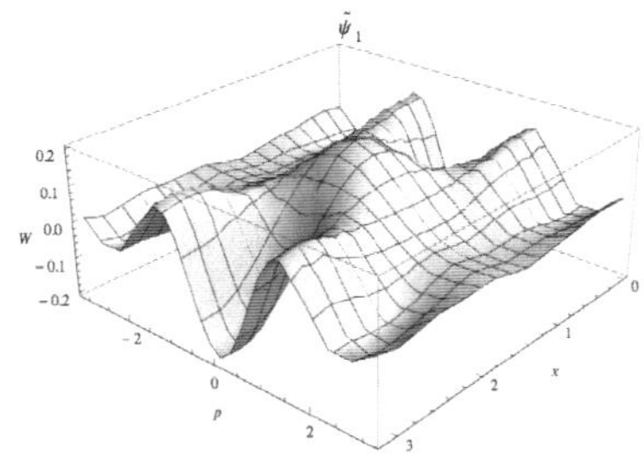

Figure 7. Wigner function of $\tilde{\psi}_1$ for the case $\tilde{V}(x; k = 3, \omega = 2)$

.

4. Conclusion

The eigenfunctions of the supersymmetric partners of the quantum infinite well fail to be completely connected through the action of its natural ladder operators. This makes difficult obtaining coherent states through the usual definitions. Thus we introduced a formal definition of such states given by formal series that could show the expected behavior of a coherent state in terms of an arbitrary positive function. Two natural choices have been considered leading to linear and quadratic coherent states. The linear ones show a better localization in the position at time $t = 0$ but the probability density spreads quickly as time evolves. We also analyzed the time evolution of the mean position and momentum. These mean values oscillate in a similar manner in both states. Finally, the Wigner functions of the linear and quadratic coherent states were found to be negative in regions of their domains and they are not displaced versions of the Wigner function of the ground state. Both of this characteristics differ from those of the usual coherent states of the harmonic oscillator.

Acknowledgements
This work has been supported in part by research grants from Natural sciences and engineering research council of Canada (NSERC).

References
[1] Schrödinger E 1926 *Naturwiss* **14** 664.
[2] Glauber R J 1963 *Phys. Rev.* **130** 2529; **131** 2766.

IOP Conf. Series: Journal of Physics: Conf. Series **839** (2017) 012017 doi:10.1088/1742-6596/839/1/012017

[3] Klauder J R and Skagerstam B S 1985 *Coherent States- Applications in Physics and Mathematical Physics* (Singapore: World Scientific).
[4] Perelomov A M 1986 *Generalized Coherent states and their applications* (Berlin: Springer).
[5] Gilmore R 1972 *Ann. Phys.* **74** 391.
[6] Gazeau J P 2009 *Coherent States in Quantum Physics* (Weinheim: Wiley-VCH).
[7] Fernańdez D J, Hussin V and Rosas-Ortiz O 2007, *J. Phys. A: Math. Theor.* **40** 6491.
[8] Fernandez D J and Salinas-Hernandez E 2003, *J. Phys. A: Math. Gen.* **36** 2537.
[9] Fiset M A and Hussin V 2015, *J. Phys.: Conf. Ser.* **624** 012016.
[10] Cooper F, Khare A and Sukhatme U 1995 *Phys. Rep.* **251** 267.
[11] Junker G and Roy P 1998 *Ann. Phys.* **270** 155.
[12] Samsonov B F 1999 *Phys.Lett. A* **263** 274.
[13] Aoyama H, Sato M and Tanaka T 2001 *Nucl. Phys. B* **619** 105.
[14] Mielnik B and Rosas-Ortiz O 2004 *J. Phys. A: Math. Gen.* **37** 10007.
[15] Fernández D J 2010 *AIP Conf. Proc.* **1287** 3.
[16] Quesne C 2011 *Mod. Phys. Lett. A* **26** 1843.
[17] Marquette I 2012 *J. Math. Phys.* **53** 012901.
[18] Andrianov A A and Ioffe M V 2012 *J. Phys. A: Math. Theor.* **45** 503001.
[19] Gómez-Ullate D, Grandati Y and Milson R 2014 *J. Phys. A: Math. Theor.* **47** 015203.
[20] Fernández D J and Hussin V 1999, *J. Phys. A* **32** 3603.
[21] Carballo J M, Fernández D J, Negro J and Nieto L M 2004 *J. Phys. A: Math. Gen.* **37** 10349.
[22] Bermudez D, Contreras-Astorga A and Fernández D J 2014 *Ann. Phys.* **350** 615.
[23] Quesne C 2015 *J. Phys.: Conf. Ser.* **597** 012064.
[24] Jalali H R and Tavassoly M K 2013, *Opt. Comm.* **298** 161.
[25] Dello Sbarba L and Hussin V 2007, *J. Math. Phys.* **48** 012110.

Exactly Solvable Time-Dependent Oscillator-Like Potentials Generated by Darboux Transformations

K Zelaya and O Rosas-Ortiz
Physics Department, Cinvestav, A.P. 14-740, 07000, México City, Mexico
E-mail: kdzelaya@fis.cinvestav.mx

Abstract. The stationary Schrödinger equation of the harmonic oscillator is deformed by a Darboux transformation to construct time-dependent potentials with the oscillator profile. The Darboux (supersymmetric or factorization) method is usually developed in the spatial variables of the Schrösdinger equation. Here we follow a variation introduced by Bagrov, Samsonov and Shekoyan to include the time-variable as a parameter of the transformation.

1. Introduction

It is well known that the Darboux transformations leave key geometric properties of certain classes of surfaces unchanged [1]. Such a property finds a diversity of applications in physics because the sets of solutions of the differential equations that represent dynamical laws can be modeled in geometric form. One of the branches of physics which uses the Darboux transformation in exhaustive form is the soliton theory, where nonlinear superposition principles hold. Unexpectedly, the models introduced in the eighties of the previous century to study (in the same picture) bosons and fermions were also associated with the Darboux transformation [2]. The term *supersymmetric quantum mechanics* came to denote the simplest case of such models and gave rise to a new branch of quantum physics which has grown stronger over the years [2–5]. In this context, most of the works dealing with the supersymmetric construction of exactly solvable potentials use the Darboux transformation of a given potential in the spatial variable of the Schrödinger equation. A notable exception is offered by the papers [6, 7], where the authors introduce a variation of the Darboux transformation that includes the time-variable as a parameter. Using such a method, one can construct time-dependent potentials which are exactly solvable.

In this paper we report a new family of exactly solvable time-dependent potentials that correspond to the Darboux-deformations of the harmonic oscillator in the approach of Refs. [6,7]. The set of solutions of the corresponding Schrödinger equation is also constructed and some of their basic properties are discussed.

2. The BSS approach

In this section we briefly summarize the approach proposed by Bagrov, Samsonov and Shekoyan (BSS) to generate time-dependent potentials by a Darboux transformation [6,7]. Consider the following pair of dimensionless Schrödinger equations

$$\left[i\partial_t + \partial_x^2 - V_0(x,t)\right]\phi(x,t) = 0, \tag{1}$$

$$\left[i\partial_t + \partial_x^2 - V_1(x,t)\right]\psi(x,t) = 0, \tag{2}$$

where $V_0(x,t)$ and $V_1(x,t)$ are real-valued potentials that depend on time in general. In the above equations ∂_t and ∂_x stand for time and position partial-derivatives respectively. It is assumed that the operator

$$L = \ell(t)\left[\beta(x,t) + \partial_x\right], \tag{3}$$

with $\ell(t)$ and $\beta(x,t)$ functions to be determined, intertwines (1) and (2) as follows

$$L\left[i\partial_t + \partial_x^2 - V_0(x,t)\right] = \left[i\partial_t + \partial_x^2 - V_1(x,t)\right]L. \tag{4}$$

Then one arrives at the set of equations

$$V_1(x,t) - V_0(x,t) = i\frac{\dot{\ell}}{\ell} + 2\beta_x, \tag{5}$$

$$\left(\beta^2\right)_x = i\dot{\beta} + \beta_{xx} + (V_0)_x, \tag{6}$$

where $\dot{f} := \partial_t f$ and $f_x := \partial_x f$. Let u be the transformation function, i.e. $\beta = -(\ln u)_x$. Therefore, (6) is simplified to the Schrödinger-like equation

$$i\dot{u} = -u_{xx} + [V_0 + c_1(t)]u, \tag{7}$$

with $c(t)$ an integration constant which can be set to zero. On the other hand, as the potentials V_0 and V_1 are real-valued, from (5) we have $i(\ln|\ell(t)|^2)_t = -2\partial_x(\beta_x - \beta_x^*)_x$. That is, the transformation function must satisfy

$$\left(\ln\frac{u}{u^*}\right)_{xxx} = 0. \tag{8}$$

Now, assuming that the function $\ell(t)$ is a real-valued we have

$$\ell(t) = \exp\left(2\int dt\,\mathrm{Im}(\ln u(x,t))_{xx}\right), \qquad V_1(x,t) - V_0(x,t) = -\left(\ln|u(x,t)|^2\right)_{xx}. \tag{9}$$

As usual in supersymmetry, the solutions $\psi(x,t)$ of (2) can be obtained from (4) whenever the solutions of (1) are given, and vice versa. Additionally, the missing state $\psi_0 = \frac{1}{\ell u^*}$ must be considered since it is also a solution of (2).

3. Time-dependent oscillators

Let $V_0(x,t) = x^2$ be the initial potential and

$$u(x,t) = B(t)e^{a(t)x^2} f(x,t) \tag{10}$$

the transformation function, with $a(t)$, $B(t)$ and $f(x,t)$ functions to be determined. The form of u in (10) is a generalization of the one used in [6] for the free-particle potential. In our case, the u-function could even be adapted to be one of the Gaussian wave packets discussed in e.g. [8–10]. At this stage, it is convenient to introduce an additional function $z(x,t) = b(t)x$ that couples the spatial variable x with the time-parameter t via the real-valued function $b(t)$. The introduction of $u(z(x,t))$ into (1) gives rise to a system which includes a differential equation in the variable z for f, a differential equation in the variable t for B, and a pair of constraints that are useful to determine the time-dependent functions a and b [11]. The straightforward calculation yields

$$f(x,t) = e^{-\frac{b^2(t)x^2}{2}}\left[k_a\,{}_1F_1\left(\nu, \frac{1}{2}; b^2(t)x^2\right) + k_b\, b(t)x\,{}_1F_1\left(\nu + \frac{1}{2}, \frac{3}{2}; b^2(t)x^2\right)\right], \tag{11}$$

with k_a and k_b arbitrary constants,

$$b(t) = \frac{c_0}{\sqrt{c_1 + \gamma \cos(4t + c_2)}}, \tag{12}$$

and

$$\begin{aligned} B(t) &= \frac{\exp\left[i\left(\nu - \frac{1}{4}\right)\frac{c_0^2}{\gamma\lambda\sqrt{1+\lambda^2}} \arctan\left(\frac{\lambda}{\sqrt{1+\lambda^2}} \tan\left(\frac{4t+c_2}{2}\right)\right)\right]}{[c_1 + \gamma\cos(4t + c_2)]^{1/4}}, \\ a(t) &= -i\frac{\gamma}{2}\left(\frac{\sin(4t + c_2)}{c_1 + \gamma\cos(4t + c_2)}\right). \end{aligned} \tag{13}$$

In the above expressions ${}_1F_1(\nu, \eta; w)$ stands for the confluent hypergeometric function in the variable w [12] and

$$\gamma^2 = c_1^2 - c_0^4, \quad \lambda^2 = \frac{1}{2}(c_1 - 1),$$

with c_0, c_1 and c_2 integration (real) constants such that $c_1 > c_0^2$. Therefore

$$\ell(t) = \sqrt{c_1 + \gamma\cos(4t + c_2)},$$

and the new potential is given by the time-dependent function

$$V_1(x, t) = x^2 + 2b^2(t) - 2\frac{\partial^2}{\partial x^2}\left[f(x, t)e^{\frac{b^2(t)x^2}{2}}\right]. \tag{14}$$

On the other hand, the well known stationary solutions of (1),

$$\phi_n(x, t) = \frac{e^{-\frac{x^2}{2} - i(2n+1)t}}{\sqrt{2^n n! \sqrt{\pi}}} H_n(x), \tag{15}$$

lead to the solutions of (2) for the new time-dependent potential

$$\begin{aligned} \psi_n(x, t) &= L\phi_{n+1}(x, t), \quad n = 0, 1, 2, \ldots, \\ \psi_0(x, t) &= \frac{1}{\ell(t) u^*(x, t)}. \end{aligned} \tag{16}$$

3.1. Examples

If $\nu = 1/2$, the time-dependent potentials (14) acquire the special form

$$V_1(x, t) = x^2 - 2b(t) - 4k_b b(t)\frac{\partial}{\partial x}\left[\frac{e^{-b^2(t)x^2}}{2k_a + \sqrt{\pi}k_b \mathrm{Erf}(b(t)x)}\right], \tag{17}$$

where $2k_a > \sqrt{\pi}k_b$ to avoid singularities. For $\gamma = 0$ the above function reproduces the results already reported in [13] for the conventional (time-independent) supersymmetric approach.

The time-dependent potentials (14) and (17) are depicted at three different times in Fig. 1 for the indicated sets of parameters. In both cases we can appreciate a very localized time-dependent deformation that increases in size and is shifted to the right as the time goes pass. On the other hand, the coupling function $z(x, t) = b(t)x$ is regulated by the time-dependent function defined in (12). The latter oscillates between the values $b_\pm = c_0/\sqrt{c_1 \pm \gamma}$, so that z

IOP Conf. Series: Journal of Physics: Conf. Series **839** (2017) 012018 doi:10.1088/1742-6596/839/1/012018

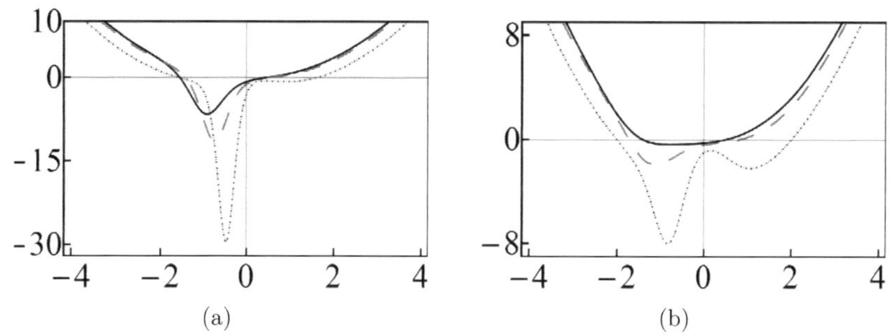

Figure 1. (Color online) The time-dependent oscillator potentials (14) for $c_0 = 1$, $c_1 = 10$, $c_2 = 0$ and (a) $k_a = 2$, $k_b = 5$, $\nu = 2$ (b) $k_a = 1.3\sqrt{\pi}$, $k_b = 2$ and $\nu = 1/2$. The analytic expression for the potentials depicted in (b) is given in Eq. (17). In both cases the potential has been evaluated at $t = 0$ (solid-black), $t = \frac{\pi}{8}$ (dashed-red) and $t = \frac{\pi}{4}$ (dotted-blue).

oscillates with the time at each value of x. The closer is x to the position of the minimum of the potential at $t = 0$, the larger is the oscillation over the time.

The squared modulus of the first three non-normalized functions (16) of the potentials (14) and (17) are shown in Fig. 2 for the same parameters and times as in Fig. 1. At $t = 0$ and $t = \frac{\pi}{4}$ all of them show a distribution of zeros which obeys the well known oscillation theorem of the stationary solutions (15). At other times like $t = \frac{\pi}{8}$, no function (16) has zeros. However, the distribution of the maxima of such functions follows the oscillation-like theorems obeyed by the wave-functions of complex-valued potentials [14]. This odd behavior deserves attention and will be analyzed elsewhere.

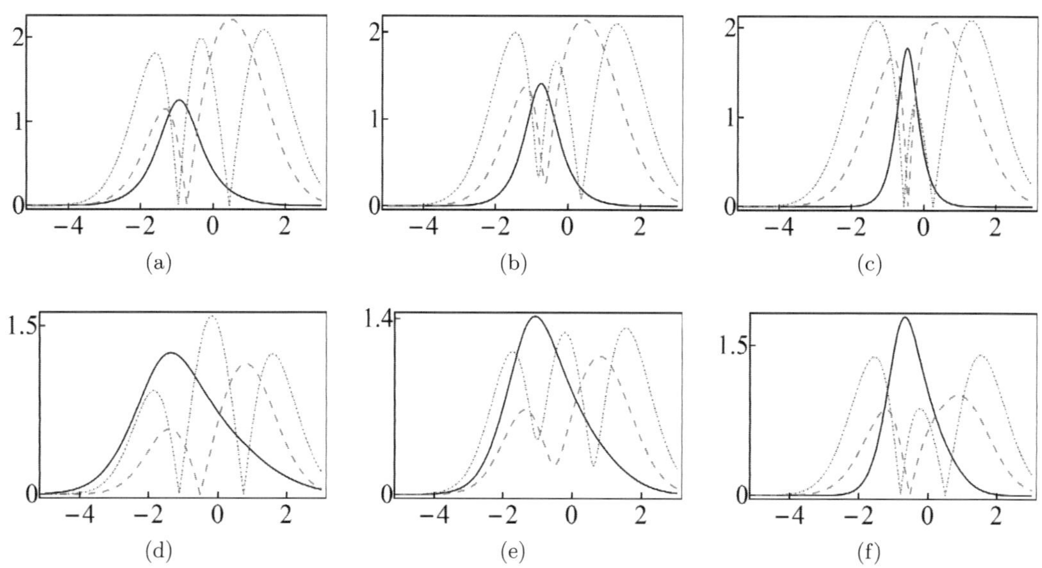

Figure 2. (Color online) Squared modulus of the non-normalized functions ψ_0 (solid-black), ψ_1 (dashed-red) and ψ_2 (dotted-blue), associated with the time-dependent oscillator potentials (14). The upper row corresponds to the potentials depicted in Fig. 1(a) and the lower row to those depicted in Fig. 1(b). From left to right the columns correspond to $t = 0$, $t = \frac{\pi}{8}$ and $t = \frac{\pi}{4}$.

4. Concluding Remarks

We have constructed time-dependent potentials which are exactly solvable and have the oscillator profile. Such potentials are Darboux-deformations of the harmonic oscillator one in the approach introduced by Bagrov, Samsonov and Shekoyan [6, 7]. We have shown that the family of oscillator-like potentials introduced by Mielnik [13] are included as a particular case of our results. In general, the oscillator-like potentials reported here are such that the deformation generated by the Darboux procedure oscillates with the time. We have found that the related solutions present an odd behavior since they lost their zeros in time by time. However, the maxima of these functions obey the oscillation theorems that are satisfied by the solutions to the Schrödinger equation of complex-valued potentials reported in [14]. The progress on the latter subject and additional results will be reported elsewhere.

Acknowledgment
KZ acknowledges the funding received through a CONACyT scholarship.

References

[1] Rogers C and Schief W K, *Bäcklund and Darboux Transformations. Geometry and Modern Applications in Soliton Theory*, Cambridge University press, Cambridge, 2002
[2] Mielnik B and Rosas-Ortiz O, *J. Phys. A: Math. Gen.* **37** (2004) 10007
[3] Bagchi B K, *Supersymmetry in Quantum and Classical Mechanics*, Chapman & Hall, Boca Raton, 2001
[4] Cooper F, Khare A and Sukhatme U, *Supersymmetry in Quantum Mechanics*, World Scientific, Singapore, 2001
[5] Gangopadhyaya A, Mallow J V and Rasinariu C, *Supersymmetric Quantum Mechanics. An Introduction*, World Scientific, Singapore, 2011
[6] Bagrov V G, Samsonov B F and Shekoyan L A, *Russ. Phys. J.* **38** (1995) 706
[7] Bagrov V G and Samsonov B F, *J. Phys. A: Math. Gen* **29** (1996) 1011
[8] Castaños O, Schuch D and Rosas-Ortiz O, *J. Phys. A: Math. Gen* **46** (2013) 218
[9] Cruz H, Schuch D, Castaños O and Rosas-Ortiz O, *Ann. Phys.* **360** (2015) 44
[10] Cruz H, Schuch D, Castaños O and Rosas-Ortiz O, *Ann. Phys.* **373** (2016) 609
[11] Zelaya K and Rosas-Ortiz O, in preparation.
[12] Olver F W J, Lozier D W, Boisvert R F and Clark C W (eds), *NIST Handbook of Mathematical Functions*, Cambridge University Press, Cambridge, 2010
[13] Mielnik B, *J. Math. Phys.* **25** (1984) 3387
[14] Jaimes-Najera A and Rosas-Ortiz O, *Ann. Phys.* **376** (2017) 126

IOP Conf. Series: Journal of Physics: Conf. Series **839** (2017) 012019 doi:10.1088/1742-6596/839/1/012019

A Time-Dependent Anharmonic Oscillator

A Contreras-Astorga

Department of Physics, Indiana University Northwest, 3400 Broadway, Gary IN 46408, USA

E-mail: `aloncont@iun.edu`

Abstract. Supersymmetric Quantum Mechanics is commonly used to generate time independent Hamiltonians with a desired spectrum. This technique can be generalized to construct time dependent potentials. In this work, the harmonic oscillator and a coherent state are taken to perform a generalized SUSY transformation in order to obtain a time dependent anharmonic oscillator.

1. Introduction

In 1984 B. Mielnik [1] constructed with a novel technique a potential with the spectrum of the quantum harmonic oscillator known as the Abraham-Moses potential [2]. Later this technique, called Supersymmetric Quantum Mechanics (SUSY), was studied in more detail and generalized in order to modify the spectrum of Hamiltonians, [3–6]. Another generalization of SUSY where the starting point is a time dependent Schrödinger equation with a potential that could depend on time is considered in [7] and [8]. In this article the SUSY technique is used to generate a time dependent anharmonic oscillator departing from the harmonic oscillator. Solutions for the new Schrödinger equation can also be obtained. The general theory is presented in Sec. 2, introducing first the simplest case known as 1-SUSY and later an iteration known as second order confluent SUSY transformation. In Sec. 3 the confluent technique is used to obtain the time-dependent anharmonic oscillator. Conclusions are presented in the last section.

2. Time-dependent supersymmetry transformation

As in the time-independent case, a supersymmetry transformation can be done relating two Schrödinger operator through an intertwining operator. In Sec. 2.1 we introduce a 1-SUSY transformation using a first order differential operator. In Sec. 2.2 we iterate this technique to generate new solvable potentials, this iteration will have a different restriction giving to this second order transformation more freedom.

2.1. Time-dependent 1-SUSY transformation

The time-dependent Schrödinger equation is given by

$$i\partial_t\psi + \partial_{xx}\psi - V_0\psi = 0, \qquad x \in (x_\ell, x_r), \quad t \in (-\infty, \infty), \tag{1}$$

where the potential $V_0 = V_0(x,t)$ is a real known function; x_ℓ and x_r are the left and right endpoints of the domain of the potential. In order to generate a new exactly solvable system consider the following intertwining relationship between two Schrödinger operators

$$S_1 L_1 = L_1 S_0, \tag{2}$$

where the Schrödinger operators S_0, S_1, and the intertwining operator L_1 are defined as

$$S_j = i\partial_t + \partial_{xx} - V_j; \quad j = 0, 1; \qquad L_1 = A_1 \left(-\partial_x + \frac{u_x}{u}\right), \tag{3}$$

where $A_1 = A_1(t)$ is in principle a complex valued function but in this article it will be considered only the case when A_1 is real-valued, the function $u = u(x,t)$ is called transformation or seed function and $u_x = \partial_x u(x,t)$. A comparison of the expansions of the left and right hand sides of the intertwining relationship (2) after some simplifications lead us to the couple of equations

$$V_1 = V_0 + i\frac{A_{1t}}{A_1} - 2\partial_{xx} \ln u, \qquad i\partial_{xt} \ln u + \partial_x \left(\frac{u_{xx}}{u} - V_0\right) = 0, \tag{4}$$

where $A_{1t} = \partial_t A_1(t)$ and $u_{xx} = \partial_{xx} u(x,t)$; integration of the last equation with respect to x simplifies the equation u needs to fulfill to

$$i\partial_t u + u_{xx} - V_0 u = 0, \tag{5}$$

as in the time-independent SUSY (see [6]), the transformation function $u(x,t)$ satisfies the Schrödinger equation, but now the time-dependent version; moreover, no separability of $u(x,t)$ is assumed. Asking V_1 to be a real potential, i.e. $\mathrm{Im}(V_1) = 0$, and using (4), a second condition the transformation function needs to fulfill is

$$\partial_{xxx} \ln \left(\frac{u}{u^*}\right) = 0. \tag{6}$$

Since we are restricting ourselves to the case where A_1 is a real valued function, it can be obtained as

$$A_1 = \exp\left\{-i \int^t \partial_{xx} \ln \left[\frac{u(x,s)}{u^*(x,s)}\right] ds\right\} = \exp\left\{2 \int^t \mathrm{Im}\left[\partial_{xx} \ln u(x,s)\right] ds\right\}, \tag{7}$$

where $\mathrm{Im}(\cdot)$ stands for the imaginary part. The second expression shows explicitly that A_1 is a real function. Substituting this last expression in (4) a simplified expression for V_1 can be obtained

$$V_1 = V_0 - 2\partial_{xx} \ln |u|. \tag{8}$$

Note that in order to obtain a potential V_1 with no more singularities than the ones already existing in V_0, it is necessary to use a transformation function u without zeros in the domain (x_ℓ, x_r) at any time.

The solutions ϕ of the new Schrödinger equation

$$i\partial_t \phi + \partial_{xx}\phi - V_1 \phi = 0, \qquad x \in (x_\ell, x_r), \quad t \in (-\infty, \infty), \tag{9}$$

can be obtained by applying the operator L_1 onto the solutions ψ of the original equation, $\phi = L_1 \psi$.

It is also important to notice that the adjoint equation of (2) give us a new intertwining relationship

$$L_1^\dagger S_1 = S_0 L_1^\dagger, \qquad \text{where} \quad L_1^\dagger = A_1 \left(\partial_x + \frac{u_x^*}{u^*}\right), \tag{10}$$

and, as can be seen from (10), the function annihilated by $L_1^\dagger$ could also be solution of $S_1\phi = 0$. To construct this solution ϕ the first step is to solve the first order differential equation $L_1^\dagger\phi = 0$ and then to ask the condition $S_1\phi = 0$, the final form of this extra solution is

$$\phi = \frac{1}{A_1 u^*}. \tag{11}$$

Thus, solutions of the equation $S_1\phi = 0$ can be obtained with the operator L_1 as $\phi = L_1\psi$ or as given by (11).

Summarizing, given a Schrödinger equation with the set of solutions ψ satisfying the boundary conditions (physical solutions) and a transformation function $u(x,t)$ fulfilling (5), (6) and $u(x,t) \neq 0$ for all $x \in (x_\ell, x_r)$ and $t \in (-\infty, \infty)$ a new time-dependent potential $V_1(x,t)$ given by (8) can be generated and the solutions of the corresponding Schrödinger equation (9) will have the form $\phi = L_1\psi$, and an extra solution can be found using (11), this is the so called missing state.

2.2. Second-order confluent supersymmetry algorithm

The time-independent confluent SUSY transformation introduced in [9–12] can be seen as an iteration of a 1-SUSY transformation. In the same way, an iteration can be performed for the time-dependent situation [8]. Let us consider the intertwining relation (2) and also

$$S_2 L_2 = L_2 S_1, \quad \text{where} \quad S_j = i\partial_t + \partial_{xx} - V_j, \quad j = 1, 2, \qquad L_2 = A_2\left(-\partial_x + \frac{v_x}{v}\right), \tag{12}$$

the intermediate transformation function $v(x,t)$ satisfies $S_1 v = 0$. If we use the function (11) as transformation function turns out that $V_2 = V_0$, i.e. we obtain the original Schrödinger operator as a result of the second transformation, so we need a more general solution v

$$v = \frac{1}{A_1 u^*}\left(\omega + \int^x |u(s,t)|^2 ds\right), \tag{13}$$

where ω is considered in this manuscript as a real constant. Now, the new potential is given by

$$V_2 = V_1 - 2\partial_{xx}\ln|v| = V_0 - 2\partial_{xx}\ln\left|\omega + \int^x |u(s,t)|^2 ds\right|. \tag{14}$$

An important remark is that every value of the constant ω defines a potential V_2, for different values different physics are modeled.

In order to produce regular potential the condition changed, we need a transformation function $u(x,t)$ such that

$$\omega + \int^x |u(s,t)|^2 ds \neq 0, \tag{15}$$

which means that any normalized function will satisfy this condition, but also some non-square-integrable functions are going to fulfill this condition. The reality condition for V_2 in terms of u can be expressed as

$$\partial_{xxx}\ln\left(\frac{v}{v^*}\right) = 0 \quad \Rightarrow \quad \partial_{xxx}\ln\left(\frac{u}{u^*}\right) = 0. \tag{16}$$

Finally, the function $A_2(t)$ appearing in the definition of the operator L_2 in (12) is

$$A_2 = \exp\left\{-i\int^t \partial_{xx}\ln\left[\frac{v(x,s)}{v^*(x,s)}\right] ds\right\} = \exp\left\{-i\int^t \partial_{xx}\ln\left[\frac{u(x,s)}{u^*(x,s)}\right] ds\right\}. \tag{17}$$

IOP Conf. Series: Journal of Physics: Conf. Series **839** (2017) 012019 doi:10.1088/1742-6596/839/1/012019

The assumption $A_2 = A_1$ will only be considered in this manuscript, (compare eqs. (7) and (17)).

Solutions ϕ of the equation $S_2\phi = 0$ where S_2 is given by (12) with V_2 as in (14) can be obtained from solutions ψ of the original equation as

$$\phi = L_2L_1\psi = A_1^2 \left\{\partial_{xx} - [\partial_x \ln(uv)]\,\partial_x + [(\partial_x \ln v)(\partial_x \ln u) - \partial_{xx} \ln u]\right\} \psi \tag{18}$$

and a missing state as

$$\phi = \frac{1}{A_1 v^*} = \frac{u(x,t)}{\omega + \int^x |u(s,t)|^2 ds}. \tag{19}$$

With this iteration u still needs to satisfy (5) and (6) but a transformation function with zeros can be used as long as the constant ω is such that (15) is satisfied.

3. A time-dependent anharmonic oscillator

The time-dependent Schrödinger equation for the harmonic oscillator is given by [13, 14]

$$\left[i\hbar\partial_t + \frac{\hbar^2}{2m}\partial_{xx} - \frac{1}{2}m\widetilde{\omega}^2 x^2\right]\psi = 0. \tag{20}$$

In this example units where $\hbar = 1$, $m = 1/2$ and $\widetilde{\omega} = 2$ will be considered, then the Schrödinger equation is

$$\left(i\partial_t + \partial_{xx} - x^2\right)\psi = 0. \tag{21}$$

Our initial potential $V_0 = x^2$ is a time-independent potential. Stationary solutions are well known and are given by

$$\psi_n(x,t) = \frac{1}{\sqrt{\pi^{1/2}2^n n!}} H_n(x) \exp\left(-\frac{x^2}{2} - iE_n t\right), \quad \text{where} \quad E_n = 2n+1, \quad n = 0,1,2,\ldots \tag{22}$$

and $H_n(x)$ are Hermite polynomials [15].

An interesting set of solutions of this equation is given by the so called coherent states, these states are Gaussian packets that maintain their shape with time and they are characterized by a complex number $z = r\exp(i\theta)$ [16]. One of this coherent states can be used as transformation function u in order to obtain SUSY partners of the harmonic oscillator, let u be given by

$$u(x,t) = \frac{1}{\pi^{1/4}} \exp\left\{-i\,t - \frac{[x - r\cos(2t)]^2}{2} + i\,\frac{r^2\cos(2t)\sin(2t)}{2} - i\,r\,x\sin(2t)\right\} \tag{23}$$

where for simplicity a parameter $z = r$ with $r \in \mathbb{R}$ is used. This election of the phase of z does not affect the generality of the results since they are periodic states, $t = 0$ is selected when $\theta = 0$. This transformation function satisfies the reality condition (6) and it is never and nowhere equal to zero. If we try to use u as given by (23) to generate new potentials with a first-order transformation we will obtain

$$V_1 = x^2 - \partial_{xx}\ln|u|^2 = V_0 + 2, \tag{24}$$

i.e., after the first order transformation we obtain a displaced version of the harmonic oscillator, this property is known in the time-independent SUSY as shape invariance and it allows to generate the solutions of the Schrödinger equation for the original potential with algebraic methods [3].

To obtain new potentials using the transformation function u we can employ the second-order confluent SUSY algorithm developed in section 2. Using the equation (14) the time-dependent SUSY partners of the harmonic oscillator are

$$V_2 = x^2 - \partial_{xx} \ln \left| \omega + \frac{1}{\pi^{1/2}} \int_{-\infty}^{x} \exp\left\{ -[s - r\cos(2t)]^2 \right\} ds \right|^2 \tag{25}$$

Note that when $t = (2m+1)\pi/4$ with $m = 0, 1, 2, \ldots$ the generated potential V_2 correspond to a Abraham-Moses potential [2], rediscovered and obtained with a first-order SUSY transformation by Mielnik [1]. It is also important to notice that to avoid singularities in V_2 the parameter ω must be in the interval $(-\infty, -1] \cup [0, \infty)$.

In Fig. 1 the time-dependent anharmonic potential $V_2(x,t)$ given by (25) with the parameters $\omega = -1.001$ and $r = 2$ is plotted at three different times, $t_0 = 0$, $t_1 = \pi/4$, $t_2 = \pi/2$. To complete a cycle note that $V_2(x, 3\pi/2) = V_2(t_1)$. The perturbation or local minimum that can be seen in $V_2(x,t_0)$ moves to the left until the position shown in $V_2(x,t_2)$ and then travels back to its initial position in the same time.

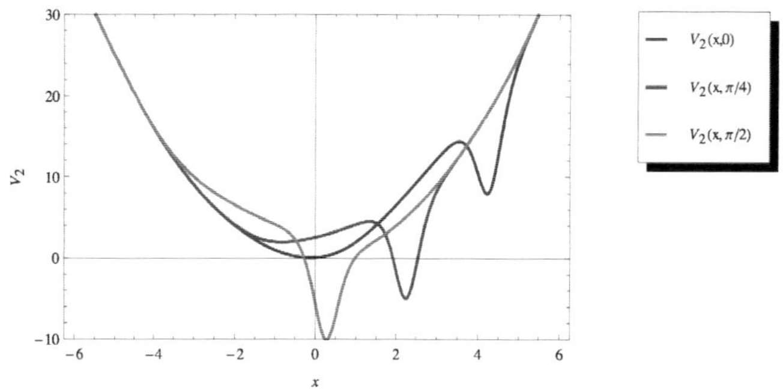

Figure 1. New potential $V_2(x,t)$ at three different times, the parameters are $\omega = -1.001$ and $r = 2$.

To obtain solutions ϕ of the new Schrödinger equation $S_2\phi = 0$ where the potential V_2 involved is given by (25) equation (18) can be used, and also (19). First step is to calculate A_1 through (7), with the selection of u in this example we can fix $A_1 = 1$. The function v is then constructed with (13). Any solution of the Harmonic Oscillator can be mapped, for example the ground state ψ_0 will be mapped to a function ϕ_0

$$\phi_0 = \frac{1}{\pi^{3/4}} r \exp\left(-\frac{x^2}{2} - 3it \right) \left\{ \frac{2\exp(-[x - r\ \cos(2t)]^2)}{\mathrm{erf}(x - r\ \cos(2t)) + 2\omega + 1} + \sqrt{\pi}\,(2x - r\ \exp(2it)) \right\}, \tag{26}$$

where, erf$(\cdot)$ is the error function [15]. On the left of figure 2 it can be seen the plot of the probabilities densities $|\phi_0(x,t)|^2$ at t_0 (blue curve), t_1 (purple curve) and t_2 (yellow curve). The missing state can be obtained using (19):

$$\phi = \frac{\exp\left\{ -i\ t - \frac{[x - r\cos(2t)]^2}{2} + i\ \frac{r^2\cos(2t)\sin(2t)}{2} - i\ r\ x\sin(2t) \right\}}{\pi^{1/4}\left\{ \frac{1}{2}[\mathrm{erf}(x - r\ \cos(2t)) + 1] + \omega \right\}}. \tag{27}$$

On the right of Fig. 2 the probability density of the missing state (27) at t_0 (blue curve), t_1 (purple curve) and t_2 (yellow curve) are shown.

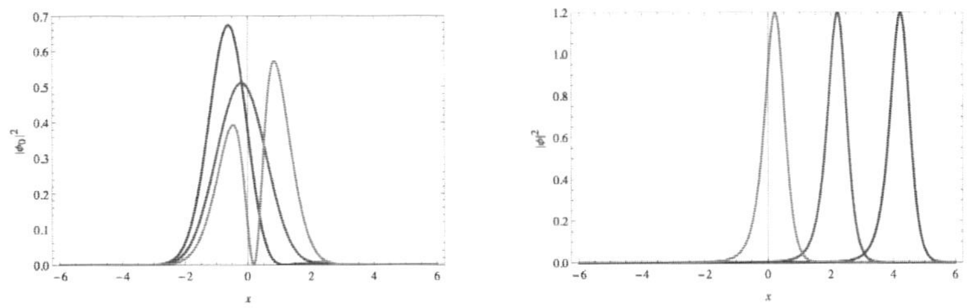

Figure 2. Probabilities densities of $\phi_0(x,t)$ (left) and the missing state $\phi(x,t)$ (right) at $t=0$ (blue curve), $t=\pi/4$ (purple curve) and $t=\pi/2$ (yellow curve). The parameters are $\omega=-1.001$ and $r=2$.

4. Conclusions

A time-dependent anharmonic oscillator was constructed using a generalized supersymmetry transformation, in particular it was shown that application of the simplest version of SUSY is not enough to generate a new potential from the traditional harmonic oscillator when a coherent state is used as transformation function. An iteration was needed to generate this family of potentials, we called this two-step procedure a confluent time-dependent SUSY transformation. Solutions of the corresponding time-dependent Schrödinger equation can be generated using any solution of the harmonic oscillator and the constructed operators L_1 and L_2, furthermore, a missing state was also found.

References

[1] Mielnik B 1984 *J. Math. Phys.* **25** 3387
[2] Abraham P B and Moses H E 1980 *Phys. Rev. A* **22** 1333
[3] Cooper F, Khare A and Sukhatme U 1995 *Phys. Rep.* **251** 267
[4] Andrianov A A and Cannata F 2004 *J. Phys A: Math. Gen.* **37** 10297
[5] Fernández C D J and Fernández-García N 2004 *AIP Conf. Proc.* **744** 236
[6] Fernández C D J 2010 *AIP Conf. Proc.* **1287** 3
[7] Matveev V and Salle M 1991 *Darboux Transformations and Solitons* (Springer-Verlag, New York)
[8] Bagrov V G, Samsonov B F and Shekoyan L A 1995 *Russ. Phys. J.* **38** 706
[9] Baye D 1993 *Phys. Rev. A* **48** 2040
[10] Stahlhofen A A 1995 *Phys. Rev. A* **51** 934
[11] Mielnik B, Nieto L M and Rosas-Ortiz O 2000 *Physi. Lett. A* **269** 70
[12] Fernández C D J and Salinas-Hernández E 2005 *Phys. Lett. A* **338** 13
[13] Flügge S 2012 *Practical quantum mechanics* (Springer, New York)
[14] Cohen-Tannoudji C, Diu B and Laloe F 1977 *Quantum mechanics* (Wiley, New York)
[15] Stegun I A and Abramowitz M 1964 *Handbook of mathematical functions with formulas, graphs, and mathematical tables Dover books on intermediate advanced mathematics* (US. Nat. Bureau Stand., New York)
[16] Schrödinger E 1926 *Die Naturwissenschaften* **28** 664

IOP Conf. Series: Journal of Physics: Conf. Series **839** (2017) 012020 doi:10.1088/1742-6596/839/1/012020

Generalized Equations and Their Solutions in the $(S,0)\oplus(0,S)$ Representations of the Lorentz Group

V V Dvoeglazov

UAF, Universidad Autónoma de Zacatecas
Apartado Postal 636, Zacatecas 98061 Zac., México

E-mail: valeri@fisica.uaz.edu.mx

Abstract.
We present three explicit examples of generalizations in relativistic quantum mechanics. First of all, we discuss the generalized spin-1/2 equations for neutrinos. They have been obtained by means of the Gersten-Sakurai method for derivations of arbitrary-spin relativistic equations. Possible physical consequences are discussed. Next, it is easy to check that both Dirac algebraic equation $Det(\hat{p}-m)=0$ and $Det(\hat{p}+m)=0$ for $u-$ and $v-$ 4-spinors have solutions with $p_0=\pm E_p=\pm\sqrt{\mathbf{p}^2+m^2}$. The same is true for higher-spin equations. Meanwhile, every book considers the equality $p_0=E_p$ for both $u-$ and $v-$ spinors of the $(1/2,0)\oplus(0,1/2))$ representation only, thus applying the Dirac-Feynman-Stueckelberg procedure for elimination of the negative-energy solutions. The recent Ziino works (and, independently, the articles of several others) show that the Fock space can be doubled. We re-consider this possibility on the quantum field level for both $S=1/2$ and higher spin particles. The third example is: we postulate the non-commutativity of 4-momenta, and we derive the mass splitting in the Dirac equation. Some applications are discussed.

1. Generalized neutrino equations

A. Gersten [1] proposed a method for derivations of massless equations of arbitrary-spin particles. In fact, his method is related to the van der Waerden-Sakurai [2] procedure for the derivation of the massive Dirac equation. I commented the derivation of the Maxwell equations in [3]. Then, I showed that the method is rather ambigious, because instead of free-space Maxwell equations one can obtain *generalized* $S=1$ equations, which connect the antisymmetric tensor field with additional scalar fields. The problem of physical significance of additional scalar chi-fields should be solved, of course, by experiment.

In the present contribution we apply the van der Waerden-Sakurai-Gersten procedure to the spin-1/2 fields. As a result we obtain equations which *generalize* the well-known Weyl equations. However, these equations are known for a long time [4]. Raspini [5, 6] analized them again in detail. I add some comments on physical contents of the generalized spin-1/2 equations.

I use the equation (4) of the Gersten paper [1a] for the two-component spinor field function:

$$(E^2-c^2\vec{\mathbf{p}}^2)I^{(2)}\psi=\left[EI^{(2)}-c\vec{\mathbf{p}}\cdot\vec{\sigma}\right]\left[EI^{(2)}+c\vec{\mathbf{p}}\cdot\vec{\sigma}\right]\psi=0\,. \qquad (1)$$

The matrix $I^{(2)}$ is the identity matrix of order 2; $\vec{\sigma}$ is the vector composed of the Pauli matrices; c is the light speed, E and $\mathbf{p}$ form the energy-momentum vector. Actually, this equation is

IOP Conf. Series: Journal of Physics: Conf. Series **839** (2017) 012020 doi:10.1088/1742-6596/839/1/012020

the massless limit of the equation which has been presented (together with the corresponding method of derivation of the Dirac equation) in the Sakurai book [2]. In the latter case one should substitute m^2c^4 into the right-hand side of Eq. (1). However, instead of the equation (3.25) of [2] one can define *ad hoc* the two-component 'right' field function

$$\phi_R = \frac{1}{m_1 c}(i\hbar\frac{\partial}{\partial x_0} - i\hbar\boldsymbol{\sigma}\cdot\boldsymbol{\nabla})\psi, \quad \phi_L = \psi \tag{2}$$

with the different mass parameter m_1. In such a way we come to the system of the first-order differential equations

$$(i\hbar\frac{\partial}{\partial x_0} + i\hbar\boldsymbol{\sigma}\cdot\boldsymbol{\nabla})\phi_R = \frac{m_2^2 c}{m_1}\phi_L\,, \tag{3}$$

$$(i\hbar\frac{\partial}{\partial x_0} - i\hbar\boldsymbol{\sigma}\cdot\boldsymbol{\nabla})\phi_L = m_1 c\phi_R\,. \tag{4}$$

It can be re-written in the 4-component form:

$$\begin{pmatrix} i\hbar(\partial/\partial x_0) & i\hbar\boldsymbol{\sigma}\cdot\boldsymbol{\nabla} \\ -i\hbar\boldsymbol{\sigma}\cdot\boldsymbol{\nabla} & -i\hbar(\partial/\partial x_0)\end{pmatrix}\begin{pmatrix}\psi_A \\ \psi_B\end{pmatrix} = \tag{5}$$
$$= \frac{c}{2}\begin{pmatrix} (m_2^2/m_1 + m_1) & (-m_2^2/m_1 + m_1) \\ (-m_2^2/m_1 + m_1) & (m_2^2/m_1 + m_1)\end{pmatrix}\begin{pmatrix}\psi_A \\ \psi_B\end{pmatrix}$$

for the function $\Psi = column(\psi_A \quad \psi_B) = column(\phi_R + \phi_L \quad \phi_R - \phi_L)$. The generalized equation (5) can be written in the covariant form.

$$\left[i\gamma^\mu\partial_\mu - \frac{m_2^2 c}{m_1 {}^-h}\frac{(1-\gamma^5)}{2} - \frac{m_1 c}{{}^-h}\frac{(1+\gamma^5)}{2}\right]\Psi = 0\,. \tag{6}$$

The standard representation of γ^μ matrices has been used here.

If $m_1 = m_2$ we can recover the standard Dirac equation. As noted in [4b] this procedure can be viewed as the simple change of the representation of γ^μ matrices. However, this is valid unless $m_2 \neq 0$ only. Otherwise, the entries in the transformation matrix become to be singular.

Furthermore, one can either repeat a similar procedure (the modified Sakurai procedure) starting from the *massless* equation (4) of [1a] or put $m_2 = 0$ in eq. (6). The *massless* equation is[1]

$$\left[i\gamma^\mu\partial_\mu - \frac{m_1 c}{\hbar}\frac{(1+\gamma^5)}{2}\right]\Psi = 0\,. \tag{7}$$

Then, we may have different physical consequences following from (7) comparing with those which follow from the Weyl equation.[2] The mathematical reason of such a possibility of different massless limits is that the corresponding change of representation of γ^μ matrices involves mass parameters m_1 and m_2 themselves.

It is interesting to note that we can also repeat this procedure for other *ad hoc* definition (or for even more general definitions);

$$\phi_L = \frac{1}{m_3 c}(i{}^-h\frac{\partial}{\partial x_0} + i{}^-h\boldsymbol{\sigma}\,\boldsymbol{\nabla})\psi, \quad \phi_R = \psi\,. \tag{8}$$

[1] It is necesary to stress that the term *'massless'* is used in the sense that $p_\mu p^\mu = 0$.

[2] Remember that the Weyl equation is obtained as $m \to 0$ limit of the usual Dirac equation.

This is due to the fact that the parity properties of the two-component spinor are undefined in the two-component equation. The resulting equation is

$$\left[i\gamma^\mu \partial_\mu - \frac{m_4^2 c}{m_3 \hbar}\frac{(1+\gamma^5)}{2} - \frac{m_3 c}{\hbar}\frac{(1-\gamma^5)}{2}\right]\tilde{\Psi} = 0\,, \tag{9}$$

which gives us yet another equation in the massless limit ($m_4 \to 0$):

$$\left[i\gamma^\mu \partial_\mu - \frac{m_3 c}{\hbar}\frac{(1-\gamma^5)}{2}\right]\tilde{\Psi} = 0\,, \tag{10}$$

differing in the sign at the γ_5 term from (7).

The above procedure can be generalized to *any* Lorentz group representations, *i. e.*, to any spins. Is the physical content of the generalized $S = 1/2$ *massless* equations the same as that of the Weyl equation? Our answer is 'no'. The excellent discussion can be found in [4a,b]. First of all, the theory does *not* have chiral invariance. Those authors call the additional parameters as the measures of the degree of chirality. Apart of this, Tokuoka introduced the concept of the gauge transformations (not to confuse with phase transformations) for the 4-spinor fields. He also found some strange properties of the anti-commutation relations (see §3 in [4a]). And finally, the equation (7) describes *four* states, two of which answer for the positive energy $p_0 = |\mathbf{p}|$, and two others answer for the negative energy $p_0 = -|\mathbf{p}|$.

I just want to add the following speculations to the discussion. The operator of the *chiral-helicity* $\hat{\eta} = (\boldsymbol{\alpha} \cdot \hat{\mathbf{p}})$ (in the spinorial representation) used in [4b] does *not* commute, *e.g.*, with the Hamiltonian of the equation (7):[3]

$$[\mathcal{H}, \boldsymbol{\alpha} \cdot \hat{\mathbf{p}}]_- = 2\frac{m_1 c}{\hbar}\frac{1-\gamma^5}{2}(\boldsymbol{\gamma} \cdot \hat{\mathbf{p}})\,. \tag{11}$$

For the eigenstates of the *chiral-helicity* the system of corresponding equations can be read ($\eta = \uparrow, \downarrow$)

$$i\gamma^\mu \partial_\mu \Psi_\eta - \frac{m_1 c}{\hbar}\frac{1+\gamma^5}{2}\Psi_{-\eta} = 0\,. \tag{12}$$

The conjugated eigenstates of the Hamiltonian $|\Psi_\uparrow + \Psi_\downarrow >$ and $|\Psi_\uparrow - \Psi_\downarrow >$ are connected, in fact, by γ^5 transformation $\Psi \to \gamma^5 \Psi \sim (\boldsymbol{\alpha} \cdot \hat{\mathbf{p}})\Psi$ (or $m_1 \to -m_1$). However, the γ^5 transformation is related to the PT ($t \to -t$ only) transformation [4b], which, in its turn, can be interpreted as $E \to -E$, if one accepts the Stueckelberg idea about antiparticles. We associate $|\Psi_\uparrow + \Psi_\downarrow >$ with the positive-energy eigenvalue of the Hamiltonian $p_0 = |\mathbf{p}|$ and $|\Psi_\uparrow - \Psi_\downarrow >$, with the negative-energy eigenvalue of the Hamiltonian ($p_0 = -|\mathbf{p}|$). Thus, the free chiral-helicity massless eigenstates may oscillate one to another with the frequency $\omega = E/\hbar$ (as the massive chiral-helicity eigenstates, see [7a] for details). Moreover, a special kind of interaction which is not symmetric with respect to the chiral-helicity states (for instance, if the left chiral-helicity eigenstates interact with the matter only) may induce changes in the oscillation frequency, like in the Wolfenstein (MSW) formalism.

Thus, we derive the same equations as in Ref. [4] beginning with the basic equation (1). The question is: how can these frameworks be connected with the Ryder method of derivation of relativistic wave equations, and with the subsequent analysis of problems of the choice of normalization and that of the choice of phase factors in the papers [7–9]? However, the conclusion may be similar to that which was achieved before: the dynamical properties of the massless particles (*e. g.*, neutrinos and photons) may differ from those defined by the well-known Weyl and Maxwell equations [10].

[3] Do not confuse with the Dirac Hamiltonian.

2. Negative energies in the Dirac equation

The recent problems of superluminal neutrinos, e. g., Ref. [11], negative mass-squared neutrinos, various schemes of oscillations including sterile neutrinos, e. g. [12], require much attention. The problem of the lepton mass splitting (e, μ, τ) has long history [13]. This suggests that something missed in the foundations of relativistic quantum theories. Modifications seem to be necessary in the Dirac sea concept, and in the even more sophisticated Stueckelberg concept of the backward propagation in time. The Dirac sea concept is intrinsically related to the Pauli principle. However, the Pauli principle is intrinsically connected with the Fermi statistics and the anticommutation relations of fermions. Recently, the concept of the *bi-orthonormality* has been proposed; the (anti) commutation relations and statistics are assumed to be different for *neutral* particles [8].

We observe some interisting things related to the negative-energy concept. The Dirac equation is:

$$[i\gamma^\mu \partial_\mu - m]\Psi(x) = 0\,. \tag{13}$$

At least, 3 methods of its derivation exist [2, 14, 15]:

- the Dirac one (the Hamiltonian should be linear in $\partial/\partial x^i$, and be compatible with $E_p^2 - \mathbf{p}^2 c^2 = m^2 c^4$);
- the Sakurai one (based on the equation $(E_p - \sigma \cdot \mathbf{p})(E_p + \sigma \cdot \mathbf{p})\phi = m^2\phi$);
- the Ryder one (the relation between 2-spinors at rest is $\phi_R(\mathbf{0}) = \pm\phi_L(\mathbf{0})$, and application of the boosts to them).

Usually, everybody uses the following definition of the field operator [16] in the pseudo-Euclidean metrics:

$$\Psi(x) = \frac{1}{(2\pi)^3} \sum_h \int \frac{d^3\mathbf{p}}{2E_p} [u_h(\mathbf{p}) a_h(\mathbf{p}) e^{-ip\cdot x} + v_h(\mathbf{p}) b_h^\dagger(\mathbf{p})] e^{+ip\cdot x}]\,, \tag{14}$$

as given *ab initio*. After actions of the Dirac operator at $\exp(\mp i p_\mu x^\mu)$ the 4-spinors ($u-$ and $v-$) satisfy the momentum-space equations: $(\hat{p} - m) u_h(p) = 0$ and $(\hat{p} + m) v_h(p) = 0$, respectively; the h is the polarization index. However, it is easy to prove from the characteristic equations $Det(\hat{p} \mp m) = (p_0^2 - \mathbf{p}^2 - m^2)^2 = 0$ that the solutions should satisfy the energy-momentum relation $p_0 = \pm E_p = \pm\sqrt{\mathbf{p}^2 + m^2}$ in both cases.

Let me remind the general scheme of construction of the field operator, which has been presented in [17]. In the case of the $(1/2, 0) \oplus (0, 1/2)$ representation we have:

$$\begin{aligned}
\Psi(x) &= \frac{1}{(2\pi)^3} \int d^4p\, \delta(p^2 - m^2) e^{-ip\cdot x} \Psi(p) = \\
&= \frac{1}{(2\pi)^3} \sum_h \int d^4p\, \delta(p_0^2 - E_p^2) e^{-ip\cdot x} u_h(p_0, \mathbf{p}) a_h(p_0, \mathbf{p}) = \\
&= \frac{1}{(2\pi)^3} \int \frac{d^4p}{2E_p} [\delta(p_0 - E_p) + \delta(p_0 + E_p)][\theta(p_0) + \theta(-p_0)] e^{-ip\cdot x} \\
&\times \sum_h u_h(p) a_h(p) = \frac{1}{(2\pi)^3} \sum_h \int \frac{d^4p}{2E_p} [\delta(p_0 - E_p) + \delta(p_0 + E_p)] \Big[\theta(p_0) u_h(p) a_h(p) e^{-ip\cdot x} + \\
&+ \theta(p_0) u_h(-p) a_h(-p) e^{+ip\cdot x}\Big] = \frac{1}{(2\pi)^3} \sum_h \int \frac{d^3\mathbf{p}}{2E_p} \theta(p_0) \Big[u_h(p) a_h(p)|_{p_0 = E_p} e^{-i(E_p t - \mathbf{p}\cdot\mathbf{x})} + \\
&+ u_h(-p) a_h(-p)|_{p_0 = E_p} e^{+i(E_p t - \mathbf{p}\cdot\mathbf{x})}\Big]
\end{aligned} \tag{15}$$

During the calculations above we had to represent $1 = \theta(p_0) + \theta(-p_0)$ in order to get positive- and negative-frequency parts. Moreover, during these calculations we did not yet assumed, which

IOP Conf. Series: Journal of Physics: Conf. Series **839** (2017) 012020 doi:10.1088/1742-6596/839/1/012020

equation this field operator (namely, the $u-$ spinor) does satisfy, with negative- or positive-mass?

In general we should transform $u_h(-p)$ to the $v(p)$. The procedure is the following one [18]. In the Dirac case we should assume the following relation in the field operator:

$$\sum_h v_h(p) b_h^\dagger(p) = \sum_h u_h(-p) a_h(-p)\,. \tag{16}$$

We know that [15][4]

$$\begin{aligned} \bar{u}_{(\mu)}(p) u_{(\lambda)}(p) &= +m\delta_{\mu\lambda}\,, & (17)\\ \bar{u}_{(\mu)}(p) u_{(\lambda)}(-p) &= 0\,, & (18)\\ \bar{v}_{(\mu)}(p) v_{(\lambda)}(p) &= -m\delta_{\mu\lambda}\,, & (19)\\ \bar{v}_{(\mu)}(p) u_{(\lambda)}(p) &= 0\,, & (20) \end{aligned}$$

but we need $\Lambda_{(\mu)(\lambda)}(p) = \bar{v}_{(\mu)}(p) u_{(\lambda)}(-p)$. By direct calculations, we find

$$-m b_{(\mu)}^\dagger(p) = \sum_\lambda \Lambda_{(\mu)(\lambda)}(p) a_{(\lambda)}(-p)\,. \tag{21}$$

Hence, $\Lambda_{(\mu)(\lambda)} = -im(\boldsymbol{\sigma}\cdot\mathbf{n})_{(\mu)(\lambda)}$, $\mathbf{n} = \mathbf{p}/|\mathbf{p}|$, and

$$b_{(\mu)}^\dagger(p) = i\sum_\lambda (\boldsymbol{\sigma}\cdot\mathbf{n})_{(\mu)(\lambda)} a_{(\lambda)}(-p)\,. \tag{22}$$

Multiplying (16) by $\bar{u}_{(\mu)}(-p)$ we obtain

$$a_{(\mu)}(-p) = -i\sum_\lambda (\boldsymbol{\sigma}\cdot\mathbf{n})_{(\mu)(\lambda)} b_{(\lambda)}^\dagger(p)\,. \tag{23}$$

The equations are self-consistent. In the $(1,0)\oplus(0,1)$ representation the similar procedure leads to somewhat different situation:

$$a_{(\mu)}(p) = [1 - 2(\mathbf{S}\cdot\mathbf{n})^2]_{(\mu)(\lambda)} a_{(\lambda)}(-p)\,. \tag{24}$$

This signifies that in order to construct the Sankaranarayanan-Good field operator, which was used by Ahluwalia, Johnson and Goldman [19], it satisfies $[\gamma_{\mu\nu}\partial_\mu\partial_\nu - \frac{(i\partial/\partial t)}{E} m^2]\Psi(x) = 0$, we need additional postulates. For instance, one can try to construct the left- and the right-hand side of the field operator separately each other [20].

However, other ways of thinking are possible. First of all to mention, we have, in fact, $u_h(E_p,\mathbf{p})$ and $u_h(-E_p,\mathbf{p})$, and $v_h(E_p,\mathbf{p})$ and $v_h(-E_p,\mathbf{p})$, originally, which may satisfy the equations:[5]

$$\left[E_p(\pm\gamma^0) - \boldsymbol{\gamma}\cdot\mathbf{p} - m\right] u_h(\pm E_p,\mathbf{p}) = 0\,. \tag{25}$$

Due to the properties $U^\dagger\gamma^0 U = -\gamma^0$, $U^\dagger\gamma^i U = +\gamma^i$ with the unitary matrix $U = \begin{pmatrix} 0 & -1\\ 1 & 0\end{pmatrix} = \gamma^0\gamma^5$ in the Weyl basis, we have

$$\left[E_p\gamma^0 - \boldsymbol{\gamma}\cdot\mathbf{p} - m\right] U^\dagger u_h(-E_p,\mathbf{p}) = 0\,. \tag{26}$$

[4] (μ) and (λ) are the polazrization indices here. According to the referee advice I use parenthesis here to stress this.

[5] Remember that, as before, we can always make the substitution $\mathbf{p}\to-\mathbf{p}$ in any of the integrands of (15).

IOP Conf. Series: Journal of Physics: Conf. Series **839** (2017) 012020 doi:10.1088/1742-6596/839/1/012020

The properties of the $U-$ matrix are opposite to those of $P^\dagger\gamma^0 P = +\gamma^0$, $P^\dagger\gamma^i P = -\gamma^i$ with the usual $P = \gamma^0$, thus giving $\left[-E_p\gamma^0 + \gamma\cdot\mathbf{p} - m\right] Pu_h(-E_p, \mathbf{p}) = -\left[\hat{p} + m\right]\tilde{v}_?(E_p, \mathbf{p}) = 0$. While, the relations of the spinors $v_h(E_p, \mathbf{p}) = \gamma_5 u_h(E_p, \mathbf{p})$ are well-known, it seems that the relations of the $v-$ spinors of the positive energy to $u-$ spinors of the negative energy are frequently forgotten, $\tilde{v}_?(E_p, \mathbf{p}) = \gamma^0 u_h(-E_p, \mathbf{p})$.

Thus, unless the unitary transformations do not change the physical content, we have that the negative-energy spinors $\gamma^5\gamma^0 u^-$ (see (26)) satisfy the accustomed "positive-energy" Dirac equation. We should then expect the same physical content. Their explicit forms $\gamma^5\gamma^0 u^-$ are different from the textbook "positive-energy" Dirac spinors. They are the following ones:[6]

$$\tilde{u}(p) = \frac{N}{\sqrt{2m(-E_p + m)}} \begin{pmatrix} -p^+ + m \\ -p_r \\ p^- - m \\ -p_r \end{pmatrix}, \tag{27}$$

$$\tilde{\tilde{u}}(p) = \frac{N}{\sqrt{2m(-E_p + m)}} \begin{pmatrix} -p_l \\ -p^- + m \\ -p_l \\ p^+ - m \end{pmatrix}. \tag{28}$$

$E_p = \sqrt{\mathbf{p}^2 + m^2} > 0$, $p_0 = \pm E_p$, $p^\pm = E \pm p_z$, $p_{r,l} = p_x \pm i p_y$. Their normalization is to $(-2N^2)$.

What about the $\tilde{v}(p) = \gamma^0 u^-$ transformed with the γ_0 matrix? Are they equal to $v_h(p) = \gamma^5 u_h(p)$? Our answer is '*no*'. Obviously, they also do not have well-known forms of the usual $v-$ spinors in the Weyl basis, differing by phase factors and in the signs at the mass terms.

Next, one can prove that the matrix

$$P = e^{i\theta}\gamma^0 = e^{i\theta}\begin{pmatrix} 0 & 1_{2\times 2} \\ 1_{2\times 2} & 0 \end{pmatrix} \tag{29}$$

can be used in the parity operator as well as in the original Weyl basis. However, if we would take the phase factor to be zero we obtain that while $u_h(p)$ have the eigenvalue $+1$ of the parity, but $(R = (\mathbf{x} \to -\mathbf{x}, \mathbf{p} \to -\mathbf{p}))$

$$PR\tilde{u}(p) = PR\gamma^5\gamma^0 u(-E_p, \mathbf{p}) = -\tilde{u}(p)\,, \tag{30}$$

$$PR\tilde{\tilde{u}}(p) = PR\gamma^5\gamma^0 u(-E_p, \mathbf{p}) = -\tilde{\tilde{u}}(p)\,. \tag{31}$$

Perhaps, one should choose the phase factor $\theta = \pi$. Thus, we again confirmed that the relative (particle-antiparticle) intrinsic parity has physical significance only.

Similar formulations have been presented in Refs. [21], and [22]. The group-theoretical basis for such doubling has been given in the papers by Gelfand, Tsetlin and Sokolik [23], who first presented the theory in the 2-dimensional representation of the inversion group in 1956 (later called as "the Bargmann-Wightman-Wigner-type quantum field theory" in 1993). M. Markov wrote *two* Dirac equations with the opposite signs at the mass term [21] long ago:

$$\left[i\gamma^\mu\partial_\mu - m\right]\Psi_1(x) = 0\,, \tag{32}$$

$$\left[i\gamma^\mu\partial_\mu + m\right]\Psi_2(x) = 0\,. \tag{33}$$

In fact, he studied all properties of this relativistic quantum model (while he did not know yet the quantum field theory in 1937). Next, he added and subtracted these equations. What did

[6] We use tildes because we do not yet know their polarization properties.

he obtain?

$$i\gamma^\mu\partial_\mu\varphi(x) - m\chi(x) = 0\,, \tag{34}$$
$$i\gamma^\mu\partial_\mu\chi(x) - m\varphi(x) = 0\,. \tag{35}$$

Thus, φ and χ solutions can be presented as some superpositions of the Dirac 4-spinors $u-$ and $v-$. These equations, of course, can be identified with the equations for the Majorana-like $\lambda-$ and $\rho-$, which were presented in Ref. [7]:[7]

$$i\gamma^\mu\partial_\mu\lambda^S(x) - m\rho^A(x) = 0\,, \tag{36}$$
$$i\gamma^\mu\partial_\mu\rho^A(x) - m\lambda^S(x) = 0\,, \tag{37}$$
$$i\gamma^\mu\partial_\mu\lambda^A(x) + m\rho^S(x) = 0\,, \tag{38}$$
$$i\gamma^\mu\partial_\mu\rho^S(x) + m\lambda^A(x) = 0\,. \tag{39}$$

Neither of them can be regarded as the Dirac equation. However, they can be written in the 8-component form as follows:

$$\left[i\Gamma^\mu\partial_\mu - m\right]\Psi_{(+)}(x) = 0\,, \tag{40}$$
$$\left[i\Gamma^\mu\partial_\mu + m\right]\Psi_{(-)}(x) = 0\,, \tag{41}$$

with

$$\Psi_{(+)}(x) = \begin{pmatrix}\rho^A(x)\\ \lambda^S(x)\end{pmatrix}, \Psi_{(-)}(x) = \begin{pmatrix}\rho^S(x)\\ \lambda^A(x)\end{pmatrix}, \Gamma^\mu = \begin{pmatrix}0 & \gamma^\mu\\ \gamma^\mu & 0\end{pmatrix}. \tag{42}$$

It is easy to find the corresponding projection operators, and the Feynman-Stueckelberg propagator.

You may say that all this is just related to the spin-parity basis rotation (unitary transformations). However, in the previous papers I explained: the connection with the Dirac spinors has been found [7, 24].[8] For instance,

$$\begin{pmatrix}\lambda^S_\uparrow(\mathbf{p})\\ \lambda^S_\downarrow(\mathbf{p})\\ \lambda^A_\uparrow(\mathbf{p})\\ \lambda^A_\downarrow(\mathbf{p})\end{pmatrix} = \frac{1}{2}\begin{pmatrix}1 & i & -1 & i\\ -i & 1 & -i & -1\\ 1 & -i & -1 & -i\\ i & 1 & i & -1\end{pmatrix}\begin{pmatrix}u_{+1/2}(\mathbf{p})\\ u_{-1/2}(\mathbf{p})\\ v_{+1/2}(\mathbf{p})\\ v_{-1/2}(\mathbf{p})\end{pmatrix}, \tag{43}$$

provided that the 4-spinors have the same physical dimension. Thus, we can see that the two 4-spinor systems are connected by the unitary transformations, and this represents itself the rotation of the spin-parity basis. However, it is usually assumed that the $\lambda-$ and $\rho-$ spinors describe the neutral particles, meanwhile $u-$ and $v-$ spinors describe the charged particles. Kirchbach [24] found the amplitudes for neutrinoless double beta decay ($00\nu\beta$) in this scheme. It is obvious from (43) that there are some additional terms comparing with the standard calculations of those amplitudes.

One can also re-write the above equations into the two-component forms. Thus, one obtains the Feynman-Gell-Mann [25] equations. As Markov wrote himself, he was expecting "new physics" from these equations.

[7] Of course, the signs at the mass terms depend on, how do we associate the positive- or negative- frequency solutions with λ and ρ.

[8] I also acknowledge personal communications from D. V. Ahluwalia on these matters.

Barut and Ziino [22] proposed yet another model. They considered γ^5 operator as the operator of the charge conjugation. Thus, the charge-conjugated Dirac equation has the different sign comparing with the ordinary formulation:

$$[i\gamma^\mu \partial_\mu + m]\Psi^c_{BZ} = 0\,, \tag{44}$$

and the so-defined charge conjugation applies to the whole system, fermion + electromagnetic field, $e \to -e$ in the covariant derivative. The superpositions of the Ψ_{BZ} and Ψ^c_{BZ} also give us the 'doubled Dirac equation', as the equations for $\lambda-$ and $\rho-$ spinors. The concept of the doubling of the Fock space has been developed in the Ziino works (cf. [23, 26]) in the framework of the quantum field theory. In their case the self/anti-self charge conjugate states are simultaneously the eigenstates of the chirality. Next, it is interesting to note that we have for the Majorana-like field operators ($a_\eta(\mathbf{p}) = b_\eta(\mathbf{p})$):

$$\left[\nu^{ML}(x^\mu) + \mathcal{C}\nu^{ML\,\dagger}(x^\mu)\right]/2 = \int \frac{d^3\mathbf{p}}{(2\pi)^3}\frac{1}{2E_p} \tag{45}$$
$$\sum_\eta \left[\begin{pmatrix} i\Theta\phi^{*\,\eta}_L(p^\mu) \\ 0\end{pmatrix} a_\eta(p^\mu)e^{-ip\cdot x} + \begin{pmatrix} 0 \\ \phi^\eta_L(p^\mu)\end{pmatrix} a^\dagger_\eta(p^\mu)e^{ip\cdot x}\right],$$

$$\left[\nu^{ML}(x^\mu) - \mathcal{C}\nu^{ML\,\dagger}(x^\mu)\right]/2 = \int \frac{d^3\mathbf{p}}{(2\pi)^3}\frac{1}{2E_p} \tag{46}$$
$$\sum_\eta \left[\begin{pmatrix} 0 \\ \phi^\eta_L(p^\mu)\end{pmatrix} a_\eta(p^\mu)e^{-ip\cdot x} + \begin{pmatrix} -i\Theta\phi^{*\,\eta}_L(p^\mu) \\ 0\end{pmatrix} a^\dagger_\eta(p^\mu)e^{ip\cdot x}\right],$$

which, thus, naturally lead to the Ziino-Barut scheme of massive chiral fields, Ref. [22].

Finally, I would like to mention that, in general, in the Weyl basis the $\gamma-$ matrices are *not* Hermitian, $\gamma^{\mu^\dagger} = \gamma^0\gamma^\mu\gamma^0$. So, $\gamma^{i^\dagger} = -\gamma^i$, $i = 1, 2, 3$, the pseudo-Hermitian matrix. The energy-momentum operator $i\partial_\mu$ is obviously Hermitian. So, the question, if the eigenvalues of the Dirac operator $i\gamma^\mu\partial_\mu$ (the mass, in fact) would be always real? The question of the complete system of the eigenvectors of the *non*-Hermitian operator deserve careful consideration [27]. Bogoliubov and Shirkov [17, p.55-56] used the scheme to construct the complete set of solutions of the relativistic equations, fixing the sign of $p_0 = +E_p$.

The main points of this Section are: there are 'negative-energy solutions' in that is previously considered as 'positive-energy solutions' of relativistic wave equations, and vice versa. Their explicit forms have been presented in the case of spin-1/2. Next, the relations to the previous works have been found. For instance, the doubling of the Fock space and the corresponding solutions of the Dirac equation obtained additional mathematical bases. Similar conclusion can be deduced for the higher-spin equations.

3. Non-commutativity in the Dirac equation.

The non-commutativity [28, 29] exibits interesting peculiarities in the Dirac case. We analized Sakurai-van der Waerden method of derivations of the Dirac (and higher-spins too) equation [30]:

$$(EI^{(4)} + \boldsymbol{\alpha}\cdot\mathbf{p} + m\beta)(EI^{(4)} - \boldsymbol{\alpha}\cdot\mathbf{p} - m\beta)\Psi_{(4)} = 0\,. \tag{47}$$

Obviously, the inverse operators of the Dirac operators of the positive- and negative- masses exist in the non-commutative case. As in the original Dirac work, we have

$$\beta^2 = 1\,, \quad \alpha^i\beta + \beta\alpha^i = 0\,, \quad \alpha^i\alpha^j + \alpha^j\alpha^i = 2\delta^{ij}\,. \tag{48}$$

For instance, their explicit forms can be chosen

$$\alpha^i = \begin{pmatrix} \sigma^i & 0 \\ 0 & -\sigma^i \end{pmatrix} , \quad \beta = \begin{pmatrix} 0 & 1_{2\times 2} \\ 1_{2\times 2} & 0 \end{pmatrix} . \tag{49}$$

We also postulate the non-commutativity relations for the components of 4-momenta:

$$[E, \mathbf{p}^i]_- = \Theta^{0i} = \theta^i , \tag{50}$$

as usual. Therefore the equation (47) will *not* lead to the well-known equation $E^2 - \mathbf{p}^2 = m^2$. Instead, we have

$$\left\{E^2 - E(\boldsymbol{\alpha}\cdot\mathbf{p}) + (\boldsymbol{\alpha}\cdot\mathbf{p})E - \mathbf{p}^2 - m^2 - i(\boldsymbol{\sigma}\otimes I_{(2)})[\mathbf{p}\times\mathbf{p}]\right\}\Psi_{(4)} = 0 \tag{51}$$

For the sake of simplicity, we may assume the last term to be zero. However, it may be used for discussion of extra physical generations. Thus, we come to

$$\left\{E^2 - \mathbf{p}^2 - m^2 - (\boldsymbol{\alpha}\cdot\boldsymbol{\theta})\right\}\Psi_{(4)} = 0 . \tag{52}$$

However, let us apply the unitary transformation. It is known [7, 31] that one can[9]

$$U_1(\boldsymbol{\sigma}\cdot\mathbf{a})U_1^{-1} = \sigma_3|\mathbf{a}| . \tag{53}$$

For $\boldsymbol{\alpha}$ matrices we re-write (53) to

$$\mathcal{U}_1(\boldsymbol{\alpha}\cdot\boldsymbol{\theta})\mathcal{U}_1^{-1} = |\boldsymbol{\theta}| \begin{pmatrix} 1 & 0 & 0 & 0 \\ 0 & -1 & 0 & 0 \\ 0 & 0 & -1 & 0 \\ 0 & 0 & 0 & 1 \end{pmatrix} = \alpha_3|\boldsymbol{\theta}| . \tag{54}$$

The explicit form of the U_1 matrix is ($a_{r,l} = a_1 \pm ia_2$, $a = |\mathbf{a}|$):

$$\begin{aligned} U_1 &= \frac{1}{\sqrt{2a(a+a_3)}} \begin{pmatrix} a+a_3 & a_l \\ -a_r & a+a_3 \end{pmatrix} = \frac{1}{\sqrt{2a(a+a_3)}} \\ &\times [a + a_3 + i\sigma_2 a_1 - i\sigma_1 a_2] , \\ \mathcal{U}_1 &= \begin{pmatrix} U_1 & 0 \\ 0 & U_1 \end{pmatrix} . \end{aligned} \tag{55}$$

Let us apply the second unitary transformation:

$$\mathcal{U}_2\alpha_3\mathcal{U}_2^\dagger = \begin{pmatrix} 1 & 0 & 0 & 0 \\ 0 & 0 & 0 & 1 \\ 0 & 0 & 1 & 0 \\ 0 & 1 & 0 & 0 \end{pmatrix} \alpha_3 \begin{pmatrix} 1 & 0 & 0 & 0 \\ 0 & 0 & 0 & 1 \\ 0 & 0 & 1 & 0 \\ 0 & 1 & 0 & 0 \end{pmatrix} = \begin{pmatrix} 1 & 0 & 0 & 0 \\ 0 & 1 & 0 & 0 \\ 0 & 0 & -1 & 0 \\ 0 & 0 & 0 & -1 \end{pmatrix} . \tag{56}$$

The final equation is

$$[E^2 - \mathbf{p}^2 - m^2 - \gamma^5_{chiral}|\boldsymbol{\theta}|]\Psi'_{(4)} = 0 . \tag{57}$$

In the physical sense this implies the mass splitting for a Dirac particle over the non-commutative space, $m_{1,2} = \pm\sqrt{m^2 \pm \theta}$. This procedure may be attractive for explanation of the mass creation and the mass splitting for fermions, because we can associate $\Delta m = |m_1 - m_2|$ to the mass difference of electron and muon, for instance.

[9] Some relations for the components of an arbitrary vector $\mathbf{a}$ should be assumed. Moreover, in our case $\boldsymbol{\theta}$ should not depend on E and $\mathbf{p}$. Otherwise, we must take the non-commutativity $[E, \mathbf{p}^i]_-$ into account again.

IOP Conf. Series: Journal of Physics: Conf. Series **839** (2017) 012020 doi:10.1088/1742-6596/839/1/012020

4. Conclusions

In this contribution we considered the generalized equation in the $(1/2, 0) \oplus (0, 1/2)$ representation. It is different from the well-known Weyl equation in the massless limit. It leads to different physical consequences. Next, we showed that the negative-energy solutions are possible in the Dirac equation. Finally, we showed that the non-commutativity parameter may give us possibility to describe fermionic mass splitting.

Acknowledgments

This talk was also presented at the 31st ICGTMP, June 19-25, 2016, CBPF, Rio de Janeiro, Brasil. I appreciate useful discussions with colleagues.

References.

[1] Gersten A 1999 *Found. Phys. Lett.* **12** 291
Gersten A2000 **13** 185
[2] Sakurai J J 1967 *Advanced Quantum Mechanics* (Addison-Wesley) §3.2.
[3] Dvoeglazov V V 2000 *J. Phys A: Math. Gen.* **33** 5011
[4] Tokuoka Z 1967 *Prog. Theor. Phys.* **37** 603
Gupta N D S 1967 *Nucl. Phys.* **B4** 147
Santhanam T S and Chandrasekaran P S 1969 *Prog. Theor. Phys.* **41** 264
Fushchich V I 1970 *Nucl. Phys.* **B21** 321
Fushchich V I 1972 *Lett. Nuovo Cim.* **4** 344
Fushchich V I and Grischenko A 1970 *Lett. Nuovo Cim.* **4** 927
Simon M T 1971 *Lett. Nuovo Cim.* **2** 616
Santhanam T S and Tekumalla A R 1972 *Lett. Nuovo Cim.* **3** 190
[5] Raspini A 1994 *Int. J. Theor. Phys.* **33** 1503
Raspini A 1996 *Fizika B* **5** 159
Raspini A 1997 *Fizika B* **6** 123
Raspini A 1998 *Fizika B* **7** 83
[6] A. Raspini 1999 in *Photon and Poincaré Group.* Dvoeglazov VV (ed.) pp. 181-188.
[7] Dvoeglazov V V 1995 *Int J. Theor. Phys.* **34** 2467
Dvoeglazov V V 1998 **37** 1909
Dvoeglazov V V 1996 *Nuovo Cim.* **B111** 483
Dvoeglazov V V 1995 *Nuovo Cim.* **A108** 1467
Dvoeglazov V V 1997 *Hadronic J.* **20** 435
Dvoeglazov V V 1997 *Fizika* **B6** 111
Dvoeglazov V V 1997 *Adv. Appl. Clifford Algebras* **7(C)** 303
Dvoeglazov V V 1999 *Adv. Appl. Clifford Algebras***9** 231
Dvoeglazov V V 1998 *Acta Physica Polon.* **B29** 619
Dvoeglazov V V 2000 *Found. Phys. Lett.* **13** 387
[8] Ahluwalia D V 1996 *Int. J. Mod. Phys.* **A11** 1855
Ahluwalia D V 1998 *Mod. Phys. Lett.* **A13** 3123
[9] Dvoeglazov V V 2000 *Czech. J. Phys.* **50** 225 in *Photon: Old Problems in Light of New Ideas.* Dvoeglazov V V (ed.) pp. 319-337.
[10] Dvoeglazov V V 2016 *Adv. Appl. Clifford Algebras* Online First (2016)
[11] Guang-jiong Ni 2004 in *Relativity, Gravitation, Cosmology* (Nova Science Pub., Huntington, NY, USA, 2004) (*Preprint* hep-ph/0306028)
[12] Bilenky S M and Giunti C 1996 *Nucl. Phys. Proc. Suppl.* **48** 381
[13] Barut a O 1978 *Phys. Lett.* B**73** 310
Barut A O 1979 *Phys. Rev. Lett.* **42** 1251
[14] Dirac P A M 1928 *Proc. Roy. Soc. Lond.* A **117** 610
[15] Ryder L H 1985 *Quantum Field Theory.* (Cambridge University Press, Cambridge)
[16] Itzykson C and Zuber J-B *Quantum Field Theory.* (McGraw-Hill Book Co.) p. 156
[17] Bogoliubov N N and Shirkov D V 1973 *Introduction to the Theory of Quantized Fields.* 2nd Edition. (Nauka, Moscow)
[18] Dvoeglazov V V 2003 *Hadronic J. Suppl.* **18** 239 (*Preprint* physics/0402094)
Dvoeglazov V V 2006 *Int. J. Mod. Phys.* B **20** 1317

[19] Ahluwalia D V, Johnson M B and Goldman J T 1993 *Phys. Lett.* **B316** 102
[20] Dvoeglazov V V 2011 *J. Phys. Conf. Ser.* **284** 012024 (*Preprint* arXiv:1008.2242)
[21] Markov M 1937 *ZhETF* **7** 579
Markov M 1937 *ZhETF* **7** 603
Markov M 1964 *Nucl. Phys.* **55** 130
[22] Barut A and Ziino G 1993 *Mod. Phys. Lett.* A **8** 1099
Ziino G 1996 *Int. J. Mod. Phys.* A **11** 2081
[23] Gelfand I M and Tsetlin M L 1956 *ZhETF* **31** 1107
Sokolik G A 1957 *ZhETF* **33** 1515
[24] Kirchbach M, Compean C and Noriega L 2004 *Eur. Phys. J.* A **22** 149
[25] Feynman R P and Gell-Mann M 1958 *Phys. Rev.* **109** 193
[26] Dvoeglazov V V 1998 *Int. J. Theor. Phys.* **37** 1915
[27] Ilyin V A 1991 *Spektralnaya Teoriya Differencialnyh Operatorov.* (Nauka, Moscow)
Budaev V D 1997 *Osnovy Teorii Nesamosopryazhennyh Differencialnyh Operatorov.* (SGMA, Smolensk)
[28] Snyder H 1947 *Phys. Rev.* **71** 38
Snyder H 1947 *Phys. Rev.* **72** 68
[29] Kempf A, Mangano G and Mann R B 1995 *Phys. Rev.* D**52** 1108
Amelino-Camelia G 2000 *Nature* **408** 661 (2000)
Amelino-Camelia G *Preprint* gr-qc/0012051
Amelino-Camelia G *Preprint* hep-th/0012238
Amelino-Camelia G *Preprint* gr-qc/0106004
Kowalski-Glikman J *Preprint* hep-th/0102098
Amelino-Camelia G and M. Arzano *Prperint* hep-th/0105120
BrunoN R, Amelino-Camelia G and Kowalski-Glikman J *Preprint* hep-th/0107039
[30] Dvoeglazov V V 2003 *Rev. Mex. Fis. Supl.* **49** 99
[31] Berg R A 1966 *Nuovo Cimento* **42**A 148

Quantum Control of the States of Light in a Mach-Zehnder Interferometer

Z Blanco-Garcia

Physics Department, Cinvestav, A.P. 14-740, 07000, México City, Mexico

E-mail: iblanco@fis.cinvestav.mx

Abstract. The experiment proposed by Elitzur and Vaidman is modified by replacing the fully absorbing obstacle (the 'bomb') with a non-linear optical medium. In this form the involved Mach-Zehnder interferometer works as a quantum beam splitter that produces NOON states which can be controlled if the input is the appropriate linear combination of two-mode Fock states.

1. Introduction

In a previous work [1] we have shown that the probabilities of detection in the well known Elitzur-Vaidman experiment [2] can be manipulated if the fully absorbing obstacle (the 'bomb') is substituted by a semitransparent object. Moreover, we found that the transparency of the obstacle determines either the particle-like or the wave-like behavior of the test photon as this occurs in the 'delayed choice' experiment proposed by Wheeler [3].

In the present contribution we follow the idea of a *quantum beam splitter* (QBS) discussed in [4] and include this in the Elitzur-Vaidman experiment to analyze the case when the obstacle is substituted by a non-linear optical medium. Our point is that the QBS works as a refinement of the Elitzur-Vaidman experiment and permits the manipulation of the photon-states that are injected into a Mach-Zehnder interferometer. Indeed, injecting N photons into one of the channels of a conventional beam splitter one obtains a superposition of two-mode Fock states at the output [5]. In contrast, a QBS produces the NOON states [6]

$$|\psi\rangle = \frac{1}{\sqrt{2}}\left(|N,0\rangle + e^{i\xi}|0,N\rangle\right), \tag{1}$$

where ξ is a phase. Remarkably, the entangled states (1) are very useful to obtain high-precision phase measurements; they have been generated up to $N = 5$ by multi-photon interference of down-converted light with a classical coherent state [7].

We are interested in the case when the following state is used as the input in a QBS

$$|\psi_{in}\rangle = |\psi_b, 0\rangle = \sum_{n=0}^{r} c_n|n,0\rangle, \tag{2}$$

where r is a non-negative integer defining the number $(r+1)$ of Fock states $|n\rangle$ involved in the superposition. Hereafter the two-mode Fock states $|n,m\rangle := |n\rangle \otimes |m\rangle$ correspond to the

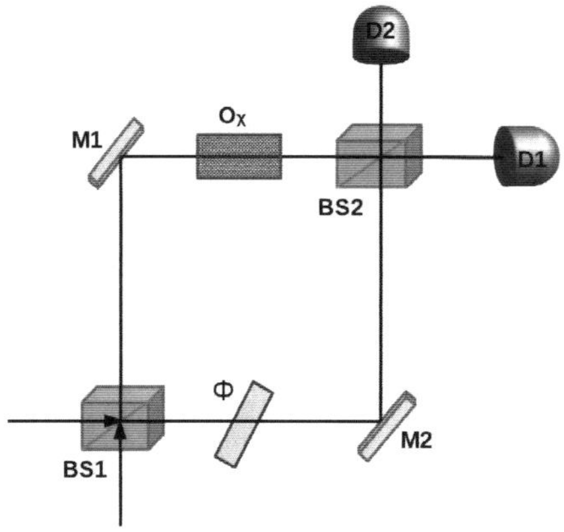

Figure 1. The Mach-Zehnder interferometer with a non-linear medium, represented by the obstacle O_χ, in one of the arms. Here BS_i, M_i and D_i, with $i = 1, 2$, stand respectively for beam-splitter, mirror and detector. The phase-shifter Φ is included to adjust the optical path difference.

biphoton states formed by n photons in the horizontal arm, and m photons in the vertical arm of the Mach-Zehnder interferometer shown in Fig. 1. For details and properties of the tensor product '$\otimes$' see, e.g. [8]. As we are going to see, the non-linear medium used instead the fully absorbing obstacle of the Elitzur-Vaidman experiment would be useful to steering the state (2) to a destination.

2. The modified Elitzur-Vaidman experiment

The setup shown in Fig. 1 represents a QBS, this is a Mach-Zehnder interferometer with a non-linear medium O_χ in its upper horizontal arm which plays the role of a refined 'bomb' in the Elitzur-Vaidman experiment. The operators representing the different devices in the optical-bench are as follows [4,9]:

$$\begin{aligned} BS &= \exp[i\theta(\mathrm{b}\otimes\mathrm{a}^\dagger + \mathrm{b}^\dagger\otimes\mathrm{a})], \qquad M = \exp\left[i\frac{\pi}{2}(\mathrm{b}\otimes\mathrm{a}^\dagger + \mathrm{b}^\dagger\otimes\mathrm{a})\right], \\ \Phi &= \exp[i\phi(\mathrm{N_H}\otimes\mathbb{I})], \qquad O_\chi = \mathrm{e}^{-i\pi(b^\dagger b)^2/2}. \end{aligned} \tag{3}$$

The straightforward calculation shows that the output state is given by

$$|\psi_{out}\rangle = \frac{\mathrm{e}^{i\pi/4}}{\sqrt{2}}\Big(|0,\psi_b'\rangle + |\psi_b'',0\rangle\Big) \tag{4}$$

where

$$|\psi_b'\rangle = \sum_{n=0}^{r} c_n i^n |n\rangle, \quad |\psi_b''\rangle = \sum_{n=0}^{r} c_n i^{2n-1} |n\rangle. \tag{5}$$

The above expressions depend explicitly on the coefficients c_n that define the initial state (2). The probability of finding the output (4) in the initial state acquires the form

$$P_{\psi_b,0} = \frac{1}{2}\Big(|c_0|^2 + \sum_{n=0}^{r}\sum_{m=0}^{r}|c_n|^2|c_m|^2(-1)^{n+m}\Big). \tag{6}$$

In turn, the probability of the transition $|\psi_{in}\rangle \to |k,0\rangle$ is given by

$$P_{k,0} = \begin{cases} |c_0|^2 & \text{if } k = 0, \\ \frac{|c_k|^2}{2} & \text{if } k \neq 0. \end{cases} \tag{7}$$

IOP Conf. Series: Journal of Physics: Conf. Series **839** (2017) 012021 doi:10.1088/1742-6596/839/1/012021

3. Applications

Let us take the coefficients c_n such that

$$|c_n|^2 = \binom{r}{n} p^n (1-p)^{r-n}, \quad 0 \leq p \leq 1, \quad n \leq r. \tag{8}$$

Then, the superposition (2) is an *optimized binomial state* [10] which is parameterized by p and r. In Fig. 2 we show the behavior of the probability (6) for this case. At $p = 0$ the probability is maximum since only the coefficient c_0 is different from zero and the initial state is the vacuum. At $p = 1$ the superposition (2) is equal to the Fock state $|r\rangle$ because only the coefficient c_r is different from zero. The the output (4) is the entangled NOON state with $N = r$.

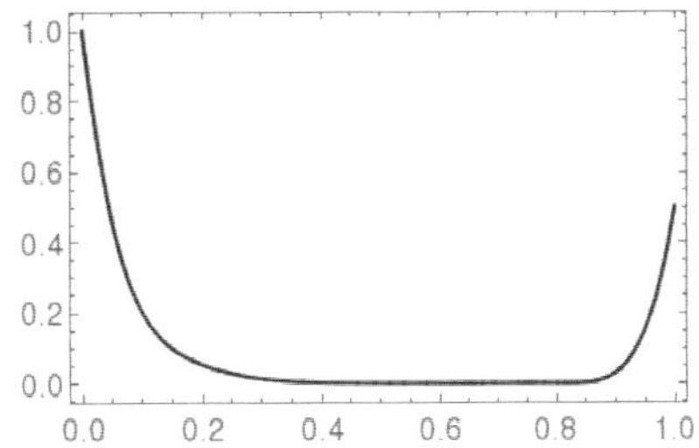

Figure 2. The probability $P_{\psi_b,0}$ defined in (6) as a function of p for the coefficients (8). The superposition (2) is an optimized binomial state.

On the other hand, the probability (7) is shown in Fig. 3 for three different values of the parameter p. As we can see, the maximum of the probability is shifted to the right as p increases and reaches its highest value in the vicinities of either $p = 0$ or $p = 1$.

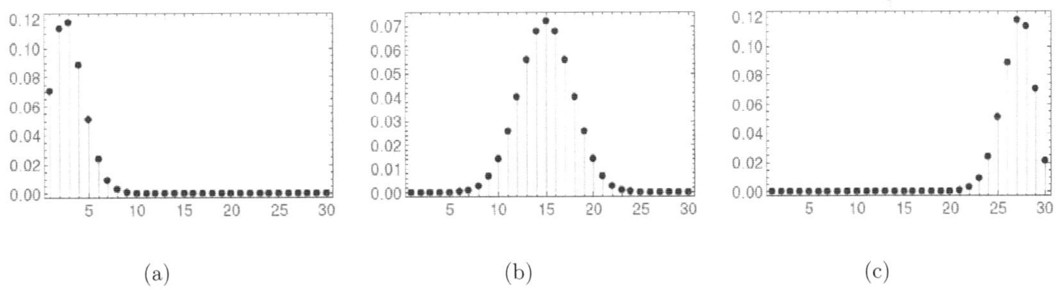

Figure 3. The probability $P_{k,0}$ of finding k photons in the horizontal arm at the output of the Mach-Zehnder interferometer for the coefficients defined in (8) with $p = 0.1$ (a), $p = 0.5$ (b) and $p = 0.9$ (c).

Another useful linear combination (2) is obtained by taking

$$c_n = \exp(-|\alpha|^2/2)\frac{\alpha^n}{\sqrt{n!}}, \quad \alpha \in \mathbb{Z}, \tag{9}$$

at the limit $r \to \infty$. Then the distribution of the Fock states in the horizontal channel is ruled by the Poisson distribution so that the input is $|\alpha, 0\rangle$, with $|\alpha\rangle$ a coherent state [10]. The output (4) is reduced to the two-mode coherent state

$$|\psi_{out}\rangle = \frac{e^{-i\pi/4}}{\sqrt{2}}|-\alpha, 0\rangle + \frac{e^{i\pi/4}}{\sqrt{2}}|0, i\alpha\rangle, \tag{10}$$

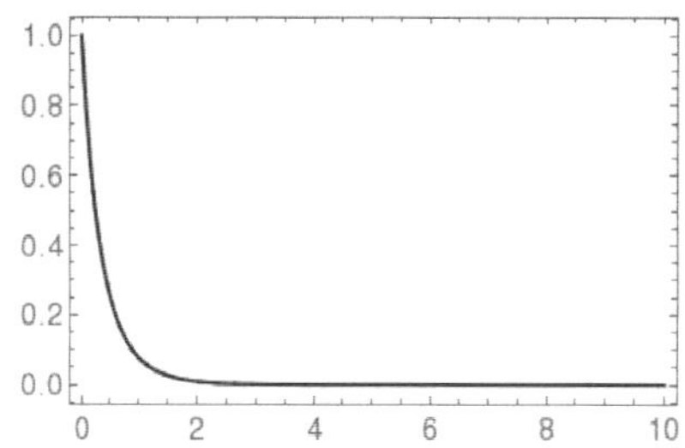

Figure 4. The probability (11) of finding the output (10) in the initial state $|\alpha, 0\rangle$ as a function of the mean number of photons $\bar{n}$.

which has been already reported in, e.g [11–13]. The probability (6) of finding the output in the initial state is now given by

$$P_{\alpha,0} = \frac{1}{2} e^{-2|\alpha|} \left(e^{-2|\alpha|} + 1 \right), \tag{11}$$

and depends on the mean number of photons involved in the initial superposition $|\alpha|^2 = \bar{n}$. Indeed, $P_{\alpha,0}$ decays exponentially as $\bar{n} \to \infty$, see Fig. 4.

In turn, the probability (7) of finding k photons in the horizontal arm of the interferometer is depicted in Fig. 5 for $\bar{n} = 10$. As expected, the probability is maximum in the vicinity of $k = 10$

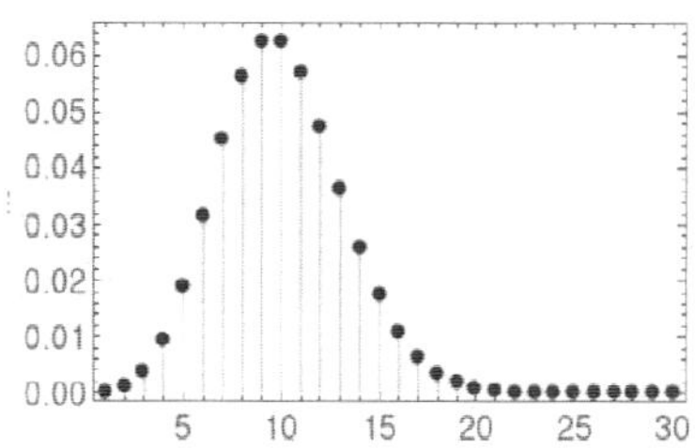

Figure 5. The probability $P_{k,0}$ of finding k photons in the horizontal arm at the output of the Mach-Zehnder interferometer for the coefficients defined in (9) with $\bar{n} = 10$.

4. Conclusions

We have shown that using different linear combinations of two-mode Fock states as the input in the Elitzur-Vaidman experiment give rise to output states which can be controlled if the fully absorbing obstacle is changed by a non-linear optical medium. The NOON states as well the two-mode coherent states already reported in the literature on the matter have been recovered as particular cases. Further insights are in progress and will be reported elsewhere.

Acknowledgments

The author is indebted to Prof. Oscar Rosas-Ortiz for illuminating discussions. The funding received through a CONACyT scholarship is acknowledged.

References

[1] Blanco-Garcia Z and Rosas-Ortiz O 2016 *J Phys Conf Ser* **698** 012013
[2] Elitzur A C and Vaidman L 1993 *Found Phys* **23** 987

[3] Wheeler J A and Zurek W H (Eds.) 1983 *Quantum Theory and Measurement* (Princeton: Princeton University Press) pp 182-213
[4] Dunningham J and Kim T 2006 *J Mod Opt* **53** 557
[5] Kim M S, Son W, Buzek V and Knight P L 2002 *Phys Rev A* **65** 032323
[6] Kok P, Boto A N, Abrams D S, et al, 2001 *Phys Rev A* 63 063407.
[7] Afek I, Ambar O and Silberberg Y 2010 *Science* **328** 879
[8] Enríquez M and Rosas-Ortiz O 2013 *Ann. Phys.* **339** 218
[9] Gerry C and Knight P 2005 *Introductory Quantum Optics* (Cambridge: Cambridge University Press)
[10] Zelaya K D and Rosas-Ortiz O 2016 *J Phys Conf Ser* **698** 012026
[11] Sanders B C 1992 *Phys Rev A* **45** 6811
[12] Gerry C C, Benmoussa A and Campos R A 2002 *Phys Rev A* **66** 013804
[13] Sanders B C 2012 *J Phys A: Math Theor* **45** 244002

IOP Conf. Series: Journal of Physics: Conf. Series **839** (2017) 012022 doi:10.1088/1742-6596/839/1/012022

Quantum and Classical Correlations in the Production of Photon-Pairs with Nonlinear Crystals

J López-Durán and O Rosas-Ortiz

Physics Department, Cinvestav, A P. 14-740, 07000, México City, Mexico

E-mail: jlopez@fis.cinvestav.mx

Abstract. The spatial distribution of photon-pairs produced by nonlinear crystals in the spontaneous parametric down conversion of type-II is analyzed. The correlations induced by the phase matching conditions are calculated by assuming that the ideal spatial distributions are of Gaussian proifile. Some predictions for the zones in the detection plane in which the entanglement of polarization is maximal are presented.

1. Introduction

It is well known that the quantum states are not separable in general. The latter is specially clear if the different parts of a given system are correlated in at least one (and the same) variable and such a correlation is preserved after the (non-destructive) manipulation of the involved variable. Such a conservation principle is the fingerprint of a *quantum correlation* usually known as *entanglement* [1, 2]. If the correlation is not maintained after the local transformation then it is merely classical. Nowadays, the most efficient system used to study quantum correlations is represented by the photon-pairs generated when a nonlinear crystal is illuminated by the appropriate light. In the crystal, one of the incoming photons, at the pump frequency ω_p, is spontaneously annihilated at the time that two new outgoing photons are created at the frequencies ω_s and ω_i. It is common to name the new photons as *signal* and *idler*, respectively. The phenomenon underlying the production of photon-pairs is known as *spontaneous parametric down conversion* (SPDC) and serves as a fundamental ingredient in many of the contemporary experiments of quantum control and quantum information [2–5].

In previous works [6, 7] we have reported some progress in the study of the spatial biphoton correlations generated in the type-I SPDC process for which the polarization of the new pair of photons is parallel to each other and orthogonal to the polarization of the pump photon. Then, the spatial distribution of the created light forms a cone that is aligned with the pump beam.

In this contribution we analyze some correlation properties of the photon-pairs generated in the SPDC process of type-II. The phenomenon is such that the polarization of the idler photons is orthogonal to the polarization of the signal ones. Then, the light produced by the nonlinear crystal forms two (not necessarily collinear) cones whose axes are symmetrically oriented with respect to the pump beam and share the same apex located somewhere in the crystal.

IOP Conf. Series: Journal of Physics: Conf. Series **839** (2017) 012022 doi:10.1088/1742-6596/839/1/012022

2. Phase matching conditions

In a first approach the SPDC occurs such that the energy and momentum are conserved. For a negative uniaxial crystal, these conservation laws are expressed as the phase matching conditions

$$\omega_p = \omega_s + \omega_i, \quad \mathbf{k}_p\left(e\right) = \mathbf{k}_s\left(o\right) + \mathbf{k}_i\left(e\right), \tag{1}$$

where ω and $\mathbf{k}$ are respectively the angular frequency and the wave number vector. If $\omega_s = \omega_i$ we say that the process is degenerated. The labels e and o stand for extraordinarily and ordinarily polarized light respectively [8]. To solve (1) we consider the wave number vectors $\mathbf{k}_p$, $\mathbf{k}_s$, $\mathbf{k}_i$, and the unitary vector $\mathbf{n}$ shown in Figure 1(a). Hereafter we take $\mathbf{n}_p$, $\mathbf{n}_s$ and $\mathbf{n}_i$ as the unitary vectors that define the direction of propagation of each one of the beams involved in the down conversion. Moreover, we shall fix $\mathbf{n} = (\sin\sigma, 0, \cos\sigma)$ as the optical axis of the nonlinear crystal.

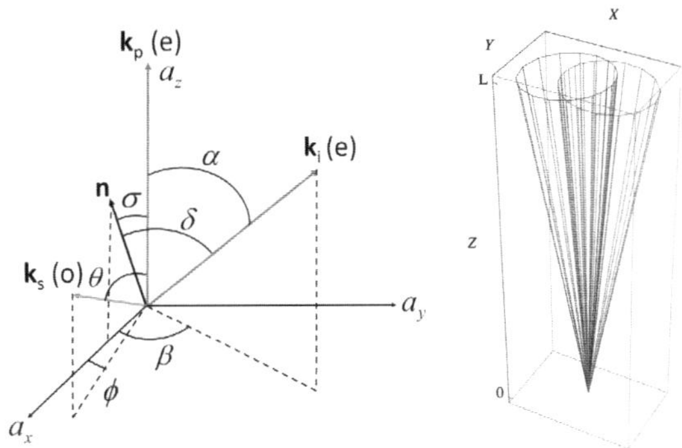

(a) (b)

Figure 1. (a) Graphic representation of the unitary vectors used to solve phase matching conditions in spherical coordinates (b) The rays of down converted light calculated from the solution of the phase matching conditions.

One can show [9] that, in the degenerated case, the solution of the phase matching conditions is given by the expressions

$$\phi = \beta + \pi, \tag{2}$$

$$\frac{n_e\left(\frac{\omega_0}{2}, \delta\right)}{n_o\left(\frac{\omega_0}{2}\right)} \sin\alpha = \sqrt{1 - \left(\frac{n_e\left(\omega_0, \sigma\right) - \frac{1}{2} n_e\left(\frac{\omega_0}{2}, \delta\right)\cos\alpha}{\frac{1}{2} n_o\left(\frac{\omega_0}{2}\right)}\right)^2}, \tag{3}$$

In Fig. 1(b) we show the distribution of rays of light that are produced in the type-II SPDC. They have been depicted by using the unitary vectors, $\mathbf{n}_s$ and $\mathbf{n}_i$, calculated from the numerical solution of the system (2)–(3).

3. Spatial correlations

Following [6, 7], we propose a Gaussian profile for the spatial distribution of the down converted photons. That is, on the detection plane (which is transversal to the propagation of the pump beam), the spatial distribution along each one of the circumferences is given by

$$F\left(r, \tau\right) = \frac{1}{2\pi\sigma_r\sigma_\tau} \exp\left(-\frac{r^2}{2\sigma_r^2} - \frac{\left(C\left(\tau\right) - d\right)^2}{2\sigma_\tau^2}\right), \qquad C\left(\tau\right) = \sqrt{d^2 - \tau^2}, \tag{4}$$

where σ_r and σ_τ are the width of the distributions with d and r the radius and the radial coordinate of the circumference. The variable τ corresponds to a coordinate perpendicular to the radial one (in a point of interest).

Using (4) one can calculate the predictions for the distribution rates of coincidences in the counting of photons at the detection zone [10]. The idler and signal channels must be measured by photo collectors located at the positions (x_i, y_i) and (x_s, y_s) on the circumferences described by a transversal cut of the cones depicted in Fig. 1(b). The collecting of photons in the laboratory should be performed by displacing the idler and/or signal detectors accordingly. In Fig. 2 we show the prediction for the horizontal-horizontal (x_i, x_s) displacements, they give rise to the ellipse-like contours oriented at 45° which are also found in the type-I SPDC production of photon-pairs [6, 7].

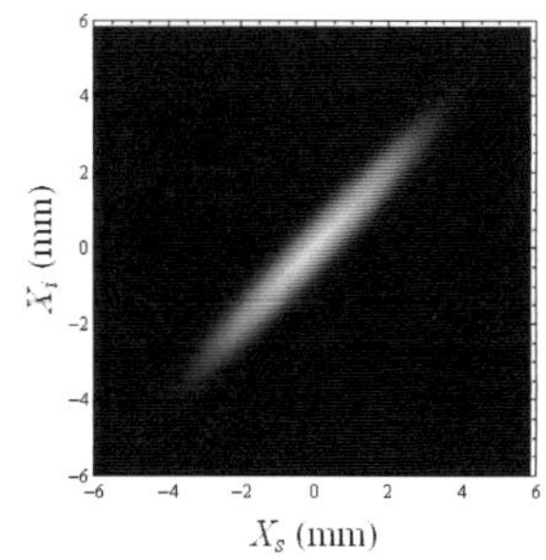

Figure 2. Distribution rate of spatial conicidences predicted for the horizontal displacements of the photon collectors in the idler and signal channels for the type-II SPDC.

4. Quantum correlations

The photon-pairs produced in the type-II SPDC are entangled in the polarization if they are found at the intersections of the cones [11]. The distribution (4) allows to visualize the behavior of the distributions at the intersections for different distances between the axes of the cones [10]. Some results are depicted in Fig. 3 for different values of the angle γ that measures the distance between the cones. The smallest zone of intersection is obtained at $\gamma = \pi/4$, see Fig. 3(c). Although this case is the closest one to the ideal photon-pair production, the photons collected in the intersection for the condition $\gamma = \pi/4$ could be not entangled with their counterparts in the other intersection. The latter because the zone of intersection in the ideal case consists of only one point on the detection plane.

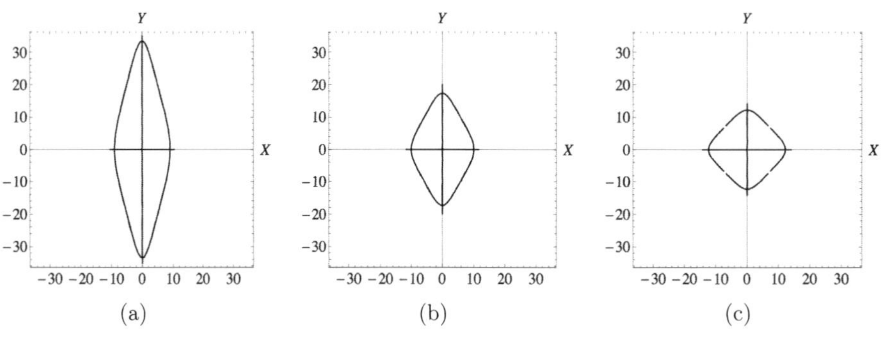

Figure 3. An intersection of the cones for different separations: (a) $\gamma = \frac{\pi}{12}$, (b) $\gamma = \frac{\pi}{6}$ and (c) $\gamma = \frac{\pi}{4}$.

IOP Conf. Series: Journal of Physics: Conf. Series **839** (2017) 012022 doi:10.1088/1742-6596/839/1/012022

Considering that both cones are identical, the entangled photons must be distinguished from a balanced mixture of extraordinarily and ordinarily polarized photons. Hence, the following density matrices must be considered.

$$\rho_{mix} = \frac{1}{2}\left(|oe\rangle\langle oe| + |eo\rangle\langle eo|\right), \tag{5}$$

$$\rho_{spdc} := |\psi_{spdc}\rangle\langle\psi_{spdc}| = \frac{1}{2}\left(|oe\rangle\langle oe| + |oe\rangle\langle eo| + |eo\rangle\langle oe| + |eo\rangle\langle eo|\right). \tag{6}$$

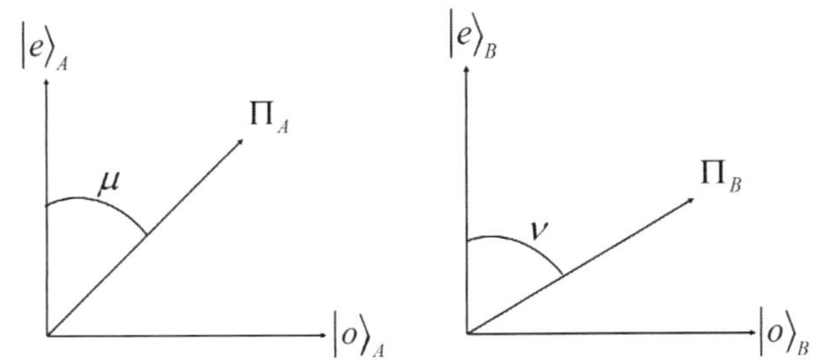

Figure 4. (Color online) Orientation of the polarizers located behind each intersection.

For the certification, a polarizer must be localized in each intersection, labelled as A and B in Fig. 4. Behind each polarizer, each one of the photodetectors will be displaced accordingly to obtain the distribution rate of spatial coincidences. The related probabilities are given by

$$P_{AB}^{spdc}(\mu,\nu) = \frac{1}{2}\sin^2(\mu+\nu), \quad P_{AB}^{mix}(\mu,\nu) = \frac{1}{2}\left[\sin^2\mu\cos^2\nu + \cos^2\mu\sin^2\nu\right], \tag{7}$$

where μ and ν give the orientation of the polarizers with respect to the extraordinary polarization (for more details see [9] and [10]).

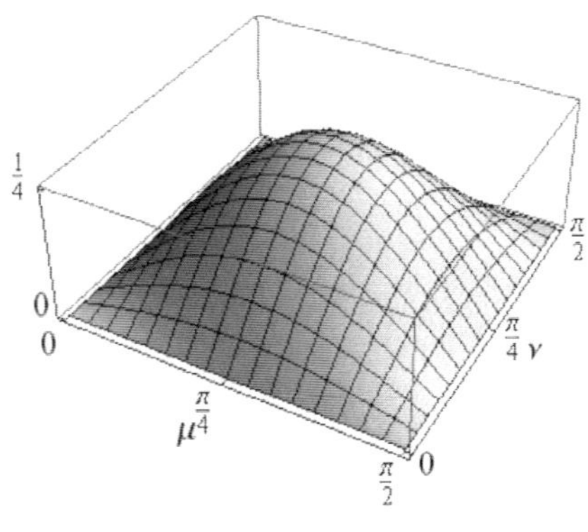

Figure 5. (Color online) Distinguishment between a statistical mixture and a pure entangled state for different polarizer configurations (μ,ν).

The quantity $\Pi = P_{AB}^{spdc}(\mu,\nu) - P_{AB}^{mix}(\mu,\nu)$ is shown in Fig. 5. As we can see, the probability of the pure state (6) is greater than that of the mixed state (5) for almost all the values of μ and ν. The exceptions are presented when μ and ν are equal to either 0 or $\pi/2$, since both probabilities are equal for the combinations of such parameters. The latter means that the probabilities will be non-distinguishable for the pairs (μ,ν) defined as either $(0,\nu)$, $(\pi/2,\nu)$, $(\mu,0)$ or $(\mu,\pi/2)$.

5. Concluding remarks

A simple geometrical model has been used to make some predictions of the spatial distribution of the photon-pairs produced in the type-II SPDC process. It has been shown that the distribution rate of spatial coincidences predicted for the horizontal displacements of the photon collectors show the ellipse-like contours oriented at 45° which are also found in the SPDC of type-I. The same model has been used to make some predictions of the zones in the detection plane where it is most probable to find the photons that are entangled in polarization. In addition, we have proposed a mechanism to distinguish between pure (entangled) and mixed states of polarization in the generated photon-pairs.

Acknowledgment

JLD acknowledges the funding received through a CONACyT scholarship.

References

[1] Aczel A, *Entanglement*, Plume, NewYork, 2003
[2] Mielnik B, Rosas-Ortiz O, *Quantum Mechanical Laws*, in Fundamentals of Physics, Vol. 1, Morán-López J L (ed), Encyclopedia of Life Support Systems (EOLSS), Developed under the Auspices of the UNESCO, Eolss Publishers: Oxford, United Kingdom; 255?326, 2009. Available from: http:// www.eolss.net.
[3] Menzel R, *Photonics*, Springer-Verlag, Berlin, 2007
[4] Bouwmeester D, Ekert A K, Zeilinger A (eds), *The Physics of Quantum Information: Quantum Cryptography, Quantum Teleportation, Quantum Computation*, Springer-Verlag, Berlin, 2000
[5] Zeilinger A, *Dance of the Photons: From Einstein to Quantum Teleportation*, Farrar, Straus and Giroux, New York, 2010
[6] Procopio L M, Rosas-Ortiz O and Velázquez V, *AIP Conf. Proc.* **1287** (2010) 1287
[7] Procopio L M, Rosas-Ortiz O and Velázquez V, *Math. Meth. Appl. Sci.* **38** (2015) 2053
[8] Fowles G R 1968 *Introduction to Modern Optics* (United States: Holt, Rinehart and Winston Inc.)
[9] López-Durán J, *Quantum entanglement and classical correlations in the photon-pairs produced by nonlinear crystals*, M.Sc. Thesis (in Spanish), Physics Department, Cinvestav, México City, 2015
[10] López-Durán J and Rosas-Ortiz O, in preparation
[11] Kwiat P G, Mattle K, Weinfurter H and Zeilinger A, *Phys. Rev. Lett.* **75** (1995) 4337

Is There a Quantum Origin in the Biological Memory?

L Cano, D Lara-Bustillos, M Hautefeuille and V Velázquez

Facultad de Ciencias, Universidad Nacional Autónoma de México, Av. Universidad 3000, CP 04510, México D.F., México

E-mail: vicvela@ciencias.unam.mx

Abstract. Under the premise that all behaviors in nature have a quantum explanation, we would like to wonder if the biological memory also possesses a quantum explanation. In this work we investigate the possibility to obtain some quantum signatures that we can recognize in processes involved in the codification of data in neurons. For this purpose, we propose to use the optical criteria in order to distinguish classical and quantum time series of photon signals. Taken into account that a neuroscientist locates the memory in the synapse configuration, we conjecture that neuronal network could be more efficient if the neuronal network functioning was considered a quantum system.

1. Introduction

Think that you are in a dark room, and that you can smell the scent of your favorite dish. A lot of molecules can lead you to your objective. What is the mechanism that allows neurons to acquire memories from classical and quantum information? Molecules from the scent, photons from images and so on. This problem is very complex. There are many people in the world investigating how neurons store one single memory. Now with the rise of technological developments, the biological processes involved in several human functions can be observed at micro and nanoscale. In such way we may rephrase the question: What does a memory mean in the classical and quantum worlds? How can we transit between them? From the theory of "Relative State" of the quantum Mechanics [1], we obtain the most elemental definition of memory. This article is devoted to memory processes that could have a quantum origin. In order to accomplish this objective, in section 1 we discuss some rules that quantum memory must follow. In section 2, we introduce a short outline of the biological memory, then in section 3 we establish the difference between a classical memory and a quantum memory. In the 4th section we present a brief description of the status of storage of information in neurons. In section 5 we propose a methodology to research the possible quantum character in the memory before we conclude.

2. Memory in the quantum world

The main characteristics of a system with quantum coherence is its reversibility. One atom can absorb energy from a photon. The electronic state changes from one quantum state with a lower energy to another with a greater one. In this case, we can say that the atom has a memory

IOP Conf. Series: Journal of Physics: Conf. Series **839** (2017) 012023 doi:10.1088/1742-6596/839/1/012023

from something that comes from its environment. But this memory has a half-life close to the decay time, which represents only a few pico or nanoseconds. The description of a joint system consisting of a quantum system S and its environment is depending of its interaction. That means, on one hand that the system is capable of absorbing the interaction carriers from the environment, for example atoms in a photon reservoir where the environment consists of a bath of photons. On another hand, the interaction is dependent also on the possibility of the system S to interact with the photon bath. Then, it is necessary that the quantum system has eigenvalues corresponding to the energies of the photons in the bath. The complexity of the interaction will be dictated by the Hilbert space dimension, because the quantity of states that can be excited in the system is proportional to this dimension, the number of photons of the system and the environment interchanges increase with this space. Each photon missing in the bath field was absorbed by the system. When the average number of photons $\bar{n}$ missing in the bath is constant over time, we can say that the system and its environment are in equilibrium. In such condition, we can say that the system is a witness of the environment. Each photon absorbed with a given energy changes the state of the system. The system can go from one eigenstate to another, passing for all possible combinations in its space. It is then easy to think that the number of states in the environment is greater than the number of states in the system. If we considerer the system as a witness of the behavior of the environment, its description is only a partial description of the reality of the environment. Now what happens if the environment disappears suddenly, is that there is a very short time interval before all system states emit all photons spontaneously. The interval between the disappearance of the environment and the event in which the last photon is out of the system is the time that lasts the memory of environment in the system.

3. Memory in the classical world

If you build a very big ensemble of atoms and/or molecules, the average potential is also an ensemble of atomic quantized potentials. The average of this potentials allows energy states which separation between neighbors are virtually unpredictable. If we increase the number of components the energy levels tends to continuous: Our classical world. In a system in which we have a very big ensemble of atoms and complex molecules, the Hilbert space dimension exploits. Now the basis state of this ensemble is not longer the basis state of a single atom or molecule, then a very complicated distribution of potential energy, allows new configurations in the distribution of all its components in such a way the memory of the interactions of this system with its environment can remain longer, for a very long time and thermodynamic equilibrium where the state that we can measure with our instruments is the most probable state, over all another possibilities. One simple example comes from the quantum optics: When we send a quantum state of n inside of a Mach-Zehnder interferometer, we have $n+1$ different interference patterns. If we are not able to separate each of them, then we will only see the average. But the average have the same pattern that the interference from one single photon [2]. This is, the more stable pattern.

3.1. The memory in a neuron

Neurons are the basic unit of the nervous system. A neuron is a specialized cell with an architectural conformation that is capable of rapid voltage changes in its membrane, which gives rise to an action potential that allows it to communicate with other cells. An action potential is generated by a sudden flow of Na ions across the respective channels in the cell membrane. When a stimulus is the most depolarizing of the membrane, the change in potential opens up the Na channels, allowing it to decrease the concentration outside of it and increase in the interior. The neuron as an object of the universe has memory, but it is a dynamic interacting memory. One photon, one single molecule can change the average potential surrounding it, then its connectivity with neighboring neurons can change, resulting in a new memory. There are studies where the individuality of both photons and cells are fundamental in some processes like the navigation sense of birds, or photosynthesis [3], but it is not clear if one memory can be registered in only one neuron [4], or this memory requires a minimal set of neurons [5]. Because a neuron is not a computer, people are looking for the mechanism that allows one neuron to store information. In [6] it is proposed that phosphorus is the only biological element with a nuclear spin able to be used in processes of quantum information. However, other neuroscientists think that memory can be related to one or more neurons but also to the neuronal network. Instead of storing memories in a single neuron, or a minimal set of neurons, the memories can be stored in the network. This means, one memory may be associated with the connectivity of one or more neurons.

4. Quantum neuronal network

If a particular memory is located in a neuron or a set of neurons, we need to investigate how neurons have to be organized to store a memory? This question allows us to think about the configuration space available in order to store memories. If one memory may be stored in a single neuron, then the possibility of memories in the brain is dependent of the number of neurons available. But if one memory is established in a particular connection of the network, then the possibilities of storage grows exponentially. For the moment, there is no clear evidences over the storage mechanism, we can only say that spin is a very interesting candidate to explain the storage of information, while random walks (quantum random walks [7]) may be the mechanism for the neuronal network, if we are looking for quantum evidences. Spins conjecture has been cited in [6]. Here we conjecture about the advantages of memory stored in the neural network. It is well known that a quantum signal that travels a given network does not need to travel all possible paths in order to take the shorter-time path in concordance with a classical signal, but quantum signals can interfere in some pathways that are not useful. If the ions of different channels of communication have an atomic mass lower than 40 amu (Na^+, K^+, mainly) the question raised above means the possibility than an individual current of ions or the potential generated by themselves can be considered as the superposition of the several options built in the network. For this mass scale there are evidences of quantum paths superpositions. Arndt and coworkers [8] have shown that the interference of ^{60}C is possible. A ball of carbon atoms can obey the rules of quantum superposition when launched at an optical grid with a period near to a micrometer. Because this molecule is more or less 50 times heavier than our ions (the speed can be considered similar), and considering that the neuronal network can be modeled as a biological grating, if the network has spatial gaps of 50-100 micrometers, then, it is possible that a single ion can travel over different paths before activating a single synapse. The consequences of this possibility are very interesting. In quantum mechanics phenomena, we know that the quantum effects are hidden in superposition instead of detection. A set of ions can activate only one synapse, but determining which synapses in particular has a dependence of the superposition. Then, there is a difference between both process, this difference comes from the quantum interference, the possibility that each ion has to go along each path producing self-interference. In the theory of optical random walks, one photon that travels through a $n \times n$ beamsplitters network has only two preferential paths [9], while the probability to find it in any output port is depending on the two imposed initial conditions: the input port selection and the pathway n-phase differences among all paths.

Then, in this conception of the memory, each ion belonging to a given ion-current has the possibility to travel for several pathways, and its probability to activate a synapse depends on their initial conditions and phases in the several pathway-options. This mechanism may be very efficient, because the interference can destroy such pathways that are not appropriate for the establishment of a new memory, or to remember it. In such case the initial conditions are related with the correlation to a particular memory, because, these conditions define the more probable paths that information must travel.

5. Quantum memory detection

In 1926, Adrian and Zotterman [10] studied the action potentials that come from neurons. They found a succession of action currents with the same form and time relationships. Most of action currents conformed to a standard size. There is a uniformity on the type of signals that keep the communication between neurons. What is very interesting is that these signals can be simulated in time and form by algorithms that take into account the electrical processes involved [11]. Apparently the quantity of ions and electrical potential associated are very well defined. This kind of uniformity in the signals represents the possibility to have quantum effects by interference in the neuronal network. If someone says that one process in particular has quantum effects, it will be considered a conjecture until the effect can be tested. Because we are conjecturing that neuronal networks may work like a quantum neuronal network, the first step to test it requires to have a criterion to measure the quantum behavior of the signals. In quantum optics, statistics of light signals is determined by the criteria of fluctuations around the average [12]. Paraphrasing Schrödinger [13], if the signals emerging from the cells have a standard deviation equal to the square root of the average in some observable, then the process involved has random behavior. Now, if the ratio between this quantity is lower than one ($f = \sigma_s/\sqrt{\bar{s}}$) the signal can be considered a quantum signal, and it is possible to observe quantum effects like quantum interference. Particularly if $f \approx 1/2$ [14] the signals can be considered to have a quantum chaotic behavior.

6. Conclusions

We have made a brief description about classical and quantum memories. Regarding the possibility of the existence of quantum effects that could explain the functioning of the memory, we have theorized about the memory alike a set of memories located in the different neuronal configurations established by the interaction with the environment. We have conjectured that if the potentials that control gates and synapses are generated by currents of ions that follow a quantum criterion of number of charges, then it is possible to have quantum effects in the neuronal network, hence, in the memory. Moreover, we propose a criteria to distinguish signals with quantum signatures using the factor f in order to know if it is possible to have quantum coherence in the communication between the environment and neurons; and communication between neurons.

References

[1] Everett III H 1957 Relative State Formulation of Quantum Mechanics *Rev. Mod. Phys.* **29** 454-462
[2] Ramírez-Cruz N, Bastarrachea-Magnani M A and Velázquez V 2016 Quantum Anatomy of the Classical Interference of n-Photon States in a Mach-Zehnder Interferometer *J. Phys.: Conf. Ser.* **698** 012015
[3] Lambert N, Chen Y-N, Cheng Y-C, Li C-M, Chen G-Y and Nori F 2013 Quantum Biology *Nature Physics* **9** 10-18
[4] Quiroga R Q, Reddy L, Kreiman G, Koch C and Fried I 2005 Invariant visual representation by single neurons in the human brain *Nature* **435** 1102-1107
[5] Liu X, Ramirez S, et. al 2012 Optogenetic stimulation of a hippocampal engram activates fear memory recall *Nature* **484** 381-387
[6] Fisher M P A 2015 Quantum cognition: The possibility of processing with nuclear spins in the brain *Ann. Phys.* **362** 593-602
[7] Aharonov Y, Davidovich L and Zagury N 1993 Quantum random walks *Phys. Rev.* A **48** 1687-1690
[8] Arndt M, Nairz O, Vos-Andreae J, Keller C, van der Zouw G and Zeilinger A 1999 Wave-particle duality of C_{60} molecules *Nature* **41** 680-682

[9] Gonzlez-Arteaga H, López-Moreno E, Grether M and V. Velázquez, to be published.
[10] Adrian E D and Zotterman Y 1926 The impulses produced by sensory nerve-endings. Part 2. The response of a single end-organ *J. Physiol.* **62** 151-71
[11] Borisyuk A and Rinzel J 2005 Understanding neuronal dynamics by geometrical dissection of minimal models, in: Chow C, Gutkin B, Hansel D, Meunier C and Dalibard J (eds.) Models and Methods in Neurophysics, Proc Les Houches Summer School 2003, (Session LXXX), Elsevier 19-72.
[12] Fox M 2006 *Introductory quantum optics* (Oxford University Press)
[13] Schrodinger E 1967 *What is life?* (Cambrige Press)
[14] Ramírez N et. al to be published.

IOP Conf. Series: Journal of Physics: Conf. Series **839** (2017) 012024 doi:10.1088/1742-6596/839/1/012024

A Note on the Off-Axis Gaussian Beams Propagation in Parabolic Media

Z Gress and S Cruz y Cruz

Instituto Politécnico Nacional, UPIITA. Av. Instituto Politécnico Nacional 2580 Col. La Laguna Ticomán, Del. Gustavo A. Madero, 07340 Ciudad de México, Mexico

E-mail: zgressm1400@alumno.ipn.mx, sgcruzc@ipn.mx

Abstract. The form of the off-axis Gaussian beam is constructed by solving the paraxial wave equation for a parabolic index medium. This construction is accomplished by reducing the problem to the solution of an Ermakov equation. It is shown that the dynamics of the system is comprised in, both, the beam width and the position of the center of the wavepacket, that are defined by the focusing properties of the material. The cases of on-axis behavior and propagation in a homogeneous medium are obtained as special cases.

1. Introduction

The description of propagation processes of paraxial beams has been a problem of main interest since old times. For instance, it is well known that Hermite-Gaussian and Laguerre-Gaussian modes are complete sets of solutions to the paraxial wave equation that have a number of applications in the description of laser resonators and waveguides modes [1] and in many other areas of physics [2]. These modes are characterized to be shape-invariant under propagation along the optical axis having transversal optical intensity distributions consistent of rectangular or axial-symmetric patterns with Gaussian envelops. In this context, the description of propagation of Gaussian beams in homogeneous, as well as in graded index media becomes relevant for all applications involving these kind of modes. The Gaussian-type solution of the paraxial wave equation in generic inhomogeneous media has been considered from different approaches [3, 4]. The particular case of a transverse quadratic refractive index is interesting because of its focusing, re-directing and collimation properties [3–6]. In the present work, we deal with the construction of off-axis Gaussian wavepacket-type solutions to the paraxial wave equation for a parabolic medium. It is shown that the problem is reduced to the solution of an Ermakov equation for the beam width together with a classical dynamical law of motion for the center of the wavepacket. Some examples are presented in order to illustrate our results.

2. Gaussian beams in quadratic-index media

Consider an inhomogeneous medium whose refractive index profile is given by

$$n(r) = n_0 \left(1 - \frac{\Omega^2}{2} r^2\right), \tag{1}$$

where r is the transversal radial coordinate. In the weakly-guiding regime we have $\Omega^2 r^2 \ll 1$ and the paraxial approximation is valid. Here n_0 is the refractive index along the optical axis, which in this case is chosen as the $z-$axis, and $\Omega \in \mathbb{R}$ is a constant parameter defining the focusing properties of the medium. In the paraxial regime the electric field amplitude of a monochromatic, TE wave has the form $E(\mathbf{r}, z) = U(\mathbf{r}, z)e^{ik_0 n_0 z}$, where $\mathbf{r} = (x, y)$ is the transversal position vector, k_0 is the wave number in free space and the function $U(\mathbf{r})$ is such that its second derivative with respect to the longitudinal coordinate can be neglected. Under this conditions U satisfies the paraxial wave equation

$$\frac{i}{k_0}\frac{\partial U}{\partial z} = \left\{-\frac{1}{2k_0^2 n_0}\left(\frac{\partial^2}{\partial x^2} + \frac{\partial^2}{\partial y^2}\right) + \frac{n_0}{2}\Omega^2 r^2\right\} U. \tag{2}$$

We are interested in finding Gaussian wavepacket-type solutions to the equation (2) which represent physical beams carrying finite transverse optical power as they propagate along the optical axis. In the off-axis general case

$$U(\mathbf{r}, z) = N(z)\exp\left\{i\left[S(z)(\mathbf{r} - \mathbf{r}_0(z))^2 + k_0\mathbf{p}_0(z)\cdot\mathbf{r} + g(z)\right]\right\}, \tag{3}$$

with $\mathbf{r}_0(z) = (x_0(z), y_0(z))$ the (variable) position of the center of the wavepacket. The normalization factor $N(z)$ and the coefficient $S(z)$ are, in general, assumed to be complex-valued functions of z, while the functions $\mathbf{p}_0(z)$ and $g(z)$ are, without loss of generality, supposed to be real-valued. Substituting (3) into (2) we find that $S(z)$ and $N(z)$ satisfy the set of equations

$$\frac{d}{dz}\left(\frac{2S}{k_0 n_0}\right) + \left(\frac{2S}{k_0 n_0}\right)^2 + \Omega^2 = 0, \tag{4}$$

$$\frac{d}{dz}\ln N(z) = -\frac{2}{k_0 n_0}S(z). \tag{5}$$

It also follows from (2) that $\mathbf{p}_0 = n_0\dot{\mathbf{r}}_0$, meaning that the center of the wavepacket obeys the dynamical law of a symmetric two-dimensional classical harmonic oscillator of frequency Ω

$$\ddot{\mathbf{r}}_0 + \Omega^2\mathbf{r}_0 = 0, \tag{6}$$

and $g(z) = \frac{k_0 n_0}{2}(\mathbf{r}_0\cdot\dot{\mathbf{r}}_0)$ (here the dot stands for derivation with respect to z).

Note that equation (4) is a complex Riccati equation. Observe also that the imaginary part of S is related to the width $w(z)$ of the wavepacket. Hence, it is convenient to write S in the form [7,8]

$$\frac{2}{k_0 n_0}S(z) = \frac{1}{R(z)} + \frac{2i}{k_0 n_0 w^2(z)}. \tag{7}$$

Actually, the function $R(z)$ turn out to be the radius of curvature of the wavefront [1,9]. As this expression is introduced in (4), the imaginary part of this equation leads us to $\frac{1}{R(z)} = \frac{d}{dz}\ln w(z)$, while the real part is transformed into the Ermakov equation

$$\frac{d^2 w}{dz^2} + \Omega^2 w = \frac{4}{k_0^2 n_0^2 w^3}. \tag{8}$$

On the other hand, the expression (7) also provide the form of the normalization factor $N(z)$ through (5). Indeed

$$N(z) = \frac{N_0}{w(z)}e^{-i\chi(z)}, \quad \dot{\chi}(z) = \frac{2}{k_0 n_0 w^2(z)}, \tag{9}$$

where N_0 is a constant to be fixed and the function χ, known as the Guoy phase, is a cumulative on-axis phase shift experienced by the wavefront in passing through a focal plane [1] which is determinant in a number of optical phenomena [10] and can be interpreted as a geometric phase [11, 12].

It is well known that the solution of the Ermakov equation $\ddot{w} = M(z)w + \alpha/w^3$, with α a constant, can be explicitly constructed provided a solution of the corresponding linear equation $\ddot{w} = M(z)w$ is known [13, 14]. Thus, in this case, after some algebra we get [8]

$$w(z) = w_0 \left[\cos^2(\Omega z) + \frac{1}{(\Omega z_R)^2}\sin^2(\Omega z)\right]^{1/2}, \qquad z_R = \frac{k_0 n_0 w_0^2}{2}, \tag{10}$$

where $w_0 = w(0)$ is the initial width and z_R, known as the Rayleigh range or collimation distance, is a parameter characterizing the divergent nature of the optical beam. On the other hand, from (6), the trajectory of the center of the wave-packet is given by

$$\mathbf{r}_0(z) = \mathbf{a}\cos(\Omega z) + \mathbf{b}\sin(\Omega z)/\Omega, \tag{11}$$

where $\mathbf{a} = (a_x, a_y)$ and $\mathbf{b} = (b_x, b_y)$ are, respectively, its initial position and velocity. In this way, the functions $w(z)$ and $\mathbf{r}_0(z)$ encode all the information about the propagation processes of the light beam.

The expression for the Gaussian mode is then (compare to [7])

$$U(\mathbf{r}) = \frac{N_0}{w(z)} e^{-i(\chi(z)+g(z))} e^{\frac{ik_0 n_0 (\mathbf{r}-\mathbf{r}_0(z))^2}{2R(z)}} e^{-\frac{(\mathbf{r}-\mathbf{r}_0(z))^2}{w^2(z)}} e^{ik_0 n_0 \dot{\mathbf{r}}_0(z)\cdot\mathbf{r}}, \tag{12}$$

with

$$R(z) = \frac{\tan(\Omega z)}{\Omega z_R\left(1 - \Omega^2 z_R^2\right)}\left[1 + \left(\frac{\Omega z_R}{\tan(\Omega z)}\right)^2\right], \quad \chi(z) = \frac{2}{k_0 n_0}\int^z \frac{dt}{w^2(t)}. \tag{13}$$

Observe the oscillating behavior of these functions inherited from the harmonic nature of w and $\mathbf{r}_0$. This fact is due to the focalization properties of the medium: the beam experiences, at the same time, the spreading effects of wave propagation and the medium confinement, resulting in a periodically self-focusing wave-packet with period π/Ω whose center follows the trajectory of a two-dimensional harmonic oscillator of frequency Ω in the transversal plane. If the center of the wavepacket is fixed on the optical axis, then $\mathbf{r}_0 = 0$ and we get the on-axis Gaussian mode

$$U(\mathbf{r}) = \frac{N_0}{w(z)} e^{-i\chi(z)} e^{\frac{ik_0 n_0 r^2}{2R(z)}} e^{-\frac{r^2}{w^2(z)}}. \tag{14}$$

In the case that $\Omega = 0$ we recover the well known width, radius of curvature and Guoy phase shift for a Gaussian beam in a homogeneous medium [1, 9]

$$w(z) = w_0\sqrt{1 + \left(\frac{z}{z_R}\right)^2}, \quad R(z) = z\left[1 + \left(\frac{z_R}{z}\right)^2\right], \tag{15}$$

and $\chi(z) = \arctan(z/z_R)$. Notice that in this particular case w_0 is the minimum width and so it is called the waist radius. This means that the transversal plane $z = 0$ is actually the focal plane of the system. The beam doubles its cross section at a distance $z = z_R$ from the focus $(w(z_R) = \sqrt{2}w_0)$. Within this distance the width remains nearly constant but out of this region it increases as a linear function of z with a total divergence angle given by $2\theta_0 = 2w_0/z_R$ [1]. On the other hand, in the case that $\Omega = 1/z_R$, the divergent nature of the beam is perfectly

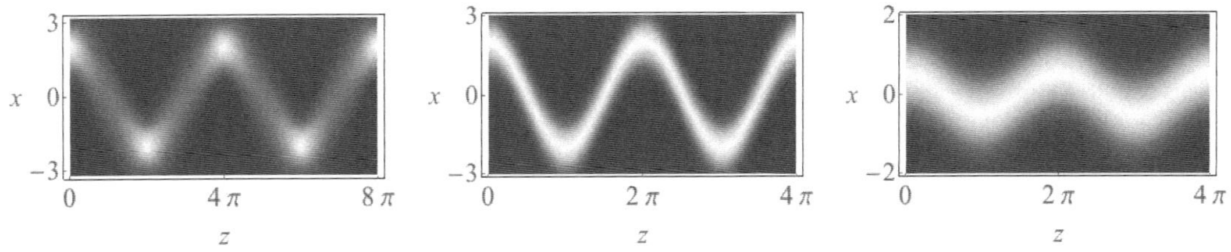

(a) $a_x = 1$, $\Omega = \frac{0.5}{z_R}$ (b) $a_x = 2$, $\Omega = \frac{1}{z_R}$ (c) $a_x = 0.5$, $\Omega = \frac{1}{z_R}$

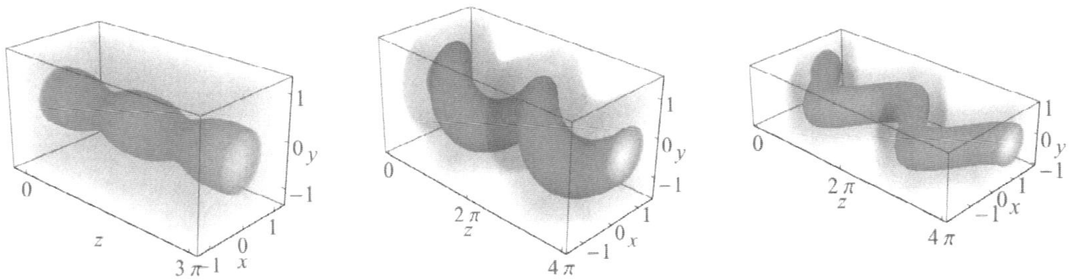

(d) $a_x = b_y = 0$, $\Omega = \frac{0.7}{z_R}$ (e) $a_x = 0.5, b_y = \frac{0.5}{\Omega}$, $\Omega = \frac{1}{z_R}$ (f) $a_x = 1$, $b_y = 0$, $\Omega = \frac{1}{z_R}$

Figure 1. Upper row: Optical intensity distributions $|U(\mathbf{r}, z)|^2$ of the Gaussian modes as functions of x and z in the longitudinal plane $y = 0$ for the parameters indicated in each case. In (a) it can be observed the combination of the off-axis and re-focusing dynamics. Lower row: Optical intensity distributions as functions of all coordinates. In (d) it is shown the on-axis behavior for the case of $\Omega = 0.7/z_R$. In (e) and (f) the maximum of the beam evolves along rotating and oscillating trajectories respectively, according to the values of a_x and b_y. The transverse and longitudinal coordinates are expressed, respectively, in units of w_0 and z_R.

balanced with the focalization properties of the medium and the beam propagates with constant width $w(z) = w_0$.

In Figure 1 we present the optical intensity distribution $|U(\mathbf{r}, z)|^2$ of the Gaussian beam for different values of the parameters. In all cases we have chosen $a_y = b_x = 0$ in order to present clear plots. This is the case in which the center of the beam departs from a point on the x-axis with a velocity oriented in the y-direction. In the upper row (Figures 1 (a),(b),(c)) it is shown the longitudinal plane $y = 0$ of the distribution for $b_y = 0$ and different values of a_x and Ω. Notice, in (a), that the period of the trajectory of the maximum of intensity is twice the period at which the beam recovers its initial width. In (b) and (c), the parameter Ω was chosen to match $1/z_R$, and so, the beam propagates with a constant width, following the oscillating dynamics of its maximum. In Figure 1(d) we have set $a_x = b_y = 0$ in order to show the on-axis self-focusing behavior of the optical intensity as a function of coordinates. In (e) and (f), as $\Omega = 1/z_R$, the beam propagates with constant width while it is redirected following, respectively, helical and sinusoidal (plane) trajectories according to the values of a_x and b_y.

3. Summary

The construction of the general off-axis Gaussian mode, in the paraxial regime, was presented for the case of a medium with a quadratic refractive index profile. This construction was accomplished by reducing the problem to the solution of an Ermakov-type equation for the width of the Gaussian beam. It was shown that the field amplitude is completely defined by the width and the trajectory of the center of the wavepacket, meaning that these two functions are determinant in the description of the propagation processes of the light beam. As special

IOP Conf. Series: Journal of Physics: Conf. Series **839** (2017) 012024 doi:10.1088/1742-6596/839/1/012024

cases, the on-axis Gaussian beam in a quadratic index profile, as well as the parameters for the propagation in a homogeneous medium were obtained for $\mathbf{r}_0 = 0$ and $\Omega \to 0$ respectively. Some examples were presented in order to illustrate our results.

Acknowledgments
The financial support from CONACyT (PhD scholarship for Z. Gress, grant 257292) and from Instituto Politécnico Nacional, Mexico through the project SIP20170233 is acknowledged. The authors are indebted to Marco Enríquez for enlightening comments.

References
[1] Siegman A E 1986 *Lasers* (California University Science Books)
[2] Andrews D L 2008 *Structured light and its applications: An intorduction to phase-structured beams and nano-scale optical forces* (London Academic Press)
[3] Kogelnik H 1965 *Appl. Opt.* **4** 1562
[4] Casperson L W 1973 *Appl. Opt.* **12** 2434
[5] Permitin G V and Smirnov A I 1996 *JETP* **82** 395-402
[6] Bornatici M and Maj O 2003 *Plasma Phys. Control Fusion* **45** 707-719
[7] Castaños O, Schuch D and Rosas-Ortiz O 2013 *J. Phys. A: Math. Theor.* **46** 075304 (20pp)
[8] Cruz y Cruz S and Gress Z 2017 Group approach to the paraxial propagation of Hermite-Gaussian modes in a parabolic medium *Preprint* IPN2017 *Submitted*
[9] Saleh B E A and Teich M C 1991 *Fundamental of Photonics* (John Wiley and Sons, New York)
[10] Feng S 2001 *Opt. Lett.* **26** 845-847
[11] Simon R and Mukunda N 1993 *Phys. Rev. Lett.* **70** 880-883
[12] Habrakem S J M and Nienhuis G 2010 *J. Math. Phys.* **51** 082702 19 pages
[13] Ermakov V P 2008 *Appl. Anal. Discrete. Math.* **2** 123-145
[14] Rosas-Ortiz O, Castaños O and Schuch D 20015 *J. Phys. A: Math. Theor.* **48** 445302 (23pp)

IOP Conf. Series: Journal of Physics: Conf. Series **839** (2017) 012025 doi:10.1088/1742-6596/839/1/012025

Using Spatial Correlations of SPDC Sources for Increasing the Signal to Noise Ratio in Images

A I Ruíz, R Caudillo, V M Velázquez and E Barrios

Facultad de Ciencias, Universidad Nacional Autónoma de México, Ciudad Universitaria, D.F. 04510, México

E-mail: e.barrios@ciencias.unam.mx

Abstract. We experimentally show that, by using spatial correlations of photon pairs produced by Spontaneous Parametric Down-Conversion, it is possible to increase the Signal to Noise Ratio in images of objects illuminated with those photons; in comparison, objects illuminated with light from a laser present a minor ratio. Our simple experimental set-up was capable to produce an average improvement in signal to noise ratio of 11dB of Parametric Down-Converted light over laser light. This simple method can be easily implemented for obtaining high contrast images of faint objects and for transmitting information with low noise.

1. Introduction

A main problem in receiving low intensity signals is the detectors noise level. It takes relevance when the signal amplitude is on the order of the detectors electronic noise because it leads to a practical limit for the detection of weak signals [1]. An example of this is the detection of optical signals at the level of photon counting. Avalanche Photodiodes (APDs) have the capacity for counting photons [2, 3]; however, they suffer from a background noise (Dark Counts), which is random in nature, is present even when there is no light impinging on the detector and typically is on the order of hundreds of counts per second. This type of noise could be removed from our measuring device if we could have knowledge (or a reference) of how it is produced and behaves. However, this turns highly difficult to do because this noise is random (lack of predictable behavior) and if the signal is weaker than the background noise. This problem can be circumvented, to a certain degree, by using the correlations produced by non-linear materials, which can give us a reference to identify the photons (signal) of interest instead of removing the noise. There have been experiments showing the effectiveness of this idea [4–7]; however, they require particular and time consuming experimental conditions [5, 6]. In this work we show how this can be simplified and still obtain a good capability for reducing the noise in a signal.

2. Spontaneous parametric down conversion

The theory behind Spontaneous Parametric Down Conversion (SPDC) has been widely studied and is based on the nonlinear interaction of some particular materials with an intense electromagnetic field [8] to produce pairs of photons with certain properties. In SPDC, a pump photon (with frequency ω_p) is annihilated to generate two new photons called signal and idler (with frequencies ω_s and ω_l respectively) according to energy and momentum conservation (phase matching condition):

$$\omega_p = \omega_s + \omega_l, \qquad \vec{k_p} = \vec{k_s} + \vec{k_l}. \tag{1}$$

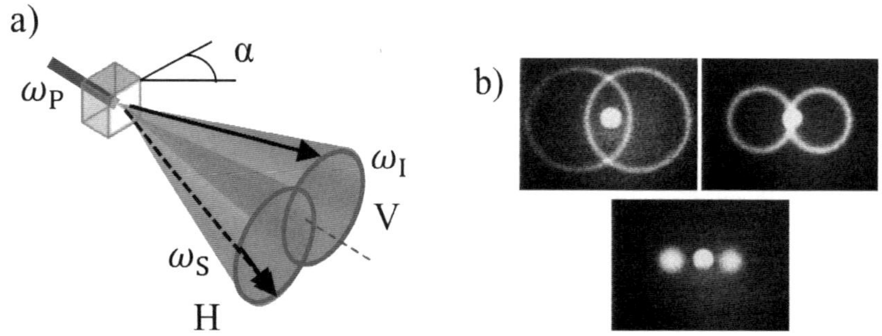

Figure 1. a) Type II SPDC geometry. Conical distributions had orthogonal polarizations; each photon of a pair is in one of the two cones. b) Transversal view of the cones for horizontal angular phase matching conditions of $\alpha = 0°$, $2°$, $4°10''$. The central spot is reminiscent of the pump beam.

In the degenerated case both photons have the same frequency ($\omega/2 = \omega_s = \omega_l$). Under Type II phase matching photons have orthogonal polarizations, horizontal photons form one cone while vertical ones form another (Figure 1a)). By changing the phase matching conditions it is possible to modify the geometry of the cones [5, 8] (Figure 1b). Due to momentum conservation, each pair of photons is produced in times and directions correlated between them according to equation (1). By placing two detectors in an anti-symmetric way, one in each cone, it will be possible to detect the simultaneous arrival of the photons of a pair (Coincidence) [9, 10], any other not anti-symmetrical position will not produce coincidences because it does not conserve momentum. This phenomenon can be interpreted as a spatial correlation. In contrast, classical light sources (lasers) will not present this correlation because the origin of each photon is not correlated with any other photon. Experimentally the degree of spatial correlation between a pair of spatial regions can be relatively quantified through the number of coincidences, where a great number of coincidences will imply a great correlation degree [10]. Then, if we use an SPDC source to transmit a signal through coincidences, we can reduce the amount of noise in its detection, because the correlations will help us to discriminate the signal from noise and because the random noise will not produce coincidences.

3. Signal to noise ratio

To quantify the advantage of SPDC light over the classical sources we use a simple Signal to Noise Ratio (SNR) measurement. This quantity is of great importance to quantify the performance of a receiver in a communication system. The SNR compares the power level of a signal with the power level of the noise in the receiver:

$$SNR_{dB} = 10log_{10}\left(\frac{P_{Signal}}{P_{Noise}}\right) = P_{Signal,dB} - P_{Noise,dB}, \tag{2}$$

where the grater the difference of the signal and noise power levels, the better the performance of the system. In our case, the power of our signal (or noise) will be given by the measured number of photons or coincidences (or dark counts): $P_{Signal} = N_{Signal}$ ($P_{Noise} = N_{Noise}$).

We calculated the SNR for two types of signals (light): a laser source, were the power of the signal will be proportional to the number of photons or coincidences detected, and an SPDC

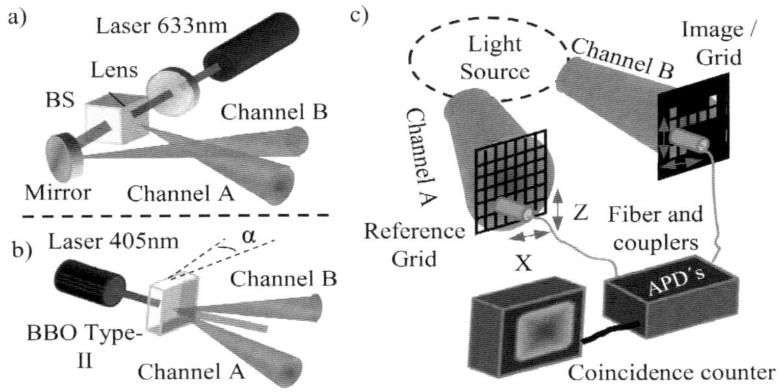

d) Channel A Grid numbering

A1	A2	A3	A4	A5	A6
A7	A8	A9	A10	A11	A12
A13	A14	A15	A16	A17	A18
A19	A20	A21	A22	A23	A24
A25	A26	A27	A28	A29	A30
A31	A32	A33	A34	A35	A36

Channel B Grid numbering

B1	B2	B3	B4	B5	B6
B7	B8	B9	B10	B11	B12
B13	B14	B15	B16	B17	B18
B19	B20	B21	B22	B23	B24
B25	B26	B27	B28	B29	B30
B31	B32	B33	B34	B35	B36

Figure 2. Set-ups for measuring spatial correlations and SNR. a) Laser light source. The beam is expanded and divided in two by a Beam Splitter (BS), the cones area cover the grids. b) SPDC source. The pairs of correlated photons are distributed in two cones (A y B). c) Detection. Composed by reference grids, the image (for noise reduction), XZ moving fiber collectors with 1mm^2 aperture, two APDs for detection and a coincidence counter. d) Grids numberings.

source, were the signal power will be given by the number of coincidences. The noise power will be given by the number of dark counts in the case of the laser, and by the number of dark (false) coincidences measured between no correlated regions or when there is no light measured, for the case of the SPDC source. It can be expected that for the SPDC source the number of coincidences (and the SNR ratio) will be greater than for the laser source, as long as we measure coincidences.

4. Experimental set-up

The experiment is divided in two sections. In the first, we quantified the degree of spatial correlation and the SNR through coincidences (or counts) for both, laser and SPDC sources. In the second part we analyzed an image illuminated with this two types of sources and quantify the SNR by using counts or coincidences.

For the laser source, we used a 633nm laser attenuated to an average of 9000 photons per second. The beam is expanded with a lens and divided in two with a beam-splitter to generate a pair of cones of light with the same dimensions of that of the SPDC cones (Figure 2a)). For the SPDC source, a 405nm laser is used to pump a BBO crystal type-II at an incident angle of $\alpha = 4°10''$ in the horizontal plane, in order to produce completely separated and solid 810nm light cones (Figure 2b)). In either case, each cone is denoted Channel A and Channel B.

The detection system is common to both sources (Figure 2c)). Photons are collected by a single mode fiber with a fiber coupler which has a 1mm circular aperture. This fibers are mounted on XZ displacers that allow collecting light from different spatial regions. For reference, in front of these light collectors we placed two numbered grids to divide the transversal area of each cone in sub-areas of $1mm^2$ (Figure 2d)). The fibers are connected to a pair of Avalanche Photodiodes (APD), which generate an electrical pulse for each photon collected. This pulses are received by a pulse counting card which also counts the number of coincidences (simultaneous arrival of photons) within a measuring window of 5ns.

In the first experiment we measured the number of photon counts in each channel at different positions and the number of coincidences between them. Collector in Channel A was placed in a particular grid position (pixel) while collector in Channel B was swept through all B grid. This procedure is done for both types of sources. To measure the SNR of the coincidences, the signal power will be the number of coincidences and the noise power the average of the background

Table 1. Spatial Correlation Graphs [a]

Laser source		SPDC source	
Conditions	Avg. Coinc.	Conditions	Avg. Coinc.
Ch A: {A8} Ch B: {B1-B36}	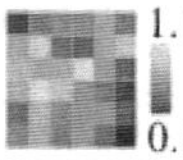	Ch A: {A8} Ch B: {B1-B36}	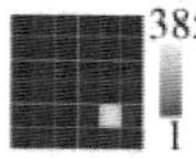
Ch A: {A29} Ch B: {B1-B36}	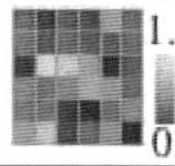	Ch A: {A29} Ch B: {B1-B36}	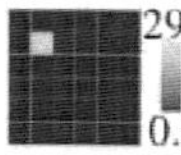
Ch A: {A26} Ch B: {B1-B36}		Ch A: {A26} Ch B: {B1-B36}	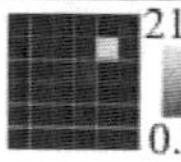
Ch A: {A11} Ch B: {B19-B36}	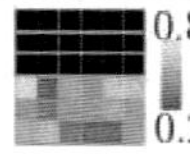	Ch A: {A11} Ch B: {B19-B36}	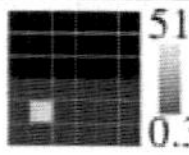

[a] Black elements means no measurement performed.

Table 2. SNR Image Analysis Graphs [a]

Case	Avg. counts or coinc.	Avg. SNR
Laser photon counts.	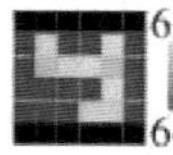	9.16dB
SPDC photon counts.		11.3dB
Laser Coinc. Counts.	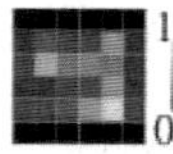	7.32dB
SPDC Coinc. Counts.		22.3dB

[a] Black elements, no measurement.

noise.

In the second experiment we showed how SDPC spatial correlations can improve the SNR of an image as long as the correlation analysis is performed in an Anti-symmetrical way to take into account the photon momentum conservation of the SPDC. For this purpose, we placed in Channel B an image (aligned with the grid) and performed simultaneous sweeps through both grids in the following order (A1,B36), (A2,B35), (A36,B1). From the obtained counts or coincidence data, the SNR of the image can be calculated using equation (2), for both laser light or with SPDC light.

5. Results

5.1. Spatial Correlation

For the laser source no spatial correlations were found. The very low number of coincidences are fortuitous and can be assumed as a background noise for coincidences. We got an average $SNR_{Coin.laser}$ of $-.285dB$. Meaning that we cannot distinguish any signal from the noise.

For the SPDC source, we found a great number of coincidences between anti-symmetrical regions in the light cones, giving an average $SNR_{Coin.SPDC}$ of $23.22dB$. Table 1 summarizes this correlations for different pairs of regions. As it can be seen, the SPDC correlation is greater than for the laser.

5.2. SNR reduction

To prove that SPDC light allows us to increase the SNR factor in an image due to spatial correlations, in Table 2 we present a comparison for the SNR obtained using a laser (or SPDC light without coincidences) against the SNR obtained using a laser or SPDC source through coincidences. In cases where there is no coincidence reference, the average SNR is similar for both the laser and the SPDC; when coincidences are used to reproduce the image, we obtained an average of 22.31dB for the SPDC and 7.32dB for the laser. In addition, it can be noticed that due to the momentum conservation, the image reconstructed through coincidences is rotated by 180°.

6. Conclusions

Our results show that, when using an SPDC sources, the production of pairs of photons according to momentum conservation generates a clear anti-symmetrical spatial correlation signal between the channels pixels above the APD noise, while for a laser source this correlation is non-existent. By taking advantage of this spatial correlations we were able to improve the SNR of an image illuminated with an SPDC source by an average of 11dB, which is translated in higher image contrast; on the contrary, when we used a laser source for illumination, the lack of correlation produces an average decrease on the SNR of image of 1.84dB, which can be interpreted as a loss on information. The increase in SNR by using an SPDC source for illumination can be very useful for applications where the object is faint or it requires low intensities of illumination in order to avoid damage [6].

Acknowledgments
We would like to thank the support of CONACYT's Red de Tecnologías Cuánticas de la Información.

References

[1] Vaseghi S V 2008 *Advances Digital Signal Processing and Noise Reduction* (London: Wiley)
[2] Renker D 2006 *Nucl. Instr. and Meth.* A **567** 48-56
[3] Vasile S, Gothoskar P, Farrell R and Sdrilla D 1998 *IEEE T. Nucl. Sci.* **45** 720-723
[4] Kolobov M I 2007 *Quantum Imaging* (New York: Springer Science)
[5] Brembilla E, Caspani L, et.al 2008 *Phys. Rev.* A **77** 053807
[6] Brida G, Genovese M and Berchera I R 2010 *Nat. Photonics* **4** 227-230
[7] Lugiato L A, Gatti A and Brambilla A 2002 *J. Opt.* B **4** S176
[8] Boyd R W 2003 *Non linear Optics* (London:Academic Press)
[9] Brambilla E, Gatti A, Bache M and Lugiato L A 2004 *Phys. Rev* A **69** 023802
[10] Procopio L M, Rosas-Ortiz O and Velázquez V 2010 *AIP Conf. Proc.* **1287** 80